J. Warnatz • U. Maas • R. W. Dibble

Combustion

Springer

Berlin
Heidelberg
New York
Barcelona
Hong Kong
London
Milan
Paris
Singapore
Tokyo

J. Warnatz • U. Maas • R. W. Dibble

Combustion

Physical and Chemical Fundamentals, Modeling and Simulation, Experiments, Pollutant Formation

3rd Edition
With 208 Figures and 14 Tables

Springer

Prof. Dr. Dr. h. c. Jürgen Warnatz
Universität Heidelberg
Interdisziplinäres Zentrum
für Wissenschaftliches Rechnen
Im Neuenheimer Feld 368
D-69120 Heidelberg, Germany

Prof. Dr. Ulrich Maas
Universität Stuttgart
Institut für Technische Verbrennung
Pfaffenwaldring 12
D-70550 Stuttgart, Germany

Prof. Dr. Robert W. Dibble
University of California
Dept. of Mechanical Engineering
Etcheverry Hall
94720 Berkeley, CA, USA

ISBN 3-540-67751-8 3rd Ed. Springer-Verlag Berlin Heidelberg New York
ISBN 3-540-65228-0 2nd Ed. Springer-Verlag Berlin Heidelberg New York

Library of Congress Cataloging-in-Publication-Data applied for

Die Deutsche Bibliothek - CIP-Einheitsaufnahme
Warnatz, Jürgen: Combustion : physical and chemical fundamentals, modeling and simulation, experiments, pollutant formation ; with 14 tables / J. Warnatz ; U. Maas ; R. W. Dibble. - 3. ed. - Berlin ; Heidelberg ; New York ; Barcelona ; Hong Kong ; London ; Milan ; Paris ; Singapore ; Tokyo : Springer, 2001
Einheitssacht.: Technische Verbrennung <engl.>
ISBN 3-540-67751-8

Springer-Verlag Berlin Heidelberg New York
a part of Springer Science+Business Media

Printed in Germany

Typesetting: cameraready by authors
Cover design: MEDIO GmbH, Berlin
Printed on acid free paper 62/3111 – 5 4 3 2 SPIN: 11331421

Preface

This book has evolved from a lecture series (of *J. Wa.*) on combustion at Stuttgart University. The lectures were intended to provide first-year graduate students (and advanced undergraduates) with a basic background in combustion. Such a course was needed since students of combustion arrive with a wide variety of backgrounds, including physics, physical chemistry, mechanical engineering, computer science and mathematics, aerodynamics, and atmospheric science. After a few years of improving printed matter distributed to the students, the lecture notes have been organized into a book, first in German, and later translated and augmented in an English version. Due to the fact that the second edition was sold out after only two years, the present third English edition solely experienced improvement with respect to some misprints, style, and face-lifting of formulae and figures.

We intend that the book provides a common basis from which research begins. Thus, the treatment of the many topics is compact with much citation to the research literature. Beyond this, the book expects that combustion engineers and researchers will increasingly rely on mathematical modeling and numerical simulation for guidance toward greater understanding, in general, and, specifically, toward producing combustion devices with ever higher efficiencies and with lower pollutant emissions. Laminar flame computer codes and data to run them are available on the internet at *http://reaflow.iwr.uni-heidelberg.de/software/.*

Because this book is a research launching point, we expect it to be updated in a timely fashion. We invite the readers to browse at our internet address at U. C. Berkeley (*http://www.me.berkeley.edu/cal/book/*) for additional comments and constructive critical comments that may be part of the next edition.

Like the students mentioned above, the authors themselves come from diverse backgrounds and owe much thanks to too many people to cite. However, we do have one place in common: We express thanks to our colleagues at the Combustion Research Facility at Sandia National Laboratories in Livermore, California; the CRF has been for us a fertile crossroads and offered continuous hospitality (to *J. Wa.*) for finishing this book, including this third edition.

Heidelberg, Stuttgart, Berkeley, in October 2000 *J. Warnatz, U. Maas, R. W. Dibble*

Preface

Table of Contents

1 Introduction, Fundamental Definitions and Phenomena

1.1 Introduction

Combustion is the oldest technology of mankind; it has been used for more than one million years. At present, about 90% of our worldwide energy support (e. g., in traffic, electrical power generation, heating) is provided by combustion; therefore it is really worthwhile studying this process.

Combustion research in the past was directed to fluid mechanics that included global heat release by chemical reaction. This heat release was often described simply with the help of thermodynamics, which assumes infinitely fast chemical reaction. This approach was useful to some extent for designing stationary combustion processes; it is not sufficient for treating transient processes like ignition and quenching or if pollutant formation shall be treated. However, pollutant formation during combustion of fossil fuels is, and will continue to be, a central topic in future.

The focus of this book is therefore to treat the coupling of chemical reaction and fluid flow; in addition, combustion-specific topics of chemistry (hydrocarbon oxidation, large reaction mechanisms, simplification of reaction mechanisms) and combustion specific topics of fluid mechanics (turbulent flow with density change by heat release, potential generation of turbulence by heat release) shall be considered.

Thus, this book will not consider in great detail the theory of chemical reaction rates and experimental methods for the determination of reaction rate coefficients (this is the task of reaction kinetics). Nor will this book discuss the details of turbulence theory and the handling of complex geometries (this is the task of fluid mechanics), although all of these topics are needed in understanding combustion.

1.2 Some Fundamental Definitions

The quantitative treatment of combustion processes requires some understanding of fundamental concepts and definitions, which shall be described in this section.

A *chemical reaction* is the exchange and/or rearrangement of atoms between colliding molecules. In the course of a chemical reaction, e. g.,

$$HCN + OH \rightarrow CN + H_2O \,,$$

the atoms (relevant in combustion: C, H, O, and N) are conserved; i. e., they are not created or destroyed. On the other hand, molecules (e. g., HCN, OH, CN, and H_2O) are not conserved. A partial list of molecules relevant in combustion is given in Table 1.1. Reactant molecules are rearranged to become product molecules, with simultaneous release of heat. A primary interest in the heat of reaction sets combustion engineering apart from chemical engineering.

Atoms and molecules are conveniently counted in terms of *amount of substance* or (worse, but everywhere used) *mole numbers* (unit: mol). 1 mol of a compound corresponds to $6.023 \cdot 10^{23}$ particles (atoms, molecules, etc.). Accordingly, the *Avogadro* constant (also called *Avogadro's* constant) is $N_A = 6.023 \cdot 10^{23} \mathrm{mol}^{-1}$. The *mole fraction* x_i of the species i denotes the ratio of the *mole number* n_i of species i to the total *mole number* $n = \Sigma n_i$ of the mixture ($x_i = n_i / n$).

The *mass m* is a fundamental property of matter (units of kg in the SI-system). The *mass fraction* w_i is the ratio of the mass m_i of the species i and the total mass $m = \Sigma m_i$ of the mixture ($w_i = m_i / m$).

The *molar mass* (obsolete: *molecular weight*) M_i (units of, e. g., g/mol) of species i is the mass of 1 mol of this species. Some examples using atomic carbon, molecular hydrogen, molecular oxygen, and methane are $M_C = 12$ g/mol, $M_{H_2} = 2$ g/mol, $M_{O_2} =$ 32 g/mol, $M_{CH_4} = 16$ g/mol. The *mean molar mass* of a mixture $\overline{M}$ (in g/mol, e. g.) denotes an average molar mass, using the mole fractions as weight ($\overline{M} = \Sigma x_i M_i$).

Frequently mass fractions and mole fractions are expressed in terms of percentages (*mole-%* or *mass-%*). The following relations hold, which can be verified by simple calculations (S denotes the number of different compounds):

$$w_i = \frac{M_i \, n_i}{\sum_{j=1}^{S} M_j \, n_j} = \frac{M_i \, x_i}{\sum_{j=1}^{S} M_j \, x_j} \,, \tag{1.1}$$

$$x_i = \frac{w_i}{M_i} \overline{M} = \frac{w_i / M_i}{\sum_{j=1}^{S} w_j / M_j} \,. \tag{1.2}$$

Densities do not depend on the size (extent) of a system. Such variables are called *intensive* properties, and are defined as the ratio of the corresponding *extensive* properties (which depend on the extent of the system) and the system volume V. Examples of intensive properties are

mass density (*density*) $\rho = m / V$ (in, e. g., kg/m^3) ,
molar density (called *concentration*) $c = n / V$ (in, e. g., mol/m^3) .

It follows (very easy to verify) that the mean molar mass is given by the expression

$$\frac{\rho}{c} = \frac{m}{n} = \overline{M} \,. \tag{1.3}$$

Tab. 1.1. List of molecules relevant for combustion processes

	FAMILY										
	Alkane	Alkene	Alkyne	Arene	Haloalkane	Alcohol	Ether	Amine	Aldehyde	Ketone	Carboxylic Acid
Specific Example	$CH_3\text{-}CH_3$	$CH_2{=}CH_2$	$HC{\equiv}CH$	C C C C C C (ring)	CH_3CH_2Cl	CH_3CH_2OH	CH_3OCH_3	CH_3NH_2	$CH_3\overset{\overset{O}{\|}}{C}H$	$CH_3\overset{\overset{O}{\|}}{C}CH_3$	$CH_3\overset{\overset{O}{\|}}{C}OH$
IUPAC Name	Ethane	Ethene	Ethyne	Benzene	Chloroethane	Ethanol	Methoxy-methane	Methylamin	Ethanal	Propanone	Ethanoic Acid
Common Name	Ethane	Ethylene	Acetylene	Benzene	Ethylchloride	Ethyl-alcohol	Dimethylether	Methyl-amine	Acetaldehyde	Acetone	Acetic Acid
General Formula	RH	$H_2C{=}CH_2$ $RCH{=}CH_2$ $RCH{=}CHR$ $R_2C{=}CHR$ $R_2C{=}CR_2$	$RC{\equiv}CH$ $RC{\equiv}CR$	ArH, ArR	RX	ROH	ROR	RNH_2 R_2NH R_3N	$R\overset{\overset{O}{\|}}{C}H$	$R\overset{\overset{O}{\|}}{C}R$	$R\overset{\overset{O}{\|}}{C}OH$
Functional Group	C–H bonds C–C bonds	$>C{=}C<$	$-C{\equiv}C-$	Aromatic Ring	$-\overset{\|}{\underset{\|}{C}}-X$	$-\overset{\|}{\underset{\|}{C}}-OH$	$-\overset{\|}{\underset{\|}{C}}-O-\overset{\|}{\underset{\|}{C}}-$	$-\overset{\|}{\underset{\|}{C}}-\overset{\|}{N}-$	$-\overset{\overset{O}{\|}}{C}-H$	$-\overset{\|}{\underset{\|}{C}}-\overset{\overset{O}{\|}}{C}-C-$	$-\overset{\overset{O}{\|}}{C}-OH$

In chemistry, concentrations c of chemical species defined in this way are usually denoted by species symbols in square brackets (e. g., $c_{H_2O} = [H_2O]$).

For the gases and gas mixtures in combustion processes, an equation of state relates temperature, pressure, and density of the gas. For many conditions it is satisfactory to use the *perfect gas equation of state*,

$$pV = nRT\,, \tag{1.4}$$

where p denotes the pressure (in units of Pa), V the volume (in m^3), n the mole number (in mol), T the absolute temperature (in K), and R the *universal gas constant* ($R = 8.314\ J{\cdot}mol^{-1}{\cdot}K^{-1}$). It follows that

$$c = \frac{p}{RT} \quad \text{and} \quad \rho = \frac{p\overline{M}}{RT} = \frac{p}{RT\sum\limits_{i=1}^{S}\frac{w_i}{M_i}}\,. \tag{1.5}$$

When temperatures are near or less than the critical temperature, or when pressures are near or above the critical pressures, the concentration or density is inadequately predicted using the perfect gas equation of state, i. e., (1.5). The system is better approximated as a *real gas*. One example of a real gas equation of state is that of *van der Waals*. Details of this and other equations of state for real gas conditions can be found in textbooks on physical chemistry (e. g., Atkins 1990).

1.3 Basic Flame Types

Tab. 1.2. Example of combustion systems ordered with respect to premixedness and flow type

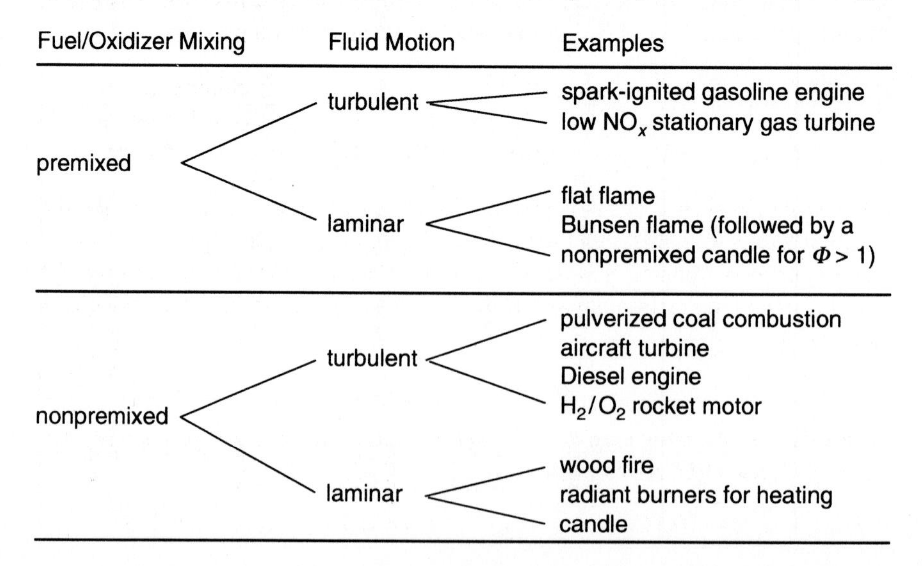

Fuel/Oxidizer Mixing	Fluid Motion	Examples
premixed	turbulent	spark-ignited gasoline engine low NO_x stationary gas turbine
	laminar	flat flame Bunsen flame (followed by a nonpremixed candle for $\Phi > 1$)
nonpremixed	turbulent	pulverized coal combustion aircraft turbine Diesel engine H_2/O_2 rocket motor
	laminar	wood fire radiant burners for heating candle

In combustion processes, fuel and oxidizer (typically air) are mixed and burned. It is useful to identify several combustion categories based upon whether the fuel and oxidizer is mixed first and burned later (*premixed*) or whether combustion and mixing occur simultaneously (*nonpremixed*). Each of these categories is further subdivided based on whether the fluid flow is laminar or turbulent. Table 1.2 shows examples of combustion systems that belong to each of these categories, which will be discussed in the following sections.

Laminar Premixed Flames: In *laminar premixed flames*, fuel and oxidizer are premixed before combustion and the flow is laminar. Examples are laminar *flat flames* and (under fuel-lean conditions) *Bunsen flames* (see Fig. 1.1).

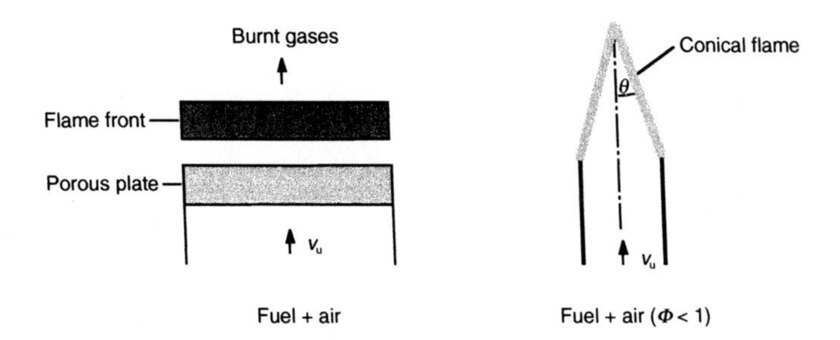

Fig. 1.1. Schematic illustration of a laminar flat flame (left) and of a Bunsen flame (right), both premixed

A premixed flame is said to be *stoichiometric*, if fuel (e. g., a hydrocarbon) and oxidizer (e. g., oxygen O_2) consume each other completely, forming only carbon dioxide (CO_2) and water (H_2O). If there is an excess of fuel, the system is called *fuel-rich*, and if there is an excess of oxygen, it is called *fuel-lean*. Examples are

$$\begin{array}{lllll} 2\,H_2 + O_2 & \rightarrow & 2\,H_2O & & \text{stoichiometric} \\ 3\,H_2 + O_2 & \rightarrow & 2\,H_2O + H_2 & & \text{rich } (H_2 \text{ left over}) \\ CH_4 + 3\,O_2 & \rightarrow & 2\,H_2O + CO_2 + O_2 & & \text{lean } (O_2 \text{ left over}). \end{array}$$

Each symbol in such a chemical reaction equation represents 1 mol. Thus, the first equation means: 2 mol H_2 react with 1 mole O_2 to form 2 mole H_2O.

If the reaction equation is written such that it describes exactly the reaction of 1 mol fuel, the mole fraction of the fuel in a stoichiometric mixture can be calculated easily to be

$$x_{\text{fuel,stoich.}} = \frac{1}{1+\nu} \,. \tag{1.6}$$

Here ν denotes the mole number of O_2 in the reaction equation for a complete reaction to CO_2 and H_2O. An example is

$$H_2 + 0.5\,O_2 \rightarrow H_2O \qquad \nu = 0.5 \qquad x_{H_2,\text{stoich.}} = 2/3.$$

If air is used as an oxidizer, it has to be taken into account that dry air contains only about 21 % oxygen (78% nitrogen, 1% noble gases). Thus, for air $x_{N_2} = 3.762\, x_{O_2}$. It follows that the mole fractions in a stoichiometric mixture with air are

$$x_{\text{fuel,stoich.}} = \frac{1}{1 + \nu \cdot 4.762}, \; x_{O_2\text{,stoich.}} = \nu \cdot x_{\text{fuel,stoich.}}, \; x_{N_2\text{,stoich.}} = 3.762 \cdot x_{O_2\text{,stoich.}} \quad (1.7)$$

ν denotes, again, the mole number of O_2 in the reaction equation for a complete reaction of 1 mol of fuel to CO_2 and H_2O. Some examples are given in Table 1.3.

Tab. 1.3. Examples of stoichiometric numbers ν and of fuel mole fractions at stoichiometric conditions $x_{\text{fuel, stoich}}$ in fuel/air mixtures

Reaction	ν	$x_{\text{fuel, stoich.}}$
$H_2 + 0.5\ O_2 + 0.5 \cdot 3.762\ N_2 \rightarrow H_2O + 0.5 \cdot 3.762\ N_2$	0.5	29.6 mol-%
$CH_4 + 2.0\ O_2 + 2.0 \cdot 3.762\ N_2 \rightarrow CO_2 + 2\ H_2O + 2.0 \cdot 3.762\ N_2$	2.0	9.50 mol-%
$C_3H_8 + 5.0\ O_2 + 5.0 \cdot 3.762\ N_2 \rightarrow 3\ CO_2 + 4\ H_2O + 5.0 \cdot 3.762\ N_2$	5.0	4.03 mol-%
$C_7H_{16} + 11.0\ O_2 + 11.0 \cdot 3.762\ N_2 \rightarrow 7\ CO_2 + 8\ H_2O + 11.0 \cdot 3.762\ N_2$	11.0	1.87 mol-%
$C_8H_{18} + 12.5\ O_2 + 12.5 \cdot 3.762\ N_2 \rightarrow 8\ CO_2 + 9\ H_2O + 12.5 \cdot 3.762\ N_2$	12.5	1.65 mol-%

Premixtures of fuel and air (the proper amount of N_2 has to be added in this case on both sides of the reaction equation; see Table 1.3) are characterized by the *air equivalence ratio* (sometimes *air number*)

$$\lambda = (x_{\text{air}}/x_{\text{fuel}}) \,/\, (x_{\text{air,stoich.}}/x_{\text{fuel, stoich.}}) = (w_{\text{air}}/w_{\text{fuel}}) \,/\, (w_{\text{air,stoich.}}/w_{\text{fuel, stoich.}})$$

The reciprocal value, the *fuel equivalence ratio* Φ ($\Phi = 1/\lambda$). This formula can be rewritten to allow the evaluation of the mole fractions in a mixture from Φ by

$$x_{\text{fuel}} = \frac{1}{1 + \frac{4.762 \cdot \nu}{\Phi}}, \quad x_{\text{air}} = 1 - x_{\text{fuel}}, \quad x_{O_2} = x_{\text{air}}/4.762, \quad x_{N_2} = x_{O_2} \cdot 3.762 .$$

Accordingly, premixed combustion processes can now be divided into three groups,

rich combustion:	$\Phi > 1$,	$\lambda < 1$
stoichiometric combustion:	$\Phi = 1$,	$\lambda = 1$
lean combustion:	$\Phi < 1$,	$\lambda > 1$.

The burning of freely burning premixed laminar flat flames into the unburnt mixture can be characterized by the *laminar burning velocity* v_L (e. g., in m/s); other names in the literature are *flame velocity* or *flame speed*. It will be shown in Chapter 8 that the burning velocity depends only on the mixture composition (Φ or λ), the pressure p, and the initial temperature T_u.

If the laminar burning velocity of a flat flame is less than the velocity v_u of the unburnt gases (see Fig. 1.1), the flame blows off. Therefore, the inequality $v_L > v_u$

has to be fulfilled for flat flames. Right before the blowoff $v_L \approx v_u$,. Thus, the inlet gas velocity at flame lift-off is a measure of the laminar burning velocity.

Higher inlet velocities are possible when the flat flame is at an angle θ to the flow. In the case of a premixed Bunsen flame attached to the exit of a round pipe, the flame front is approximately flat (the flame thickness is small compared to the curvature). It follows (see Fig. 1.1) that

$$v_L = v_u \sin \theta \,. \tag{1.8}$$

Thus, a measurement of θ, perhaps from a photograph, and of the inlet gas velocity v_u will lead to a measure of v_L. Problems connected with the determination of v_L and better experimental methods are discussed by Vagelopoulos and Egolfopoulos (1998).

Turbulent Premixed Flames: As Table 1.2 indicates, other examples of premixed flames include the ubiquitous spark-ignited engine (Otto engine) where the flow is seldom laminar. In this case, premixed flame fronts burn and propagate into a turbulent fluid flow. If the turbulence intensity is not too high, curved laminar premixed flame fronts are formed. The turbulent flame can then be viewed as an ensemble of many premixed laminar flames. This so-called *flamelet* concept will be discussed in detail in Chapter 14.

The advantage of premixed combustion is that much greater control of the combustion is possible. By premixing at lean conditions, high temperatures are avoided and hence combustion with low production of the pollutant nitric oxide (NO) is accomplished. In addition, only a very small amount of soot is formed at these circumstances as soot is largely a product of rich combustion (see Chapters 17 and 18).

Despite the advantages, premixed combustion is not widely used because of the potential for accidental collection of large volumes of premixed reactants, which could burn in an uncontrolled explosion.

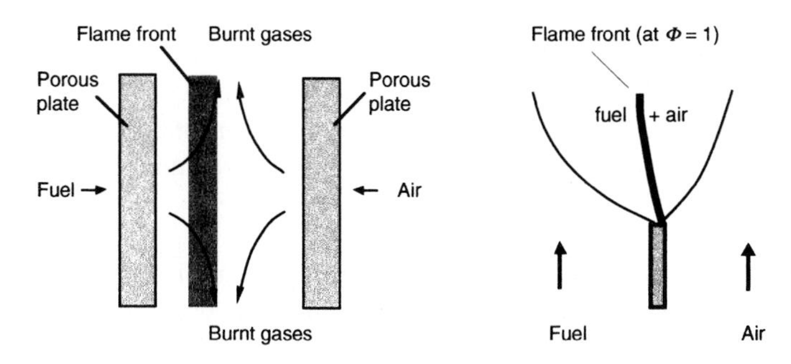

Fig. 1.2. Schematic illustration of a laminar counterflow nonpremixed flame (left) and a laminar coflow nonpremixed flame (right)

Laminar Nonpremixed Flames: In *laminar nonpremixed flames* (obsolete: *laminar diffusion flames*), fuel and oxidizer are mixed during the combustion process itself. The flow is laminar. As Table 1.2 indicates, examples include candles, oil lamps, and campfires. For research purposes, two important configurations are used: *laminar counterflow* and *laminar coflow nonpremixed flames* as shown in Fig. 1.2.

Nonpremixed flames include more complex chemistry than premixed ones, because the equivalence ratio Φ covers the whole range from 0 (air) to ∞ (pure fuel). Rich combustion occurs on the fuel side, lean combustion on the air side. The flame front, which is usually characterized by intense luminescence, is fixed to regions near the location of the stoichiometric composition, since, as it will be shown later, this is where the temperature is highest. Thus, unlike premixed flames, nonpremixed flames do not propagate and, therefore, cannot be characterized by a laminar flame speed.

Turbulent Nonpremixed Flames: In this case, nonpremixed flames burn in a turbulent flow field, and for low turbulence intensities the so-called flamelet concept can be used again (see Chapter 13). Because of the safety considerations mentioned above, nonpremixed flames are mostly used in industrial furnaces and burners. Unless very sophisticated mixing techniques are used, nonpremixed flames show a yellow luminescence, caused by glowing soot particles formed by fuel-rich chemical reactions in the rich domains of the nonpremixed flames.

Hybrid Premixed-Nonpremixed Laminar Flames: Home heating and heating of water is often done with flames of this type. Fuel and air are premixed slightly rich to $\Phi = 1.4$ so that sooting is suppressed. The products of this rich flame subsequently burn in a nonpremixed flame with the surrounding air. There are millions of such burners in service, collectively accounting for 30 % of natural gas consumption.

1.4 Exercises

Exercise 1.1. (a) How much O_2 is needed for a stoichiometric combustion of methane CH_4 and octane C_8H_{18} respectively (molar ratio and mass ratio)? (b) Which values of mole and mass fractions do stoichiometric mixtures of CH_4 and of C_8H_{18} have with air? (c) How much air is needed for the preparation of a C_8H_{18}-air mixture with an equivalence ratio of $\Phi = 2/3$, assuming 1 mol of mixture?

Exercise 1.2. A safe is to be blown up. The safe, which has a volume of 100 dm^3, is filled with 5 dm^3 of hydrogen in air ($T = 298$ K). In order to avoid noise, the safe is sunk in a cold ($T = 280$ K) lake and then the mixture is ignited. The pressure in the safe is 1 bar prior to ignition. The reaction can be taken to be isochoric (constant volume); assume that the safe did not burst. (a) How many moles of gas are in the safe prior to ignition? What are the values of the mole fractions of hydrogen, oxygen and nitrogen? What is the value of the mean molar mass? (b) How many moles of gas are left after the combustion, if the hydrogen is consumed completely and the resulting water condenses? (c) What is the pressure long after the ignition? What is the mean molar mass now? Is the safe heavier or lighter after the ignition?

2 Experimental Investigation of Flames

Computer simulations (which are treated in detail in some of the following chapters) are increasingly part of the discovery and design process. One can expect this growing trend to continue. Because of (not in spite of) the increasing level of sophistication of the numerical simulations, guidance from increasingly sophisticated experiment is a necessity. Such guidance is needed for several reasons:

- A first reason for comparison with experiment is that the discovery of previously unknown chemical reactions or physics may emerge. It is through this iterative comparison between simulations and experiment that progress is made (*investigation*).
- Secondly, in the interest of obtaining approximate solutions in an acceptable time, the simulations must be done on *modelled* equation sets where one has knowingly left out or simplified terms in the equations. With experience, one learns what terms may safely be neglected on the basis that they contribute little to the features of interest. This experience is obtained through comparison of numerical predictions with experiment (*validation*).

In general, the experiments supply measurements that critically test an aspect of what the simulations predict, namely the velocity, temperature, and concentrations of species. In the past, progress was made by inferences made from intrusive probe measurements. As the models are steadily improving, the details demanded from the simulations steadily increase. Accordingly, increasingly detailed experiments are needed for the understanding of the fundamental physics and chemistry that is then embodied into the model.

The experiments that challenge the models today are largely based upon optical diagnostics. In particular, the development of *laser spectroscopic methods* has lead to considerable progress in the field of combustion. This progress is addressed in the texts by Eckbreth (1996) and Thorne (1988) and in special editions of journals (see, e. g., Wolfrum 1986, 1998, Kompa et al. 1993). Application of the laser techniques requires specific knowledge of molecular physics and spectroscopic methods that are beyond the scope of this book. The intent of this chapter is to show how diagnostics lead to the observations that are compared with the simulations.

The state of a reacting gas mixture at one spatial location is described completely when the velocity $\vec{v}$, temperature T, pressure p, density ρ, and the mole fractions x_i

or mass fractions w_i are known. Examples will be given of how modern techniques often have high spatial and temporal resolution, such that even two-dimensional (or, in the future, three-dimensional) fields of these variables can be determined. Furthermore, development tends towards *nonintrusive* optical methods which (in contrast to conventional methods like sampling) do not disturb the reaction system.

2.1 Velocity Measurements

The measurement of velocities in gaseous flows is usually called *anemometry*. A simple device for measuring velocities is the *hot-wire anemometer*. In this technique the axis of a thin platinum wire is suspended normal to the flow. The wire temperature is sustained above the gas temperature by electrical heating. The heat transfer from the wire is related to the approach velocity. A disadvantage of the hot wire anemometry is that changes or fluctuations in the temperature or composition of the gas are interpreted as changes in the velocity. At higher temperatures, the wire can act catalytically with fuel-air mixtures. In spite of these limitations, the hot wire has been a major tool for air velocity measurements upstream of a flame. When used inside of a pipe, the hot wire technique is the basis for the large commercial industry of electronic mass flow controllers of fuel, air, and other gases.

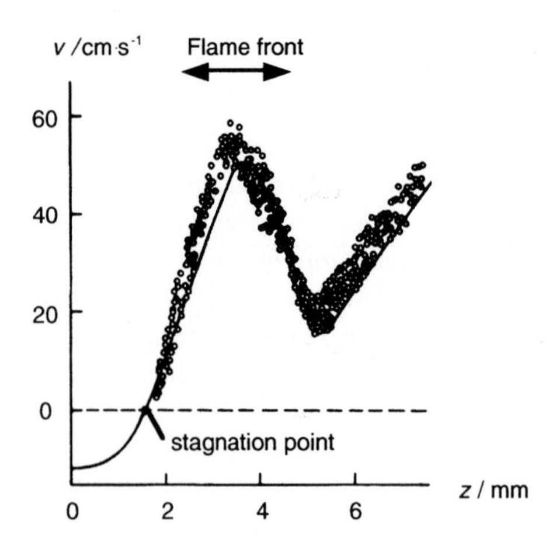

Fig. 2.1. Particle tracking velocity measurements (points) and calculated velocities in a counterflow nonpremixed flame; fuel supply is at $z = 0$, air supply at $z = \infty$

In the *laser-Doppler-anemometry, LDA* (also called *laser-Doppler-velocimetry, LDV*), particles are added ("seeded") to the flow. As with all particle scattering of light, conservation of momentum leads to a Doppler effect, i. e., slight shift of frequency of the scattered light. When the scattered light is mixed with the laser light, the slight difference in frequency is easily detected, and it is proportional to the speed of the particle. By using intersecting laser beams, the direction, speed, and hence the velocity, is determined at the zone of intersection, leading to a high spatial resolution. As with all particle techniques, the LDA measures the velocity of the *particle.* The gas and particle velocity are nearly equal as long as the particle drag/inertia ratio ($C_d\, d^2/\rho{\cdot}d^3$) is large enough; this requirement favors the addition of submicron particles, which is at odds with the light scattering intensity being proportional to d^2 in this *Mie scattering* regime ($d/\lambda > 1$). One-micrometer-diameter particles have proven to be a satisfactory compromise for subsonic flows.

In the *particle tracking* method, particles with a size on the order of 1 μm are added to the flow. The traces of the particles are determined by taking pictures with sufficiently long times of exposure. In this way velocities and even velocity fields can be measured. Fig. 2.1 shows a comparison between velocity measurements by particle tracking (Tsuji and Yamaoka 1971) and calculated velocities (Dixon-Lewis et al. 1985, see Chapter 9) in a counterflow nonpremixed flame. Although there is a considerable scatter in the experimental results, the figure shows that this method allows a reliable measurement of velocities. A variation of this technique, called *PIV* (for *particle image velocimetry*), uses the thin planar sheet of light from a pulsed laser to illuminate the particles. Due to the multiple pulses of laser light, the particle images are dots displaced in space, and the displacement can be related to the velocity. Mungal et al. (1995) review the PIV technique and demonstrate its use in turbulent reactive flows.

2.2 Density Measurements

Density ρ (or concentration $c = \rho/M$) is usually determined from the ideal gas equation of state combined with a measurement of temperature and pressure.

A direct measurement of density can be inferred from the extinction (caused by either absorption or scattering) of a laser beam as it penetrates a medium; in accordance with the Lambert-Beer law (Atkins 1996) the adsorbance A is given by

$$A = \log\left(\frac{I_{\text{ext}}}{I_{\text{laser}}}\right) = l \cdot c_i \cdot \sigma_{i,\text{ext}} ,$$

where l = path length in the sample of concentration c_i with the extinction coefficient $\sigma_{i,\text{ext}}$, and $I_{\text{ext}}/I_{\text{laser}}$ is the ratio of the initial light intensity and that after path length l. Accurate measurements are difficult when this ratio is near to 1 (transparent medium) or near to 0 (opaque medium) for a given path length. However, when one can adjust either the path length l or the concentration c_i, a measurable absorbance allows a tomographic reconstruction of the concentration (Nguyen et al. 1993) or development of, for example, a fuel-to-air ratio probe (Mongia et al. 1996).

A modern optical method for density measurements is based on *laser Rayleigh-scattering*. The Rayleigh limit of light scattering, which compliments the Mie limit used for LDA, occurs when the particle diameter d is small compared to the laser wavelength λ ($d/\lambda < 1$). With visible lasers, the Rayleigh limit extends from submicron particles (tobacco smoke, fog, and soot) to molecules. The ability to see a laser beam as it passes through air is a consequence of the small amount of elastic light scattering. Experience shows that a considerable part of the observed light is due to Mie scattering from suspended particulates while the remainder is Rayleigh scattered (i. e., elastically scattered) light from molecules. The intensity $I_{\text{scattered}}$ of the scatter light is simply proportional to the number density (or concentration) of scatterers,

$$I_{\text{scattered}} \propto I_{\text{laser}} \cdot l \cdot \Omega \cdot \eta \cdot \sum_i c_i \sigma_{i,\text{scatt}} , \tag{2.1}$$

where I_{laser} is the initial light intensity, l = length of laser beam observed (typically ~ 1 mm), Ω = solid angle of detection, η = quantum efficiency of the detector (typically ~ 0.1), and $\sigma_{i,\text{scatt}}$ = scatter cross-section of the particles or molecules. For typical conditions (see Fig. 2.3), the ratio of unburnt and burnt gas temperatures is 1/7, leading to a decrease of c and, hence, $I_{\text{scattered}}$ by a factor of 7.

In compressible flows, where the temperature and pressure may not be known, Rayleigh scattering still measures density, which can be used to challenge numerical simulations. The Rayleigh scattering cross-section of each gas is unique. Methane scatters 2.4 times more than air; therefore, the fuel-to-air ratio of methane with air can be determined using Rayleigh scattering. For studies such as these, both air and fuel are filtered so that Mie scattering from (unwanted) particles does not obfuscate the interpretation of temperature from the Rayleigh signal. Scattering from soot will also compete with (or even overwhelm) the scattering from gas molecules.

As with Mie scattering used for LDV, the Rayleigh scattered frequency is slightly shifted from the laser frequency due to the motion of the gas molecules or particulates. Because the molecules are in random motion, the net result is that the line width of the scattered light can be related to the random velocity and hence temperature (and density) of the molecules. Further, if the gas (or particle) cloud is in motion, the average Doppler shift can be detected, which leads to velocity.

Great progress has been made by tuning the laser so that it is coincident with an atomic absorption line. Then, non-Doppler-shifted scattered light is absorbed by the atomic line filter, while the frequency-shifted light passes through the atomic line filter and is detected. Early research in this area used iodine vapor as the filter; more recent examples of this *filter Rayleigh scattering* (FRS) are presented by Grimstead et al. (1996), by Shirley and Winter (1993), and by Mach and Varghese (1998).

2.3 Concentration Measurements

Probe Samples: A method often used in combustion diagnostics to measure the composition is the sampling by means of *probes*. A suction tube is introduced into the system. The walls of the probe are cooled so that further chemical reaction of the sample flowing inside the tube is halted, called *freezing* of the reaction. The frozen sample is then analyzed for its chemical constituents using a wide variety of techniques. The chemical analysis is rarely controversial. However, great controversy exists as to how representative the cooled sample is of the gas near the probe if the probe were absent.

Often it is expected that important radical species such as OH, O, and H stop reacting inside the tube. As it is shown in later chapters, reactions with radical species such as these have low activation energies and, hence, cooling the sample mixture (without pressure reduction) has little effect on the reactions. Furthermore, the reaction times are often much shorter than the cooling times. Thus, large differences between optical diagnostics and probe sampling can result even for stable species (see, e. g., Nguyen et al. 1993). In any case, the increasing demand on numerical simulations demands measurement of reactive species as they exist in the flame zone. For such measurements, laser based techniques will, therefore, be the method of choice.

Raman-Spectroscopy: As indicated in Fig. 2.2, it is profitable to view the scattering process as an absorption of a laser photon that results in the molecule being excited to an upper *virtual state*, which has a short half-life of ca. 10 femtoseconds. A photon is re-emitted as the molecule returns to the original state in the case of Rayleigh scattered light. From the virtual state, the molecule can return to a state other that the original one. When this happens, the emitted photon has less (or more) energy than the absorbed laser photon. This inelastic process is termed *Stokes* (or *anti-Stokes*) *Raman-scattering*. In either event, the energy difference between the laser photon and the emitted photon is proportional to the spacing of the vibrational energy level, E_i. The vibrational energy level spacing is unique for each molecule; thus, the Raman scattered light is at a slightly different wavelength for each molecule. It is straightforward for a spectrometer to isolate each wavelength and measure the amount of Raman scattered light. As with Rayleigh scattering, the amount of Raman scattered light is proportional to the concentration (cross sections are given in Eckbreth 1996).

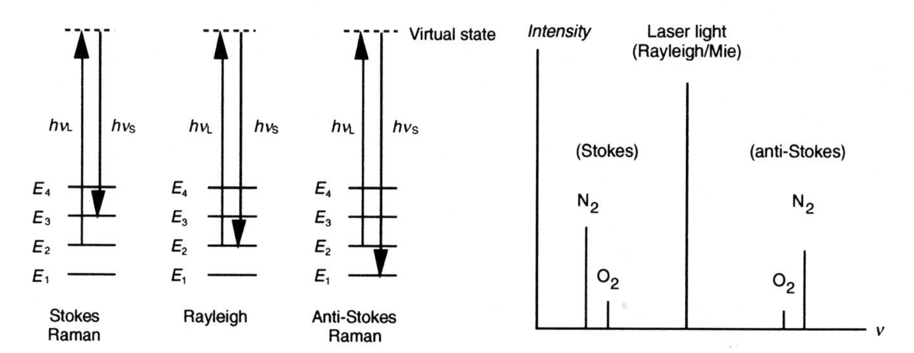

Fig. 2.2. Illustration of the basic processes in Rayleigh and Raman spectroscopy and resulting Raman spectrum, E_i = vibrational levels of the molecule considered, $h\nu_L$ = energy of the irradiated laser light, $h\nu_S$ = energy of the scattered light (h = Planck's constant, ν = light frequency)

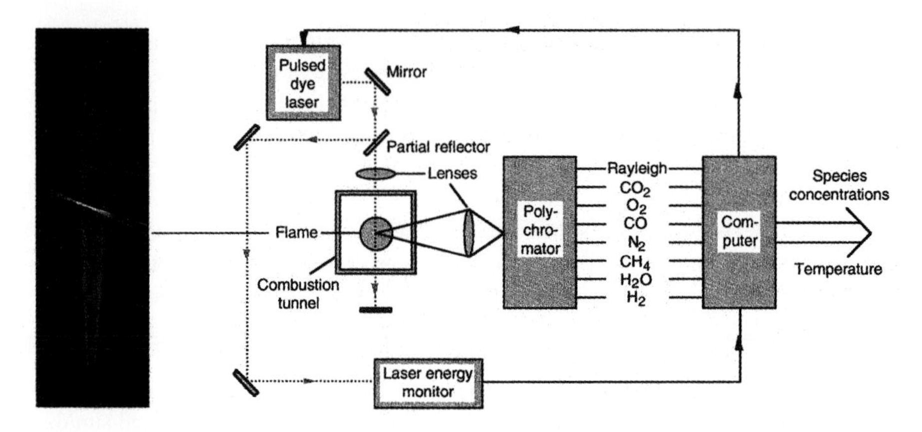

Fig. 2.3. Experimental setup for the determination of concentration and temperature profiles by Raman and Rayleigh spectroscopy (Masri et al. 1988)

Even simultaneous measurements of different species are possible (e. g., N_2, CO_2, O_2, CO, CH_4, H_2O, H_2, Dibble et al. 1987). Two-dimensional measurements (e. g., in turbulent flames) are possible, too (Long et al. 1985).

The Raman process would be the method of choice in nearly all combustion research were it not for the basic physical fact that the *Raman effect* is very weak, i. e., the Raman scattering cross section is small (~2000 times smaller than the Rayleigh scattering cross section). Only powerful lasers allow the use of this effect in combustion diagnostics, and, even then, only species with concentrations above 1 percent have been routinely studied. In spite of these limitations, much improved understanding of combustion has occurred through the application of Raman scattering to carefully contrived experiments (at non-sooting conditions).

CARS Spectroscopy: Closely related to Raman spectroscopy is the *CARS* spectroscopy (*coherent anti-Stokes Raman-spectroscopy*). In this method, additional light (Stokes light, see Fig. 2.4) with a frequency ν_S is irradiated besides the so-called pump laser (frequency ν_P). Interaction of the laser light with the molecules generates light with the frequency ν_{CARS}, which is given by $\nu_{CARS} = 2\ \nu_P - \nu_S$ (see Fig. 2.4).

The physical processes are too complex to be treated here in detail (see, e. g., Eckbreth 1996, Thorne 1988). The frequency of the pump laser is usually left constant, and the Stokes-laser frequency is varied such that the different energy levels of the molecules can be scanned. From the spectral shape the temperature is reliably inferred, while from the spectral intensities concentration is inferred.

A major advantage of CARS spectroscopy is the fact that three laser beams (two pump beams and one Stokes beam) have to coincide. Special geometrical configurations allow high *spatial resolution* (e. g., Eckbreth 1996) and high *temporal resolution* can be attained by use of pulsed lasers.

A second major advantage of CARS spectroscopy is the strong signal strength of the generated CARS signal. This signal, which itself is a coherent laser beam, is easily detected in particulate- or droplet-laden flows, sooting flows, and flows with high radiant background such as flame-formed diamond films (Bertagnolli and Lucht 1996).

In spite of the strong signal, a disadvantage of CARS spectroscopy is the nontrivial evaluation of the measured CARS spectrum as well as the high costs.

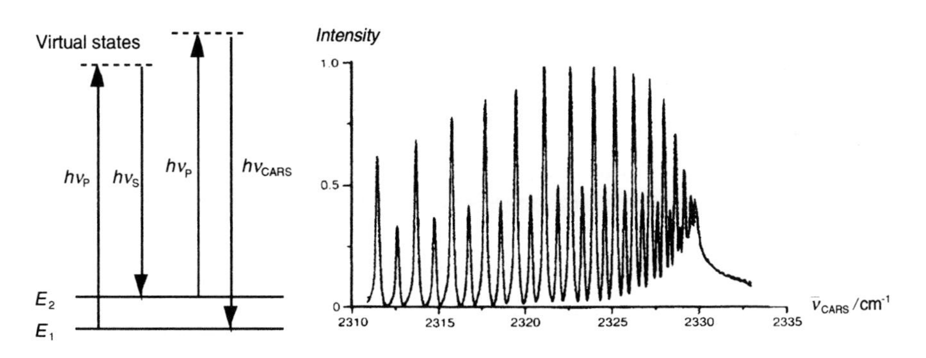

Fig. 2.4. Illustration of a CARS process (left); the experimental spectrum $I = f(\bar{\nu}_{CARS})$ of N_2 at $T = 1826$ K was obtained by scanning ν_S; $\bar{\nu} = \nu/c$ (Farrow et al. 1982)

Laser Induced Fluorescence (LIF): Laser light of tunable frequency is used in this nonintrusive method to excite selectively electronic states in molecules (see, e. g., Wolfrum 1986). During the transition the electronic structure of the molecules is changed. The difference in energy between ground state and excited states usually is relatively large. Thus, energetic light (UV-light, ultraviolet light) has to be used. The excited states return to states with lower energy, accompanied by a radiation of light (*fluorescence*, see Fig. 2.5) emitted with different energies and, thus, frequencies ν_{LIF}.

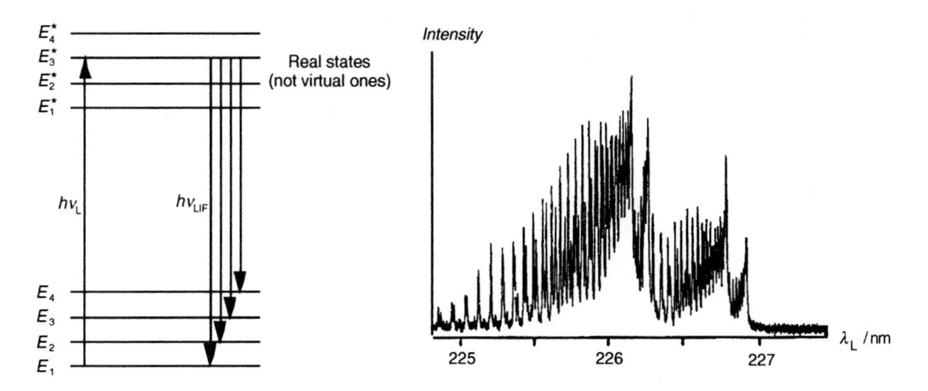

Fig. 2.5. Basic processes in laser induced fluorescence (left), E_i^* = vibrational levels of the electronically excited molecules, E_i = vibrational levels in the electronic ground state ν_L = laser frequency, ν_{LIF} = fluorescence light frequencies; the excitation spectrum $I = f(\lambda_L)$ of NO at $T = 300$ K was obtained by scanning ν_L; $\lambda = c/\nu$ (Westblom and Aldén 1989)

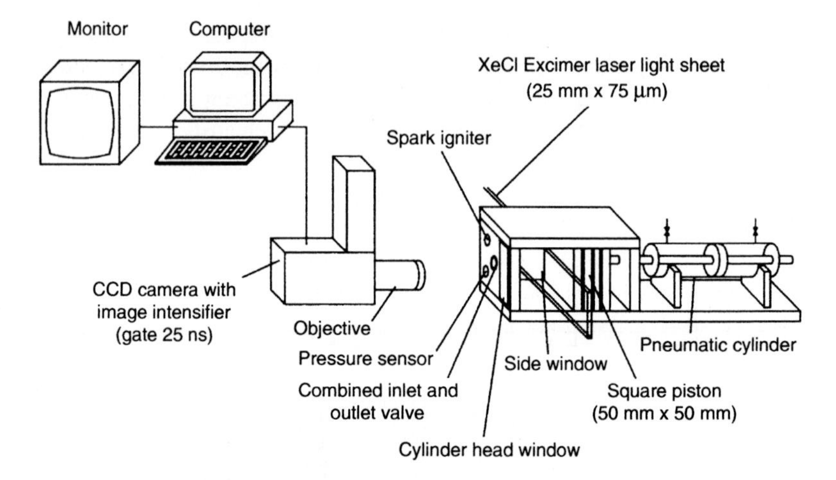

Fig. 2.6. Experimental setup for LIF-spectroscopy with a two-dimensional light sheet (Becker et al. 1991). The light sheet passes the side windows of a square piston engine simulator, LIF is measured at the cylinder head.

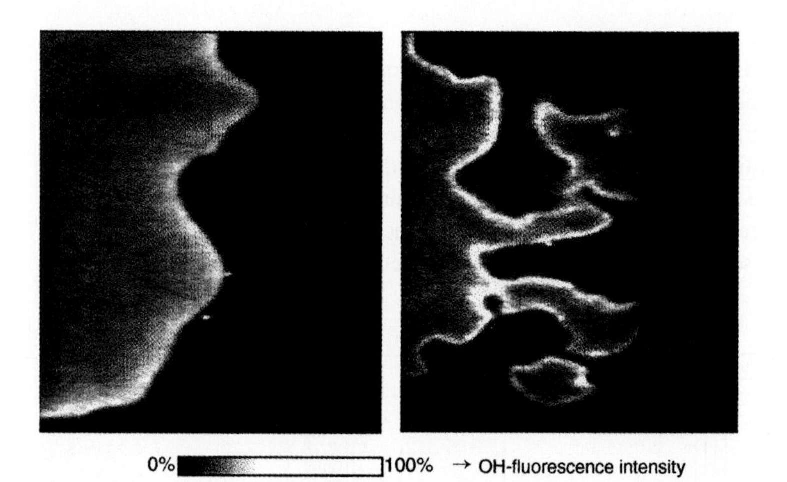

Fig. 2.7. Two-dimensional laser light sheet (about 50 µm thickness) measurements (OH-LIF; see Fig. 2.6) in an Otto test-engine at lower (left hand side) and higher (right hand side) turbulence; the higher turbulence case is leading to distinct flame quenching which is responsible for unburnt hydrocarbon emission (Becker et al. 1991)

With an apparatus such as depicted in Fig. 2.3, the overall fluorescence can be recorded while the pump laser is tuned (*excitation spectra*) or the spectrally resolved fluorescence can be recorded while holding the pump-laser frequency fixed (*fluorescence spectra*). Advantages of LIF are the high sensitivity and selectivity, because the fluorescence scattering cross-section is typically a million times larger than the

Rayleigh scattering cross-section. A barrier is that laser-accessable electronic transitions have to exist in the molecules or atoms. Due to the high sensitivity of LIF, many intermediate species, which only occur in low concentrations, can be measured (e. g., H, O, N, C, OH, CH, CN, NH, HNO, SH, SO, CH_3O etc.).

The possibility of spatially two-dimensional measurements is advantageous. In this case a thin *light sheet* is used, leading to good spatial resolution (even three-dimensional measurements are possible by using several light sheets). The fluorescence is measured by a two-dimensional detector (*photo diode array*), stored electronically, and the data are evaluated subsequently (see Figs. 2.6 and 2.7); see Hanson (1986).

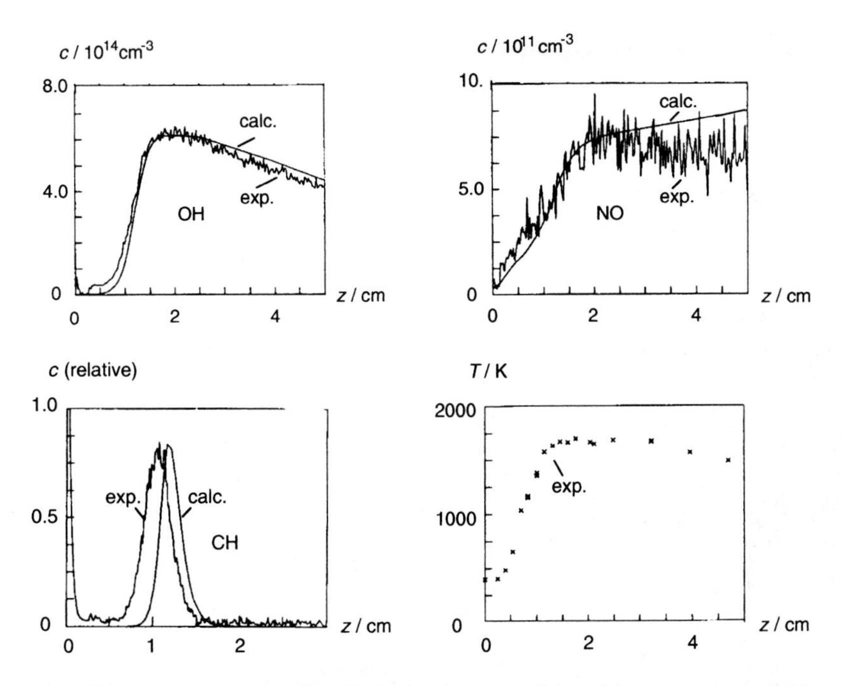

Fig. 2.8. LIF measurements of profiles of temperature, OH and NO absolute particle concentrations, and CH relative particle concentration (no calibration available for CH) in a premixed laminar flat CH_4-air flame at $p = 40$ mbar (Heard et al. 1992). The calculations are done with a reaction mechanism similar to that presented in Chapter 6.

Special fields of application are instantaneous measurements of turbulent flames (see Chapters 12-14) with short light pulses (high temporal resolution). Figure 2.7 shows two-dimensional laser light sheet measurements (OH-LIF) in an Otto test-engine (Becker et al. 1991), which clearly show the turbulent nature of the combustion. If fluorescent molecules (NO_2, NO, CO, acetone, acetaldehyde) are added to the fuel (or air), additional information about regions with or without fuel, temperatures, etc. can be obtained (e. g., Paul et al. 1990; Tait and Greenhalgh 1992; Lozano et al. 1992; Bazil and Stepowski 1995, Wolfrum 1998)

Usually, these measurements are qualitative in nature, because a calibration of concentration measurements is very difficult due to collisional quenching (relaxation of the molecules after collisions with other molecules, see, e. g., Wolfrum 1986). A lot of calibration data has to be provided for a quantitative evaluation. Nevertheless, quantitative measurements can be done in some cases, as demonstrated in Fig. 2.8. Shown are absolute concentrations of OH and NO, and relative concentrations of CH. Even the relative concentration profile provides valuable information on the shape and position of the CH-radical profile.

2.4 Temperature Measurements

Thermocouples: Temperature fields can be measured very easily by thermocouples which are pairs of junctions between different metals. A voltage, which is approximately proportional to the temperature difference between the two junctions, is induced (*thermoelectric effect*). Usually different metal combinations are used depending on the temperature range (for instance, platinum/platinum-rhodium or tungsten/ tungsten-molybdenum).

The major disadvantage is that the method measures the temperature of the metal-metal junction, which then has to be related to the surrounding gas temperature. Thus, the temperature measured (i. e., that of the metal) can be very different (hundreds of K) from that of the gas. An energy balance will include catalytic reaction at surface of the thermocouple, conductive heat losses via the wires to the (usually) ceramic support, radiation from the wire, and conduction and convection from the gas phase to the wire (see Fristrom 1995). Nevertheless, the method is easy to use and inexpensive and can be applied for qualitative measurements. A comparison between thermocouple and optical thermometry is presented in Rumminger et. al (1996).

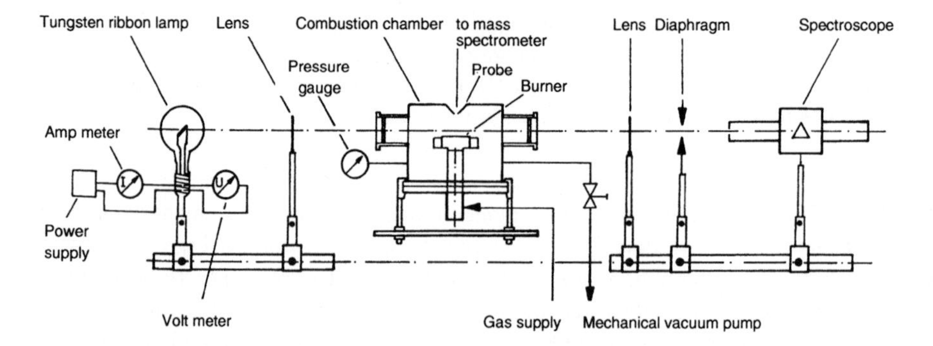

Fig. 2.9. Experimental setup for the measurements of temperatures (Na line-reversal) and concentrations (probe sampling, mass spectroscopy, OH-absorption) in a laminar flat premixed low pressure flame (Warnatz et al. 1983)

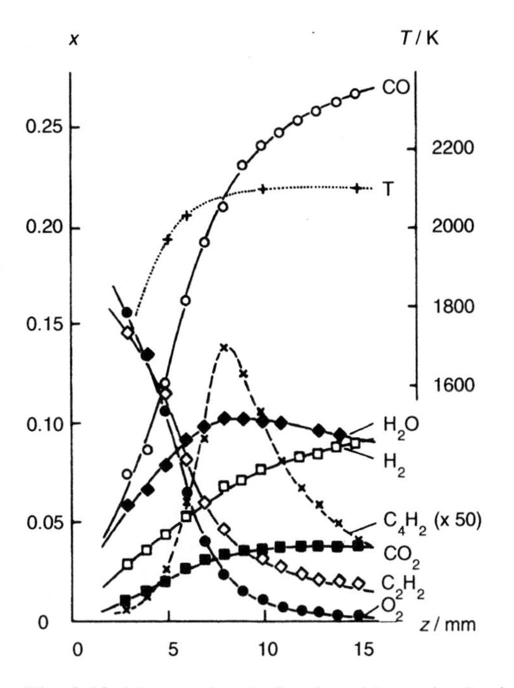

Fig. 2.10. Measured mole fractions (determined using mass spectroscopy) and temperatures (determined using Na-line reversal) in a laminar flat premixed low-pressure acetylene/oxygen/argon flame (Warnatz et al. 1983)

Na-Line-Reversal-Method: Here sodium-containing compounds are added to the reactants. Sodium atoms can absorb or (at elevated temperatures) radiate yellow light. The yellow Na-emission of added Na in comparison with a *black body* disappears if the Na-atoms have exactly the same temperature as the black body. If the temperature is higher, more light is emitted than absorbed; if the temperature is lower, more light is absorbed than emitted. The advent of diode array detectors greatly improved the detection of adsorption or emission (see Gaydon and Wolfhard 1979). Fig. 2.9 shows an experimental setup, and Fig. 2.10 shows temperature measurements using this method together with concentration measurements, obtained by mass spectroscopy in a rich laminar premixed acetylene-oxygen-argon low-pressure flame.

CARS Spectroscopy: Besides species concentrations, temperatures can be measured by CARS spectroscopy even more accurately. High resolution spectra are compared with simulated spectra which can be derived theoretically from molecular properties of the species under consideration. Temperature and concentration are varied in the simulation until best agreement with measurement.

Advantages are, again, the high spatial (about 1 mm^3) and temporal (about 10 ns) resolution, disadvantages are the high costs and the complicated evaluation of the data, which is mainly due to the nonlinear dependence of the signal on laser intensity and on species concentration (Sick et al. 1991).

Laser-Induced Fluorescence: The selective excitation of different energy levels in molecules (e. g., in OH-radicals) can be used to determine the population distribution of the energy levels. Assuming a Boltzmann distribution, temperatures can be determined (see, e. g., Eckbreth 1996, Thorne 1988). Care has to be taken to account for loss of energy in the laser beam as well as *self adsorption* of the fluorescent light as it emerges from the zone of laser excitation (Rumminger et al. 1996).

In some cases, fluorescent compounds are added to the flow. Seitzmann et al. (1985) added NO and performed LIF thermometry on this molecule. They showed that the NO had negligible effect on the combustion process.

2.5 Pressure Measurements

In unconfined, subsonic-flow flames (such as candles, torches, and flares) the pressure is nearly constant. In confined flows (such as occurs in combustion chambers and furnaces) the pressure is often steady with a slight gradient of pressure maintaining the subsonic flow and accounts for acceleration of the flow. These average pressures are suitably measured with conventional liquid or electronic *manometers*.

Quite often in combustion, the pressure varies in time; examples include piston engines and pulse combustors. These changing pressures are usually measured with wall-mounted *piezoelectric pressure transducers*. These are quartz crystals which change their electrical properties depending on the deformation by pressure variations. The implicit assumption is that the pressure measured at the wall is close to that away from the wall. This assumption is suitable when the time for the pressure changes is long compared to that for a sound wave to traverse the chamber. Even though the pressure is uniform, there can be large variations in density, because the temperature behind a flame front is typical much larger than the temperature ahead of the flame front. In spite of these large variations in density, the flow is still incompressible as long as the square of the Mach number M is small compared to unity, i. e., a flow with $M^2 << 1$ is incompressible, even if there are large variations in density.

When the Mach number approaches or exceeds unity, the flow is *compressible*. Optical measurement of pressure is desirable in compressible flows since spatial and temporal variations in pressure, velocity, temperature, and density are substantial and probes greatly perturb the flow. Pressure is often inferred from measurement of temperature and density, using the ideal gas equation of state (for example, see Hanson et al. 1990, and McMillin et al. 1993).

2.6 Measurement of Particle Sizes

In multi-phase combustion systems (spray combustion, coal combustion, soot formation etc.) not only velocity, temperature, and concentrations, but also the size and the distribution of fuel particles (coal particles or droplets) are desired. In this case, too, laser spectroscopic methods can be applied. Usually Mie scattering (see Section 1.1) is used, i. e., the scattering of light by particles, which are larger than the wavelength of the light (Arnold et al. 1990a, Subramanian et al. 1995). Fuel sprays and coal particles are typically in the range of 1 to 100 microns diameter. The diffraction and scattering of a beam of laser light by a cloud of such particles can be used to infer a size distribution. This technique is called *ensemble diffraction* and is the basis of successful commercial instruments (e. g. Harville and Holve 1997).

For liquids, a technique based on the refraction of light through the droplet (a lens effect) is called the *phase Doppler technique*. A review of the technique is given by Brena de la Rosa et al. (1992) and by Bachalo (1995). This technique examines single particle events rather than ensembles. Another single particle counter that works with both liquids and particles is based on the near-forward scattering of light as particles transit a laser beam (Holve and Self 1979a,b). Furthermore, special LIF-based techniques can be taken to measure size and distribution of droplets (Brown and Kent 1985).

When particle size is less than 1 μm, as it is the case for soot, cigarette smoke, dust, viruses, one usually has to use collective scattering since scattering from a single particle is weak and, in the Rayleigh limit, is proportional to d^6. A change in particle diameter from 0.1 μm to 10 nm leads to a decrease in scattering intensity by 1 million. A reliable measure of the particulate volume fraction is obtained from *line-of-sight extinction* of a laser beam as it penetrates through a cloud of particles (Hodkinson 1963, Flower and Bowman 1986). Also, as in filtered Rayleigh scattering from gases, the broadening of the scattered light, caused by the Brownian motion of the particles, can be used to infer particle size (*Dynamic light scattering* or *Diffusion Broadening Spectroscopy*, Penner et al. 1976a,b).

2.7 Simultaneous Diagnostics

The introduction to this chapter stressed that interaction between experiment and numerical simulation is optimal for model building. This interaction appears unbalanced. The numerical simulations predicts many things while, by comparison, experiments measure only a few physical properties. Experience has shown that a model can usually be improved to obtain precise agreement with a few measurements of a single parameter. Far more challenging is the comparison of simulations with several parameters, even if the measured accuracy of each parameter is moderate. For example, a model of a nonpremixed jet flame is more challenged by moderate accuracy profiles of velocity (say by LDV) and density (say by laser Rayleigh scattering) than by a highly accurate measure of either profile by itself.

The apparatus depicted in Fig 2.11 combines Raman and fluorescent scattering of simultaneous measurement of major and minor species in flames. The arrays of multiple scalars obtained from this apparatus have been the basis for much model improvements (Barlow 1998).

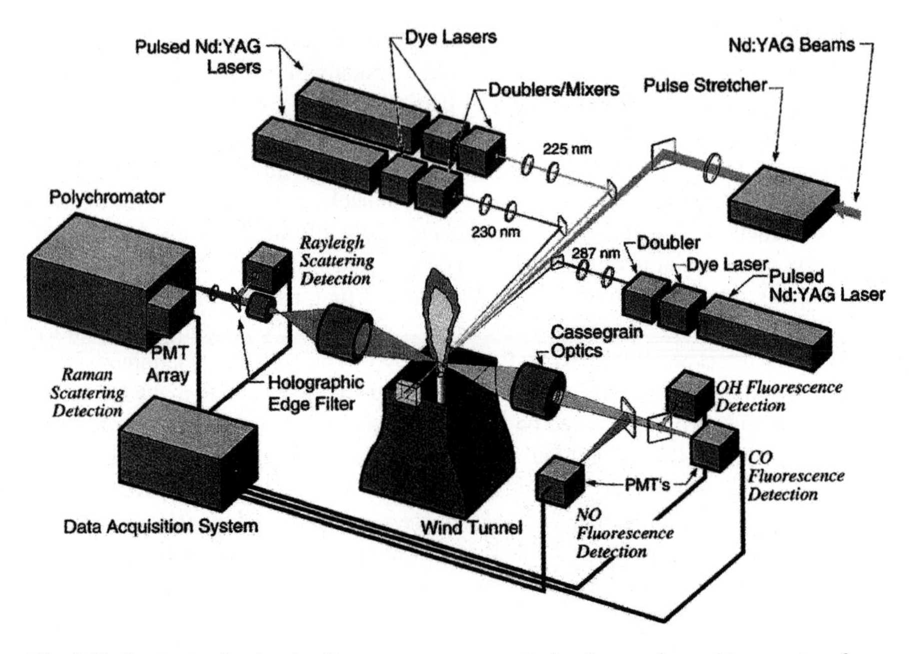

Fig. 2.11. Apparatus for the simultaneous measurement of major species and temperature from Raman scattered light pumped at 532 nm (see Fig. 2.3), of CO LIF pumped at 230 nm, of NO LIF pumped at 225 nm, and OH LIF pumped at 287 nm (Barlow 1998)

2.8 Exercises

Exercise 2.1. The Chief Scientist of a small company is asked about feasibility of using laser Raman scattering to determine the content of air entering and leaving a patient on an operating table in a hospital. The initial proposal was to use infrared absorption by O_2, N_2, CO_2, and (laughing gas) N_2O used as an anesthesia (mole fractions: 0.79, 0.20, 0.01, and 100 ppm, respectively). He points out that oxygen and nitrogen are homonuclear and hence have no infrared absorption, but they are Raman-active. Looking at the vibrational Raman cross sections, he is tempted to use an ultraviolet laser, but consultation with the company lawyers suggests additional safety (read: liability) problems since laser beams in ultraviolet cannot be seen. He decides on a "reliable" solid-state laser, the doubled Nd:YAG delivering 8 W at 532 nm (green light). Recalling that Magre and Dibble (1988) obtained 3000 photoelectrons per Joule of light from nitrogen in room air at normal conditions (298 K, 1 bar), the laser-beam length observed was 1 mm. Estimate the time required to measure, with a standard deviation $\sigma = 3.16$ %, the following species. Assume the vibrational Raman scattering cross sections for all species to be the same as for N_2. Species: N_2, O_2, CO_2, and N_2O. Assume that spatial resolution can be relaxed to 8 mm of beam length and that the noise is Poisson distributed (*shot noise*), hence for n photo electrons $\sigma = \sqrt{n}$.

3 Mathematical Description of Premixed Laminar Flat Flames

If a chemically reacting flow is considered, the system at each point in space and time is completely described by specification of pressure, density, temperature, velocity of the flow, and concentration of each species. These properties can be changing in time and space. These changes are the result of fluid flow (called *convection*), chemical reaction, molecular transport (e. g., heat conduction, diffusion, and viscosity), and radiation. A mathematical description of flames therefore has to account for each of these processes (Hirschfelder et al. 1964).

Some properties in reacting flows are characterized by the fact that they are conserved. Such properties are the energy, the mass, and the momentum. Summation over all the processes that change the *conserved properties* leads to the conservation equations, which describe the changes in reacting flow; accordingly, these equations are often called the *equations of change*. These equations of change, which are discussed in detail in Chapter 11, are the general starting point for mathematical descriptions of chemically reacting flows. Because all systems are described by the conservation equations, the main difference from one system to another are the boundary conditions and physicochemical conditions.

For the purpose of exposing the concepts embodied in the equations, this chapter will develop these conservation equations for the relatively simple, but instructive, system of the laminar premixed flat flame (Hirschfelder and Curtiss 1949; Warnatz 1978a,b). The development will introduce concepts of thermodynamics, molecular transport, and reaction kinetics; these subjects are the topics of Chapters 4, 5, and 6.

3.1 Conservation Equations for Laminar Flat Premixed Flames

Laminar premixed flames on a flat burner constitute an instructive example for the mathematical treatment of combustion processes. As Fig. 3.1 illustrates, the burner is usually a porous disk, ~10 cm in diameter, through which premixed fuel and air flow. The gases emerge from the disk and flow into the flame, which appears as a luminous disk levitating a few mm above the porous disk.

If one assumes that the burner diameter is sufficiently large, effects at the edge of the burner can be neglected as an approximation. Well within the edges, a flat flame front is observed. The properties in this flame (e. g., temperature and gas composition) depend only on the distance from the burner, i. e., only one spatial coordinate (z) is needed for the description. The conservation equations for this flame shall now be derived.

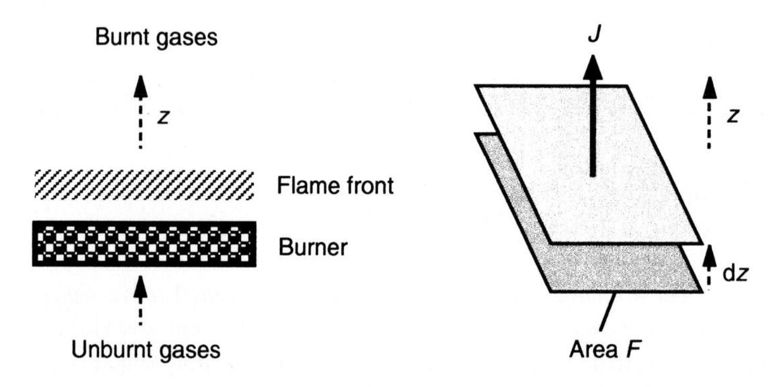

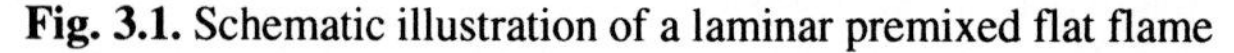

Fig. 3.1. Schematic illustration of a laminar premixed flat flame

The following assumptions will be made in order to simplify the treatment intended:

- The perfect gas law can be used ($p = c\,R\,T$; see Section 1.1).
- External forces (e. g., gravitation) are negligible.
- The system is continuous; the mean free path of the molecules is small compared to the flame thickness (a good assumption for most combustion problems).
- The pressure is constant (spatial or temporal pressure fluctuations are weak).
- The kinetic energy of the gas flow is negligible compared to other terms in the energy conservation equation (e. g., shock waves are not considered).
- The reciprocal thermal diffusion effect (*Dufour effect*) can be neglected (see below).
- Heat flux caused by radiation of gases and particles is negligible (this assumption is acceptable here when the flame is non-sooting).
- The system is in local thermal equilibrium.
- The flame is stationary, i. e., there are no temporal changes. (Formally, instationary equations are solved into stationarity for numerical reasons in the following.)

As will be seen below, these assumptions lead to reasonable predictions for laminar flat flames.

For any conserved variable E (z = spatial coordinate, t = time) in a one-dimensional system (see Fig. 3.1), the general relation

$$\frac{\partial W}{\partial t} + \frac{\partial J}{\partial z} = Q \tag{3.1}$$

holds, where W denotes the *density* of the conserved variable (= E/volume; in [E]/m^3), J a *flux* (more precisely *flux density*) of the conserved variable (= E/(surface·time); in [E]/(m^2·s) , and Q a *source* (or *sink*) of the conserved variable (= E/(volume·time)); in [E]/(m^3·s)). (3.1) is a statement that accumulation can be accomplished by influx (or outflux) and by a source (or sink). It will be shown in the following how the general equation (3.1) appears in the specific cases of conservation of mass, species, and enthalpy.

Overall mass *m* of the mixture: In the conservation of total mass, the density W in the conservation equation is given by the *total mass density* ρ (in kg/m^3). The flux J describes the movement of mass and is given as the product of density and the *mean mass velocity* (velocity of the center of mass,also called *flow velocity*), i. e., $J = \rho v$ (in kg/(m^2·s)). The source term in the mass conservation equation is zero, because chemical reactions neither create nor destroy mass ($Q = 0$). Insertion into (3.1) leads to

$$\frac{\partial \rho}{\partial t} + \frac{\partial(\rho v)}{\partial z} = 0 \,. \tag{3.2}$$

This equation is also called the *continuity equation* (here for one-dimensional systems).

Mass m_i of species *i*: Here the density W is given by the *partial density* ρ_i of species i, which denotes the mass of species i per unit volume ($\rho_i = m_i/V = (m_i/m)(m/V) = w_i\,\rho$). The flux J is given by the product of the partial density and the mass velocity v_i of the species i ($J = \rho_i\, v_i = w_i\, \rho\, v_i$) and has units of kg/(m^2·s).

In contrast to the conservation equation for the total mass (see above), this equation has a source term which describes the formation or consumption of species i in chemical reactions. This term is given by $Q = M_i\,(\partial c_i/\partial t)_{\text{chem}} = r_i$, where M_i denotes the molar mass of species i (in kg/mol), $(\partial c_i/\partial t)_{\text{chem}}$ the *chemical rate of production* of species i in chemical reactions (molar scale, units of mol/(m^3·s)), and r_i the chemical rate of production (mass scale, in kg/(m^3·s)). Together with (3.1) this leads to

$$\frac{\partial(\rho w_i)}{\partial t} + \frac{\partial(\rho w_i v_i)}{\partial z} = r_i \,. \tag{3.3}$$

The mass velocity v_i of the species i is composed of the mean mass velocity v of the center of mass of the mixture and a *diffusion velocity* V_i (relative to the center of mass), which is caused by molecular transport due to concentration gradients of the species i (discussed in Section 3.2 and Chapter 5),

$$v_i = v + V_i \,. \tag{3.4}$$

Simple transformations (product law for differentiation) of (3.3) lead then to

$$w_i \frac{\partial \rho}{\partial t} + \rho \frac{\partial w_i}{\partial t} + \rho v \frac{\partial w_i}{\partial z} + w_i \frac{\partial(\rho v)}{\partial z} + \frac{\partial j_i}{\partial z} = r_i \,,$$

where the symbol j_i denotes the *diffusion flux* of species i (in the center of mass system),

$$j_i = \rho w_i V_i = \rho_i V_i .$$

Together with (3.2), this equation simplifies to the species mass conservation equation

$$\rho \frac{\partial w_i}{\partial t} + \rho v \frac{\partial w_i}{\partial z} + \frac{\partial j_i}{\partial z} = r_i . \tag{3.5}$$

Enthalpy h of the mixture: In this case, the different terms in (3.1) are given by

$$\begin{aligned}
W &= \sum_j \rho_j h_j &&= \sum_j \rho w_j h_j && \text{J/m}^3 \\
J &= \sum_j \rho_j v_j h_j + j_q &&= \sum_j \rho v_j w_j h_j + j_q && \text{J/(m}^2\text{s)} \\
Q &= 0 && && \text{(energy conservation)} .
\end{aligned}$$

Here h_j denotes the *specific enthalpy* of species j (in J/kg) and j_q a *heat flux*, which corresponds to the diffusion flux j_i introduced above and is caused by transport of energy due to temperature gradients (see below). The term $\Sigma \rho_j v_j h_j$ describes the change of enthalpy due to the flow of species (composed of the mean mass velocity v and the diffusion velocity V_j). Insertion into (3.1) using $v_j = v + V_j$ yields

$$\sum_j \frac{\partial}{\partial z}(\rho v w_j h_j) + \sum_j \frac{\partial}{\partial z}(\rho V_j w_j h_j) + \frac{\partial j_q}{\partial z} + \sum_j \frac{\partial}{\partial t}(\rho w_j h_j) = 0 .$$

Using (3.3) and (3.4) one obtains for the first and fourth summands (T_1, T_4) that

$$\begin{aligned}
T_1 + T_4 &= \sum_j \left[\rho v w_j \frac{\partial h_j}{\partial z} + h_j \frac{\partial(\rho v w_j)}{\partial z} \right] + \sum_j \left[\rho w_j \frac{\partial h_j}{\partial t} + h_j \frac{\partial(\rho w_j)}{\partial t} \right] \\
&= \rho v \sum_j w_j \frac{\partial h_j}{\partial z} + \rho \sum_j w_j \frac{\partial h_j}{\partial t} + \sum_j h_j \left[\frac{\partial(\rho v w_j)}{\partial z} + \frac{\partial(\rho w_j)}{\partial t} \right] \\
&= \rho v \sum_j w_j \frac{\partial h_j}{\partial z} + \rho \sum_j w_j \frac{\partial h_j}{\partial t} + \sum_j h_j r_j - \sum_j h_j \frac{\partial j_j}{\partial z} .
\end{aligned}$$

For the second term (T_2) in the equation above, simple transformations lead to

$$T_2 = \sum_j \rho w_j V_j \frac{\partial h_j}{\partial z} + \sum_j h_j \frac{\partial(\rho w_j V_j)}{\partial z} .$$

Summation over all terms using $j_j = \rho w_j V_j$ yields the relation

$$\rho v \sum_j w_j \frac{\partial h_j}{\partial z} + \rho \sum_j w_j \frac{\partial h_j}{\partial t} + \sum_j h_j r_j + \sum_j j_j \frac{\partial h_j}{\partial z} + \frac{\partial j_q}{\partial z} = 0 . \tag{3.6}$$

The values for j_i and j_q (diffusion and heat flux) still have to be specified with respect to the properties of the mixture (pressure, temperature, composition). The empirical laws used to describe these relations are discussed in the next section.

3.2 Heat and Mass Transport

Empirical observations have established that concentration gradients lead to mass transport called *diffusion* and temperature gradients lead to heat transport called *heat conduction*. These empirical observations were later explained by the theory of irreversible thermodynamics (Hirschfelder et al. 1964). For the sake of brevity only the empirical laws are discussed here.

For the heat flux j_q, numerous measurements support the empirical *law of Fourier* in the form

$$j_q = -\lambda \frac{\partial T}{\partial z} \qquad \text{J/(m}^2\cdot\text{s)}, \qquad (3.7)$$

where λ denotes the *heat conductivity* of the mixture (in J/(K·m·s)). For the mass flux j_i one obtains an extended form of the *law of Fick* (which includes the first term only)

$$j_i = \frac{c^2}{\rho} M_i \sum_j M_j D_{ij} \frac{\partial x_j}{\partial z} - \frac{D_i^{\mathrm{T}}}{T} \frac{\partial T}{\partial z} \qquad \text{kg/(m}^2\cdot\text{s)}, \qquad (3.8)$$

where c denotes the molar concentration in mol/m^3; D_{ij} (units of m^2/s) are *multi-component diffusion coefficients*, x_j mole fractions, and D_i^{T} the *thermal diffusion coefficient* (in kg/(m·s)) of the species i based on the temperature gradient. Species transport caused by a temperature gradient (*thermal diffusion*) is also called the *Soret effect*. For many practical applications the simplified formula

$$j_i = -D_i^{\mathrm{M}} \rho \frac{w_i}{x_i} \frac{\partial x_i}{\partial z} - \frac{D_i^{\mathrm{T}}}{T} \frac{\partial T}{\partial z} \qquad (3.9)$$

is sufficiently accurate for the mass flux j_i. Here D_i^{M} denotes the diffusion coefficient for species i into the mixture of the other species (discussed in Chapter 5). For binary mixtures and for trace species ($w_i \rightarrow 0$) this simplified formulation is equivalent to (3.8). This assumption of strong dilution is reasonable if the oxidizer is air, because nitrogen is in excess in this case.

3.3 The Description of a Laminar Premixed Flat Flame Front

As stated above, for a complete description of laminar flat premixed flame fronts (Warnatz 1978a,b), the temperature T, the pressure p, the velocity υ and the partial densities ρ_i (i = 1,...,S for S species) or the overall density ρ and the S-1 linearly independent mass fractions $w_1,...,w_{S-1}$ ($w_S = 1 - w_1 - ... - w_{S-1}$) have to be known as functions of the spatial coordinate z. The following equations are available:

The pressure is assumed to be constant (see Section 3.1) and equal to the surrounding pressure. The density ρ can be calculated from temperature, pressure and composition using the perfect gas law (1.4).

The velocity υ is obtained from the continuity equation (3.2). Because the flame is assumed to be stationary (no temporal dependence), (3.2) reduces to

$$\partial(\rho\upsilon)/\partial z = 0 \quad \text{(therefore } \rho\upsilon = \text{const.)} \,. \tag{3.10}$$

Using the given mass flux $(\rho\upsilon)_u$ of the unburnt gases, υ can be calculated at each point in the flame.

The mass fractions w_i (i = 1 , ... , S) are determined by solving the S-1 species conservation equations, combined with the constraint that the mass factions sum to unity. Thermal diffusion, which is important for species with a small molar mass (H, H_2, He), is safely neglected here because the concentration of these species is rarely significant for this process to contribute. Then, introducing the diffusional mass flux $j_i = -D_i^M \rho(\partial w/\partial z)$ (simplified form of Eq. 3.9 for constant mean molar mass $\overline{M}$) into the species conservation (3.5) leads to

$$\rho\frac{\partial w_i}{\partial t} = \frac{\partial}{\partial z}\left(D_i^M \rho \frac{\partial w_i}{\partial z}\right) - \rho\upsilon\frac{\partial w_i}{\partial z} + r_i \,. \tag{3.11}$$

The temperature can be calculated from the energy conservation equation. Inserting the heat flux j_q (3.7) and using $c_{p,j}\mathrm{d}T = \mathrm{d}h_j$; $c_p = \Sigma w_j c_{p,j}$ (specific heat capacity of the mixture at constant pressure (in J/(kg·K); see Chapter 4) yields

$$\rho c_p \frac{\partial T}{\partial t} = \frac{\partial}{\partial z}\left(\lambda\frac{\partial T}{\partial z}\right) - \left(\rho\upsilon c_p + \sum_j j_j c_{p,j}\right)\frac{\partial T}{\partial z} - \sum_j h_j r_j \,. \tag{3.12}$$

Now all the equations required to solve the problem are given. After rearranging, they yield a partial differential equation system of the general form

$$\frac{\partial Y}{\partial t} = A\frac{\partial^2 Y}{\partial z^2} + B\frac{\partial Y}{\partial z} + C \,.$$

The numerical solution of this equation system will be discussed in Chapter 8 with special emphasis on the important consequences that the source term C (i. e., the reaction rates r_i) has for the solution method.

The terms in (3.11) and (3.12) are now discussed in detail. The term $\partial Y/\partial t$ denotes the temporal change of the variables Y at the spatial location z, the second derivatives describe the molecular transport (diffusion, heat conduction), the first derivatives describe the flow (in (3.12) $\Sigma j_j c_{p,j}$ is a correction, which accounts for transport of heat by diffusion of species), and the terms without derivatives describe the local changes due to chemical reaction (discussed in Chapter 7). The influence of the different terms can be best seen if selected simplified systems are considered, where some of the terms are negligible.

Example 3.1: Let a system be at rest without any chemical reactions (motionless inert mixture; $B = C = 0$). In this case the flow and the chemical production term vanish. If it is assumed that λ and $D_i^{\mathrm{M}} \cdot \rho$ do not depend on the location z, the result are the simple equations ($\partial^2 x_i / \partial z^2 = \partial^2 w_i / \partial z^2$ is assumed here, i. e., constant mean molar mass $\overline{M}$)

$$\frac{\partial w_i}{\partial t} = D_i^{\mathrm{M}} \frac{\partial^2 w_i}{\partial z^2} \quad \text{and} \quad \frac{\partial T}{\partial t} = \frac{\lambda}{\rho c_p} \frac{\partial^2 T}{\partial z^2} . \tag{3.13}$$

(Fick's second law) *(Fourier's second law)*

Both laws describe the broadening of profiles by diffusive processes, where the temporal change is proportional to the curvature (= 2nd derivative) of the profiles and finally leads to an equidistribution. Equations of the form (3.13) can be solved exactly (see Braun 1988). A particular solution of the diffusion equation, which demonstrates the broadening of the profiles, is shown in Fig. 3.2 and given by ($w_i^0 = w_i(t=0)$)

$$w_i(z,t) = w_i^0 \cdot \frac{1}{\sqrt{4\pi D t}} \cdot \exp\left(-\frac{z^2}{4 D t}\right) . \tag{3.14}$$

A concentration peak, which is located at $z = 0$ at time $t = 0$, distributes slowly over the whole space. In this example the profiles always have a Gaussian shape with a mean square intrusion length of $\overline{z^2} = 2\,Dt$.

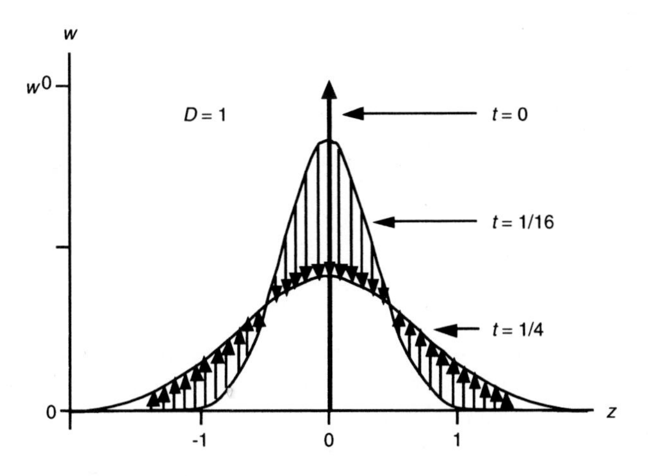

Fig. 3.2. Typical diffusion process (in dimensionless units)

Example 3.2: If no chemical reaction and no molecular transport is present in a system ($A = C = 0$), (3.11) or (3.12) yield

$$\frac{\partial Y}{\partial t} = -\upsilon \frac{\partial Y}{\partial z} \qquad (Y = w_i, T) \ . \tag{3.15}$$

This equation describes convection with a velocity υ. The temporal change at every point of the profile is proportional to its slope (= 1st derivative). This equation also can be solved analytically (see, e. g., John 1981), where the solution is given by the expression

$$Y(z,t) = Y(z-\upsilon{\cdot}t, 0) \ .$$

This means that within the time t the profile moves by a distance $\upsilon{\cdot}t$. The shape of the profile (describing a *travelling wave*) does not change during this process (Fig. 3.3).

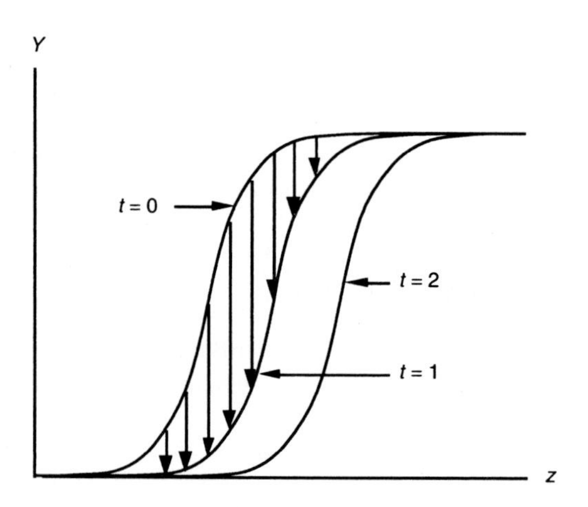

Fig. 3.3. Illustration of a convective process

Example 3.3: A third simplified case results if a system at rest without any transport processes is treated ($A = B = 0$). One obtains

$$\frac{\mathrm{d}w_i}{\mathrm{d}t} = \frac{r_i}{\rho} \quad \text{and} \quad \frac{\mathrm{d}T}{\mathrm{d}t} = -\frac{\sum h_j r_j}{\rho c_p} \ . \tag{3.16}$$

These *rate laws* of chemical kinetics describe the concentration changes during chemical reactions and the resulting heat of reaction. In order to solve these ordinary differential equation systems, the properties of r_i have to be known. This demands availability of data on the large number of chemical processes occurring and is discussed in detail in Chapter 8. In chemical engineering, this limiting case is known as a *batch reactor* or, if time is transformed to distance by a constant velocity, a *plug-flow reactor*.

Example 3.4: An important problem that has far-reaching implications toward improved understanding of flows with both convection and diffusion, and to turbulent flows in particular, is the opposed flow demonstrated in Fig. 3.4a. Flow from the left impinges onto flow from the right. Both flows splay out to the top and to the bottom; the geometry can be two-dimensional planar or axisymmetric. For the purpose of illustration, assume that the flow is nonreactive ($C = 0$) with constant density, diffusivity, and temperature, and with concentration c^+ from the right and c^- from the left. Along the line of symmetry from right to left (r=0, z), the lack of scalar gradients in the r-direction reduces the problem to one dimension that is described by (3.11) and (3.12), with the reaction term and the time-dependent term set to zero,

$$D\frac{\partial^2 c}{\partial z^2} - \upsilon\frac{\partial c}{\partial z} = 0 \ . \tag{3.17}$$

In the steady-state solution, convection across the layer is balanced by diffusion. If u and υ represent the velocity in the r- and z-direction, respectively, then for frictionless, steady-state, plane potential flow of an incompressible fluid, the velocity distribution is $u = a \cdot r$ and $\upsilon = -a \cdot z$. There is an analytic solution to (3.17),

$$c(z) = c^- + \frac{c^+ - c^-}{2}\left\{1 + \mathrm{erf}\left(z\sqrt{a/(2D)}\right)\right\} , \tag{3.18}$$

which is plotted in Fig. 3.4b. Note that increasing the ratio $a/2D$ results in a steeper gradient, grad c, with attendant increase in the mixing rate.

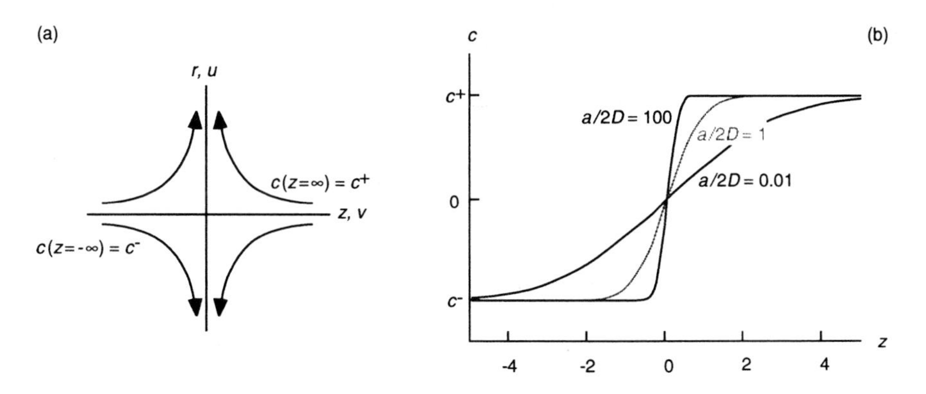

Fig. 3.4. (a) Schematic description of the 1D opposed-jet flow and (b) the solution (3.18)

Example 3.5: The conservation equations given above allow (together with the necessary transport coefficients, chemical reaction data, and thermodynamic properties) a complete description of temperature and concentration profiles in flat flames as functions of the distance z from the burner (shown, e. g., in Fig. 3.1). Such calculated profiles can be compared with corresponding experimental results such as those discussed in Chapter 2). A typical example of a very rich acetylene (C_2H_2)-oxygen flame at reduced pressure is shown in Fig. 3.5.

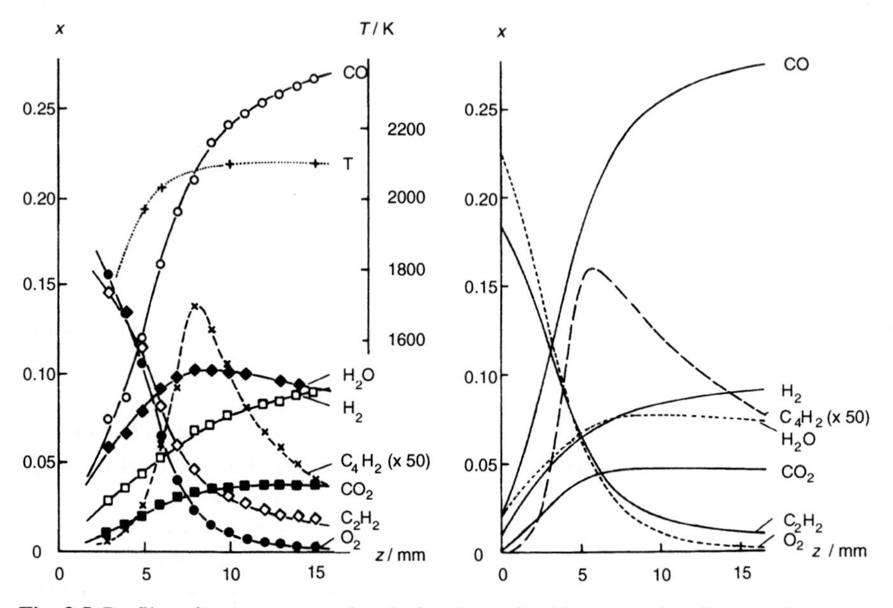

Fig. 3.5. Profiles of temperature and mole fractions of stable species in a flat acetylene-oxygen flame (diluted with argon) at low pressure. Left: experimental results (see Chapter 2); right: calculated profiles (details discussed in Chapter 8). In these calculations the energy conservation equation has not been solved; instead, the experimentally determined temperature profile has been used (Warnatz et al. 1983).

3.4 Exercises

Exercise 3.1: Determine the position of the flame front of an ethanol droplet with a constant diameter of 30 μm burning in the surrounding air. Use the following assumptions: The reaction is infinitely fast, i. e., the flame front is infinitely thin and is located at $\Phi = 1$, the stoichiometric contour somewhere between fuel and air. The reaction is given by the equation

$$C_2H_5OH + 3\,O_2 \rightarrow 2\,CO_2 + 3\,H_2O\,.$$

The diffusion coefficients and the densities are the same for all species. At the droplet surface the mass fraction of ethanol is $w_E = 0.988$.

Exercise 3.2: Calculate the velocity of the hot burnt gases of an adiabatic (no heat losses) flat premixed laminar flame, where the velocity of the unburnt gas is 35 cm/s, and the unburnt temperature is 25° C. Assume the temperature of the burnt gases to be 1700° C and the mole number to be invariant (as would be the case for a fuel-lean methane/air flame).

4 Thermodynamics of Combustion Processes

In Chapter 3, the example of a flat one-dimensional flame was used to show that several ingredients are needed for the solution of the conservation equations. One of these ingredients is the thermodynamic properties of each species, i. e., enthalpy H, entropy S, and heat capacities c_p of each species as a function of temperature and pressure. In this chapter it will be shown how H and S are generated and used. For example, one can predict the final temperature of a flame and the species composition at this final temperature using thermodynamics.

The science of thermodynamics largely emerged from 18th and 19th century efforts aimed at improving steam engines. From these many observations, three *fundamental laws of thermodynamics* were established. Although these laws were discovered through observations on engines and machines, they are of a fundamental nature that transcends their mechanical origins. The laws have widespread application, for example, in chemistry, biology, and cosmic physics. The purpose of this chapter is to review chemical thermodynamics as far as it applies to combustion; no complete treatment is attempted.

4.1 The First Law of Thermodynamics

The *first law of thermodynamics* is largely based on *Joule's experiments* (ca. 1860). Joule showed that mechanical work, applied to a thermally insulated system of water, would raise the temperature of the system under consideration. The temperature rise was the same for several ways of applying the same amount of mechanical work, such as stirring, metal-to-metal friction, and gas compression.

Thus, Joule showed that the unit of heat transfer was proportional to the unit of work. Through either route, the energy of the system is increased. The same energy unit for heat transfer and for work (the *Joule*) is used today in honor of these pioneering experiments.

The general formulation of the *first law of thermodynamics* dates back to Hermann von Helmholtz (ca. 1850). This law states that

> the sum of all energies is constant in an *isolated system* (no mass transfer and no energy transfer).

The change of the *internal energy* $\mathrm{d}U$ of a system is given by the sum of the heat δQ, transferred and the work δW done to the system,

$$\mathrm{d}U = \delta Q + \delta W\,. \tag{4.1}$$

Here the widely adopted and logical convention is used that energy added to the system is taken to be positive and energy taken from the system to be negative (Atkins 1996). Consequently, work done by the system is *negative* which is different from the awkward convention used in several textbooks where (4.1) is written in the form $\mathrm{d}U = \delta Q - \delta W$.

The two symbols d and δ for infinitesimal changes in (4.1) have special meanings: The symbol d denotes a differential change of a *state variable* Z, which depends only on the state of the system, but not on how the state was arrived at. For a state variable Z the cyclic integral along any path is identically zero,

$$\oint \mathrm{d}Z = 0\,. \tag{4.2}$$

For heat transfer or work, the cyclic integral may or may not be zero; therefore, the integral in general is path-dependent, and the symbol δ has to be used for a differential change in this case.

Work can be added to a system in various ways. A few examples are the following:

- *electrical work*, i. e., the work $e{\cdot}\mathrm{d}q$ required for a change of the charge $\mathrm{d}q$ in an electrical field e,
- *surface work*, i. e., the work $\sigma{\cdot}\mathrm{d}A$ required for a change of the surface $\mathrm{d}A$ with a surface tension σ,
- *lifting work* in a gravitational field, i. e., the energy $mg{\cdot}\mathrm{d}x$ needed to lift a mass m, by a height $\mathrm{d}x$ (with g = acceleration due to gravity),
- *compression* (or *expansion*) *work*, i. e., the energy $-p{\cdot}\mathrm{d}V$ needed to change the volume by $\mathrm{d}V$ of a gas with a pressure p (sometimes called *boundary work*).

The further treatment shall be limited to compression work only (no other kind of work is used in the following). The first law then reads

$$\mathrm{d}U = \delta Q - p\mathrm{d}V\,, \tag{4.3}$$

or

$$\mathrm{d}U = \delta Q \qquad \text{for } V = \text{const.} \tag{4.4}$$

Therefore, the change of the internal energy U equals the heat transferred at constant volume.

Very often chemical processes take place at constant pressure. It is convenient to define a state function, called the *enthalpy* H, as

$$H = U + pV \tag{4.5}$$

or

$$\mathrm{d}H = \mathrm{d}U + p\mathrm{d}V + V\mathrm{d}p\,. \tag{4.6}$$

Insertion of (4.3) leads to

$$dH = \delta Q + Vdp \tag{4.7}$$

or

$$dH = \delta Q \qquad \text{for } p = \text{const.} \tag{4.8}$$

4.2 Standard Enthalpies of Formation

Internal energy changes – and thus enthalpy changes; see (4.6) – can be measured according to (4.4) in *calorimeters*, e. g., in a *combustion bomb*. This is a *closed* (no mass transfer) constant volume tank; see Fig. 4.1. A chemical compound is mixed with oxygen and then burnt, usually at high pressure to guarantee complete reaction. The combustion bomb is placed in a large water bath which is thermally isolated with respect to the surroundings. From the heat δQ transferred to the water bath during the reaction, the internal energy change dU can be determined (calibration by electrical heating; see (4.13) and (4.14) below). Only changes in internal energy (and enthalpy), not absolute values, can be measured.

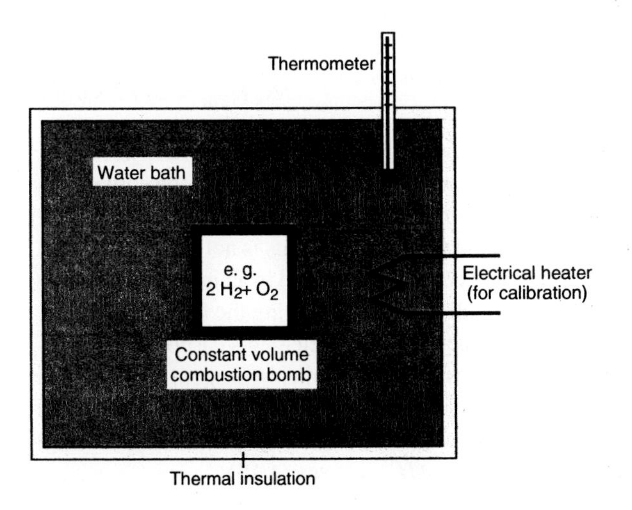

Fig. 4.1. Schematic illustration of a combustion bomb

The reaction $2\,H_2 + O_2 = 2\,H_2O$, e. g., can be rewritten as $2\,H_2O - 2\,H_2 - O_2 = 0$. In a general formulation, a chemical reaction is given by

$$\nu_1 A_1 + \nu_2 A_2 + ... + \nu_S A_S = 0 \qquad \text{or} \qquad \sum_{i=1}^{S} \nu_i A_i = 0\,, \tag{4.9}$$

where A_i denote the chemical symbols and ν_i the *stoichiometric coefficients* ($\nu_i > 0$

for products, $\nu_i < 0$ for reactants). For the reaction $2\ H_2 + O_2 = 2\ H_2O$, e. g., one obtains the symbols and stoichiometric numbers

$$A_1 = H_2,\ A_2 = O_2,\ A_3 = H_2O,\ \nu_1 = -2,\ \nu_2 = -1,\ \nu_3 = +2\ .$$

The change of the internal energy or the enthalpy in a chemical reaction (*reaction energy* and *reaction enthalpy*, respectively) is given by the sum of the internal energies or the enthalpies times the corresponding stoichiometric coefficients,

$$\Delta_R H = \Sigma \nu_i \cdot H_i \tag{4.10}$$

$$\Delta_R U = \Sigma \nu_i \cdot U_i\ . \tag{4.11}$$

Although absolute values of H and U cannot be determined in this way, one could determine the enthalpy of any species relative to the atomic elements through (4.10). One could define these atomic elements to have zero enthalpy, but it is convention to define the reference state of zero enthalpy to each of the pure elements in their most stable state at $T = 298.15$ K and $p = 1$ bar (the *standard state*).

A definition is necessary for each chemical element because one element cannot be transformed by chemical reaction into another. Using the convention given above, absolute enthalpies can be introduced for every chemical compound by the definition: The *standard enthalpy of formation* $\Delta \overline{H}^0_{f,298}$ of a compound is the reaction enthalpy $\Delta_R \overline{H}^0_{298}$ of its formation reaction from the pure elements in their most stable state at the temperature $T = 298.15$ K and the pressure $p = 1$ bar (signified by "o").

Example: $1/2\ O_2(g) \rightarrow O(g)$ $\Delta_R \overline{H}^0_{298} = 249.2$ kJ/mol

From the definition of the standard enthalpy of formation it follows: $\Delta \overline{H}^0_{f,298}(O,g) =$ 249.2 kJ/mol. The bar usually denotes molar values, i. e., in this case, the enthalpy of one mole of oxygen atoms (see next section).

Usually, direct formation of a compound from the elements is not practicable, but, because the enthalpy is a state function, it can be determined in an indirect way. This method dates back to Hess (1840) and shall be explained using the formation of ethylene (C_2H_4). Ethylene is not easily produced from the elements carbon and hydrogen, but the heats of reaction of the combustion of graphite, hydrogen, and ethylene can be determined very easily. If the three reaction equations and the enthalpies of formation are added, the standard enthalpy of formation of ethylene $\Delta \overline{H}^0_{f,298}(C_2H_4,g)$ = 52.1 kJ/mol is obtained by $\Delta_R \overline{H}^0_{298} = \sum \nu_i \Delta \overline{H}^0_{f,298,i}$; see (4.10):

No.	Reaction					$\Delta_R \overline{H}^0_{298}$ (kJ/mol)
(1)	2 C(Graphite)+	2 O_2(g)	=	2 CO_2(g)		-787.4
(2)	2 H_2(g)	+ O_2(g)	=	2 H_2O(l)		-571.5
(3)	2 CO_2(g)	+ 2 H_2O(l)	=	C_2H_4(g)	+ 3 O_2(g)	+1411.0
(1)+(2)+(3)	2 C(Graphite)+	2 H_2(g)	=	C_2H_4(g)		+52.1

The symbols in brackets denote the state of aggregation, where g means gaseous and l liquid (the enthalpies of these aggregation states differ by the heat of vaporization). Examples for the standard enthalpies of formation of some compounds are in Tab. 4.1.

Tab. 4.1. Standard enthalpies of formation and standard entropies (discussed below) of some compounds (Stull and Prophet 1971, Kee et al. 1987, Burcat 1984):

Compound		$\Delta\overline{H}^0_{f,298}$ (kJ/mol)	$\overline{S}^0_{298}$ (J/mol·K)
Oxygen	O_2(g)	0	205.04
Oxygen atoms	O(g)	249.2	160.95
Ozone	O_3(g)	142.4	238.8
Hydrogen	H_2(g)	0	130.57
Hydrogen atoms	H(g)	218.00	114.60
Water vapor	H_2O(g)	-241.81	188.72
Water	H_2O(l)	-285.83	69.95
Hydroxyl radicals	OH(g)	39.3	183.6
Nitrogen	N_2(g)	0	191.50
Nitrogen atoms	N(g)	472.68	153.19
Nitrogen monoxide	NO(g)	90.29	210.66
Nitrogen dioxide	NO_2(g)	33.1	239.91
Graphite	C(s,Graphite)	0	5.74
Diamond	C(s,Diamond)	1.895	2.38
Carbon	C(g)	716.6	157.99
Carbon monoxide	CO(g)	-110.53	197.56
Carbon dioxide	CO_2(g)	-393.5	213.68
Methane	CH_4(g)	-74.85	186.10
Ethane	C_2H_6(g)	-84.68	229.49
Ethylene	C_2H_4(g)	52.10	219.45
Acetylene	C_2H_2(g)	226.73	200.83
Propane	C_3H_8(g)	-103.85	269.91
Benzene	C_6H_6(g)	82.93	269.20
Methanol	CH_3OH(g)	-200.66	239.70
Ethanol	C_2H_5OH(g)	-235.31	282.00
Dimethylether	CH_3OCH_3(g)	-183.97	266.68

4.3 Heat Capacities

When heat is transferred to a system, the temperature changes. The *heat capacity* C of a system describes the temperature change $\mathrm{d}T$ resulting from the heat transfer δQ via

$$C = \delta Q/\mathrm{d}T\,. \tag{4.12}$$

The heat capacity C depends on the conditions during the heat addition. If the system is at constant pressure, heat transfer not only increases the temperature, but also increases the volume energy pV through expansion of the system boundary. Consequently, the heat capacity at constant pressure C_p is larger than the heat capacity at constant volume C_V.

Using the special cases (4.4) and (4.8) of the 1st law of thermodynamics one obtains

$$V = \text{const.:} \qquad \mathrm{d}U = \delta Q = C_V \mathrm{d}T, \tag{4.13}$$

$$p = \text{const.:} \qquad \mathrm{d}H = \delta Q = C_p \mathrm{d}T. \tag{4.14}$$

At a given temperature and pressure, C_V and C_p can be measured by heat transfer to an insulated system (e. g., electrically heating a wire inside the system) and measuring the resulting temperature change. In some cases, C_V, and hence C_p, can be calculated using statistical thermodynamics (e. g., discussed by Atkins 1996); these calculations are a subject outside the scope of this book. A typical example of heat capacity vs. temperature plot is shown in Fig. 4.2, which will be discussed later in Section 4.10.

With the knowledge of C_V or C_p ($C_p = C_V + nR$ for ideal gases by combination of 4.5, 4.13, 4.14 and the ideal gas law), one can generate U and H at every temperature using (4.13) and (4.14). Integration yields

$$V = \text{const.:} \qquad U_T = U_{298} + \int_{298\,\mathrm{K}}^{T} C_V \mathrm{d}T' \tag{4.15}$$

$$p = \text{const.:} \qquad H_T = H_{298} + \int_{298\,\mathrm{K}}^{T} C_p \mathrm{d}T' . \tag{4.16}$$

U, H and C depend on the amount of substance (or mole number); they are extensive values. However, it is generally advantageous to calculate in terms of intensive values. Therefore, molar and specific values are defined. *Molar values* describe internal energy, enthalpy, heat capacity, etc. per mole. They are characterized by an overbar,

$$\overline{C} = C/n; \qquad \overline{U} = U/n; \qquad \overline{H} = H/n \qquad \text{etc.}$$

Specific values describe heat capacity, internal energy, enthalpy, etc. per unit mass (e. g., 1kg). They will usually be characterized by small letters (m = total mass of the system),

$$c = C/m; \qquad u = U/m; \qquad h = H/m \qquad \text{etc.}$$

4.4 The Second Law of Thermodynamics

Many physicochemical processes do not violate the first law of thermodynamics, but never occur in nature. Two bodies with different temperatures will relax to a common temperature, if heat exchange between them is allowed. The opposite process is not possible; it will never happen that one body increases its temperature while the other cools down. The *second law of thermodynamics* results from the observation that

> a process which only withdraws heat from a cold body and transfers it to a warmer body is impossible.

Another (equivalent) formulation of the second law states that mechanical energy can be transformed completely into heat, but heat cannot be transformed completely into mechanical energy. Thus, the second law of thermodynamics contains information about the direction of thermodynamic processes and also puts a limit on the ultimate efficiency of heat engines.

A thermodynamic process is called *reversible* if the system can return to its initial state without any change of the system environment. For such processes it is necessary and sufficient that the system is in local equilibrium (examples are vaporization and condensation). In *irreversible* processes the system can only return to its initial state if the environment of the system changes (e.g., in combustion processes).

Because heat transfer Q to a system depends on the path of transfer, Q is not a state variable. However, there exists (this is another form of the second law of thermodynamics) an extensive state variable, namely the *entropy* S, with the properties

$$\mathrm{d}S = \frac{\delta Q_{\mathrm{rev.}}}{T} \quad \text{and} \quad \mathrm{d}S > \frac{\delta Q_{\mathrm{irrev.}}}{T} . \tag{4.17}$$

Here the index "rev." denotes a reversible and "irrev." an irreversible process. This relation is another equivalent formulation of the second law of thermodynamics. It follows for closed, thermally insulated systems ($\delta Q = 0$)

$$(\mathrm{d}S)_{\mathrm{rev.}} = 0 \qquad \text{or} \qquad (\mathrm{d}S)_{\mathrm{irrev.}} > 0 . \tag{4.18}$$

The change of entropy of a reversible process is obtained by integration of (4.17),

$$S_2 - S_1 = \int_1^2 \frac{\delta Q_{\mathrm{rev.}}}{T} . \tag{4.19}$$

When viewed with the advantage of statistical thermodynamics, entropy is seen as a measure of molecular chaos. Details can be found in textbooks on this topic (e. g., Atkins 1996, Tien and Lienhard 1971).

4.5 The Third Law of Thermodynamics

The second law (Section 4.4) describes the change of entropy in thermodynamic processes. Entropies, in contrast to enthalpies, have a zero point dictated by nature. The *third law of thermodynamics* identifies this zero point of entropy by

$$\lim_{T \to 0} S = 0 \quad \text{for ideal crystals of pure compounds} . \tag{4.20}$$

As in the case of internal energies and enthalpies, *standard entropies* S^0 are defined as entropies at the standard pressure. *Entropies of reaction* $\Delta_{\mathrm{R}} S$ are defined analogously to enthalpies of reaction by

$$\Delta_{\mathrm{R}} S = \sum_i \nu_i S_i . \tag{4.21}$$

For the temperature dependence of the entropy (4.13), (4.14), and (4.17) yield then

$$\mathrm{d}S = \frac{C_V}{T}\mathrm{d}T \quad \text{or} \quad S_T = S_{298\,\mathrm{K}} + \int_{298\,\mathrm{K}}^{T} \frac{C_V}{T'}\mathrm{d}T' \qquad (\text{rev.}, V=\text{const.}) \qquad (4.22)$$

$$\mathrm{d}S = \frac{C_p}{T}\mathrm{d}T \quad \text{or} \quad S_T = S_{298\,\mathrm{K}} + \int_{298\,\mathrm{K}}^{T} \frac{C_p}{T'}\mathrm{d}T' \qquad (\text{rev.}, p=\text{const.})\ . \qquad (4.23)$$

Tabulated values of $\overline{S}_{298}^0$ are shown in Table 4.1 to enable the determination of reaction enthalpies and reaction entropies.

4.6 Equilibrium Criteria and Thermodynamic Variables

If the heat transfer Q in (4.3) is replaced by the expression for the entropy (4.17), the inequality

$$\mathrm{d}U + p\mathrm{d}V - T\,\mathrm{d}S \le 0 \qquad (4.24)$$

is obtained as a combination of 1st and 2nd law of thermodynamics, where the equal sign corresponds to reversible and the "less than" sign to irreversible processes.

A reversible process can be imagined when a chemical system is at equilibrium, e. g.,

$$\mathrm{A + B + \ldots = C + D + \ldots}$$

The addition of an infinitesimal amount of A shifts the equilibrium to the right hand side, taking away an infinitesimal amount of A shifts the equilibrium to the left hand side (this is the *principle of le Chatelier*). One obtains the equilibrium condition

$$\mathrm{d}U + p\mathrm{d}V - T\mathrm{d}S = 0 \qquad (4.25)$$

or

$$(\mathrm{d}U)_{V,S} = 0\ . \qquad (4.26)$$

This equilibrium condition is inconvenient to use in practical applications, because the condition S = constant is difficult to achieve since entropy cannot be measured directly.

For the formulation of convenient equilibrium criteria, new thermodynamic state variables (analogous to enthalpy in Section 4.1) are generated from combination of state variables. If, in (4.25), $T\mathrm{d}S$ is replaced by $T\mathrm{d}S = \mathrm{d}(TS) - S\mathrm{d}T$, transformations lead directly to

$$\mathrm{d}(U - TS) + p\mathrm{d}V + S\mathrm{d}T = 0 \qquad (4.27)$$

or, defining $A = U - TS$, which is called *free energy* or *Helmholtz function*, to

$$(\mathrm{d}A)_{V,T} = 0\ . \qquad (4.28)$$

In a similar way, another useful state variable can be introduced by the transformation

$$d(U - TS + pV) - Vdp + SdT = 0 \tag{4.29}$$

such that the *free enthalpy* (or *Gibbs function*) $G = A + pV = H - TS$ enables a very convenient formulation for chemical equilibrium, namely

$$(dG)_{p,T} = 0\,. \tag{4.30}$$

4.7 Equilibrium in Gas Mixtures; Chemical Potential

The *chemical potential* μ_i of a compound i in a mixture is defined as the partial derivative of the free energy G with respect to the amount of substance (mole number) of the compound n_i,

$$\mu_i = \left(\frac{\partial G}{\partial n_i}\right)_{p,T,n_j}\,. \tag{4.31}$$

Here the indices indicate that p, T and all n_j (except for n_i) are held constant. For a pure compound one has of course the simple relation

$$\mu = \left(\frac{\partial G}{\partial n}\right)_{p,T} = \left(\frac{\partial (n\overline{G})}{\partial n}\right)_{p,T} = \overline{G}\,. \tag{4.32}$$

Now a specific expression for the chemical potential of a compound i in a gas mixture shall be derived. For T = const., Eq. (4.29) yields for the Gibbs energy

$$(dG)_T = Vdp\,. \tag{4.33}$$

Integration using the ideal gas law leads to ("0" again refers to the standard pressure)

$$G(T,p) = G^0(T) + \int_{p^0}^{p} V\mathrm{d}p' = G^0(T) + \int_{p^0}^{p} nRT\frac{\mathrm{d}p'}{p'} = G^0(T) + nRT\cdot\ln\frac{p}{p^0}\,. \tag{4.34}$$

Differentiation with respect to the amount of the compound n yields

$$\mu = \mu^0(T) + RT\cdot\ln(p/p^0)\,. \tag{4.35}$$

For an ideal mixture of gases one has accordingly (this will not be derived here in detail)

$$\mu_i = \mu_i^0(T) + RT\cdot\ln(p_i/p^0)\,. \tag{4.36}$$

Generalizing now the *total differential* (4.29) of the Gibbs energy of a pure compound

$$dG = Vdp - SdT\,,$$

to a mixture and using the definition for the chemical potential in an ideal gas mixture, one obtains

$$\mathrm{d}G = V\mathrm{d}p - S\mathrm{d}T + \sum_i \mu_i \mathrm{d}n_i \ . \tag{4.37}$$

Now a chemical reaction $\Sigma \nu_i \mathrm{A}_i = 0$ in the gas mixture is considered and the *reaction progress variable* ξ introduced (which does not depend on the total amount) by using the relation $\mathrm{d}n_i = \nu_i \mathrm{d}\xi$ (e. g., $\Delta\xi = 1$ for complete conversion according to the reaction equation). Then (4.37) at constant T and p in the equilibrium case ($\mathrm{d}G = 0$) leads to the result that

$$\Sigma \nu_i \mu_i = 0 \ . \tag{4.38}$$

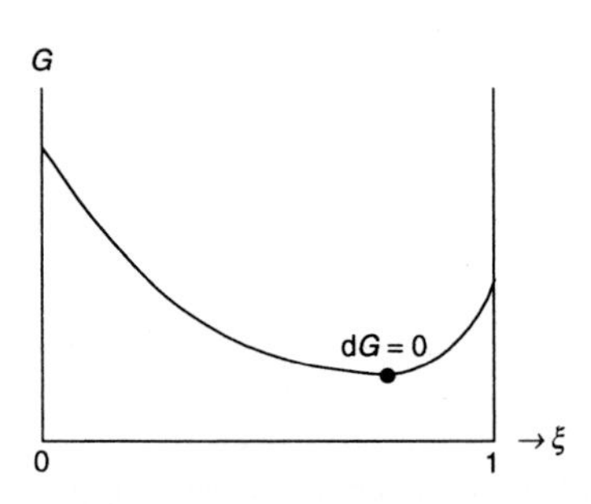

Fig. 4.2. Schematic illustration of the relationship between G and ξ

For a reacting gas mixture in chemical equilibrium, (4.35) can be substituted for the chemical potentials in (4.38); the result is

$$\sum_i \nu_i \mu_i^0 + RT \sum_i \nu_i \ln \frac{p_i}{p^0} = 0 \tag{4.39}$$

or

$$\sum_i \nu_i \mu_i^0 + RT \ln \prod_i \left(\frac{p_i}{p^0}\right)^{\nu_i} = 0 \ . \tag{4.40}$$

Noticing that, for an ideal gas, $\sum \nu_i \mu_i^0 = \sum \nu_i \overline{G}_i^0 = \Delta_\mathrm{R} \overline{G}^0 = \Delta_\mathrm{R} \overline{H}^0 - T \cdot \Delta_\mathrm{R} \overline{S}^0$ is the molar Gibbs energy of the chemical reaction considered, and introducing the *equilibrium constants* K_p and K_c, respectively, of the reaction as

$$K_p = \prod_i \left(\frac{p_i}{p^0}\right)^{\nu_i} \quad \text{and} \quad K_c = \prod_i \left(\frac{c_i}{c^0}\right)^{\nu_i} , \tag{4.41}$$

one obtains two thermodynamic relations very important for the following considerations, i. e.,

$$K_p = \exp(-\Delta_\mathrm{R} \overline{G}^0 / RT) \ \text{ and } \ K_c = \exp(-\Delta_\mathrm{R} \overline{A}^0 / RT) \ . \tag{4.42}$$

Quantitative statements about the equilibrium composition of gas mixtures are now possible using (4.41). (4.42) gives an expression allowing calculation of the equilibrium constant from thermodynamic data ($\overline{A}^0$ = free energy corresponding to $\overline{G}^0$).

4.8 Determination of Equilibrium Compositions in Gases

In this section the calculation of equilibrium compositions of the burnt gases in combustion processes will be described. Ethylene-oxygen combustion will be used as a representative example. In principle, a minimization of G has to be done; software to solve this problem is available (Gordon and McBride 1971, Reynolds 1986).

Choice of the Chemical System: First, the number S of different compounds in the reaction system must be determined. All species relevant to the system have to be considered. If necessary, the system can be extended to describe trace species as well.

Example: In order to describe the relevant species in the burnt gas of a stoichiometric C_2H_4-O_2 mixture one needs (if trace species are neglected) the compounds

$$CO_2, CO, H_2O, H_2, O_2, O, H, OH \qquad (S = 8) .$$

An adiabatic flame temperature $T_b = 2973$ K will be assumed (its determination is discussed in Section 4.9). Hydrocarbons (e. g., the fuel C_2H_4) are present in only very small amounts in the burnt gas of a stoichiometric mixture (to be verified later). In fuel-rich mixtures only CH_4 (methane) and C_2H_2 (acetylene) have to be accounted for. If the oxidizer is air, N_2 and (if desired) pollutants like NO and HCN can be added.

Determination of the Components of the System: Each mixture of S species (compounds) has a certain small number K of *components* (e. g., the chemical elements). These components are conserved and cannot be changed into each other by chemical reaction. A set of components is a minimum set of species to build up the species system under consideration.

Example: In the C_2H_4-O_2 system there are $K = 3$ different elements, i. e., components: C, H, and O. Nevertheless, for the example considered here, CO will be used as carbon-containing and H_2 and O_2 as hydrogen- and oxygen-containing components (the reason for doing so is that the corresponding mole or mass fractions have reasonable values in contrast to that for the elements).

It should be noted that care has to be taken in the choice of the components (e. g., CO_2 cannot be used as a component for carbon and oxygen at the same time, because in this case the amounts of those components cannot be varied independently).

Determination of the Independent Reactions: The compounds in the system which have not been chosen as components can change due to chemical reactions. Therefore, exactly $R = S - K$ independent chemical equilibrium conditions like (4.38) have to be specified,

$$\sum_{i=1}^{S} \nu_{ij}\, \mu_i = 0 \qquad ; \qquad j = 1,\ldots, R \; . \tag{4.43}$$

If less than R reactions are given or if some reactions are not independent, the system is underdetermined. If more than R reactions are specified, the system is overdetermined. The number of linearly independent reactions in the system (4.43) has to equal R exactly. This number corresponds to the rank of the matrix with the elements ν_{ij}.

Example: $R = S - K = 5$; the following equation system is linearly independent because the bold species occur only in one equation, respectively:

$$\begin{array}{lllll} & CO_2 & = \mathbf{CO} & + 1/2\, O_2 & K_{p,1} \\ H_2 & + 1/2\, O_2 & = \mathbf{H_2O} & & K_{p,2} \\ 1/2\, H_2 & + 1/2\, O_2 & = \mathbf{OH} & & K_{p,3} \\ & 1/2\, H_2 & = \mathbf{H} & & K_{p,4} \\ & 1/2\, O_2 & = \mathbf{O} & & K_{p,5} \end{array}$$

Formulation of the Equation System: For a given temperature and total pressure, the system is described by the S partial pressures p_i for which S equations are needed. The first one is obtained from the fact that the total pressure is the sum of the partial pressures,

$$\sum_{i=1}^{S} p_i = p \,. \tag{4.44}$$

Furthermore, the element composition of the K elements in the mixture is constant. Therefore the $K - 1$ ratios between the amounts of the different elements N_2/N_1, N_3/N_1, ..., N_K/N_1 are constant and equal to those in the initial mixture $c_{2/1}$, $c_{3/1}$, ..., $c_{K/1}$,

$$N_i/N_1 = c_{i/1} \quad ; \quad i = 2, ..., K \,. \tag{4.45}$$

The conditions (4.44) and (4.45) form a set of K linear equations (see below); the remaining R equations necessary for solution are the equilibrium conditions (4.43),

$$p_j = K^*_{p,j} \cdot \prod_{i=1}^{K} p_i^{\nu_{ij}} \quad ; \quad j = K+1, ..., S \,, \tag{4.46}$$

where $K^*_{p,j}$ denotes an equilibrium constant or its reciprocal value. The equations (4.46) are usually nonlinear.

Example: For (4.43-4.46) one obtains

$$N_H/N_{CO} = c_{H/CO} : 2\,p_{H_2O} + 2\,p_{H_2} + p_{OH} + p_H - c_{H,CO}(p_{CO_2} + p_{CO}) = 0$$

$$N_O/N_{CO} = c_{O/CO} : 2\,p_{CO_2} + 2\,p_{O_2} + p_{CO} + p_{H_2O} + p_{OH} + p_O - c_{O,CO}(p_{CO_2} + p_{CO}) = 0$$

$$\text{total pressure} : p_{CO_2} + p_{CO} + p_{H_2O} + p_{O_2} + p_{H_2} + p_{OH} + p_H + p_O - p_{total} = 0$$

equilibrium conditions:

$$p_{CO_2} = K_{p,1}^{-1} \cdot p_{CO}\sqrt{p_{O_2}} \qquad p_{H_2O} = K_{p,2} \cdot p_{H_2}\sqrt{p_{O_2}} \qquad p_{OH} = K_{p,3}\sqrt{p_{O_2}\, p_{H_2}}$$

$$p_H = K_{p,4}\sqrt{p_{H_2}} \qquad p_O = K_{p,5}\sqrt{p_{O_2}} \,.$$

Solution of the Equation system: Nonlinear equation systems are usually solved by Newton's method (see textbooks on numerical mathematics).

Example: $T = 2700\ ^0C$; $p = 1$ bar; $c_{H/C} = 2$; $c_{O/C} = 3$; $K_{p,1} = 0.498$; $K_{p,2} = 5{,}00$; $K_{p,3} = 1.06$; $K_{p,4} = 0.363$; $K_{p,5} = 0.304$. Starting estimates are $p_{CO}^{(0)} = p_{O_2}^{(0)} = p_{H_2}^{(0)} = 0.1$ bar. The solution then is: $p_{H_2}^{(5)} = 0.092$ bar, $p_{O_2}^{(5)} = 0.103$ bar, and $p_{CO}^{(5)} = 0.210$ bar.

4.9 Determination of Adiabatic Flame Temperatures

In a closed, adiabatic system ($\delta Q = 0$) at constant pressure, the first law of thermodynamics yields $\mathrm{d}H = 0$. Furthermore, the overall mass is constant.

Therefore, the unburnt gases (denoted by the index u) and the burnt gases (index b) have the same specific enthalpy. The molar enthalpies of the burnt and unburnt gases often differ, because the amount of molecules usually changes in a chemical reaction. Thus,

$$h^{(\mathrm{u})} = \sum_{j=1}^{S} w_j^{(\mathrm{u})} h_j^{(\mathrm{u})} = \sum_{j=1}^{S} w_j^{(\mathrm{b})} h_j^{(\mathrm{b})} = h^{(\mathrm{b})} . \tag{4.47}$$

Analogously to (4.16), for constant pressure the relation

$$h_j^{(\mathrm{b})} = h_j^{(\mathrm{u})} + \int_{T_\mathrm{u}}^{T_\mathrm{b}} c_{p,j} \, \mathrm{d}T \tag{4.48}$$

holds. Using this equation, the *adiabatic flame temperature* T_b can be determined, i. e., the temperature resulting after combustion provided that heat losses to the surrounding are negligible. T_b can be computed easily by a simple iteration method:

First the equilibrium composition and the enthalpies $h^{(1)}$, $h^{(2)}$ are calculated for two temperatures which are lower and higher, respectively, than the flame temperature expected ($T_1 < T_\mathrm{b}$ and $T_2 > T_\mathrm{b}$). Then the composition and the specific enthalpy $h^{(\mathrm{m})}$ at the mean temperature $T_\mathrm{m} = (T_2 + T_1)/2$ are calculated. If $h^{(\mathrm{u})}$ is between $h^{(1)}$ and $h^{(\mathrm{m})}$, one assigns $T_2 = T_\mathrm{m}$; otherwise the assignment $T_1 = T_\mathrm{m}$ is taken. This interval halving is continued until the result is sufficiently accurate. The reason for the applicability of this method is the monotonic dependence of the enthalpy on the temperature.

Examples of adiabatic flame temperatures T_b and the corresponding compositions x_b are given in Table 4.2 (Gaydon and Wolfhard 1979). Remark: The numbers refer to $T_\mathrm{u} = 298$ K, $p = 1$ bar; equilibrium concentrations can be negligible, but never zero.

Tab. 4.2. Adiabatic flame temperatures T_b and compositions (x_i) in stoichiometric mixtures

Mixture	:	H_2/air	H_2/O_2	CH_4/air	C_2H_2/air	C_2N_2/O_2
T_b [K]	:	2380.	3083.	2222.	2523.	4850.
H_2O		0.320	0.570	0.180	0.070	---
CO_2	:	---	---	0.085	0.120	0.000
CO	:	---	---	0.009	0.040	0.660
O_2	:	0.004	0.050	0.004	0.020	0.000
H_2	:	0.017	0.160	0.004	0.000	---
OH	:	0.010	0.100	0.003	0.010	---
H	:	0.002	0.080	.0004	0.000	---
O	:	.0005	0.040	.0002	0.000	0.008
NO	:	.0005	---	0.002	0.010	.0003
N_2	:	0.650	---	0.709	0.730	0.320

4.10 Tabulation of Thermodynamic Data

Thermodynamic data of a large number of species are tabulated as functions of the temperature (Stull and Prophet 1971, Kee et al. 1987, Burcat 1984). In the JANAF-Tables (Stull and Prophet 1971) the values of $\overline{C}_p^0$, $\overline{S}^0$, $-(\overline{G}^0 - \overline{H}_{298}^0)/T$, $\overline{H}^0 - \overline{H}_{298}^0$, $\Delta\overline{H}_f^0$, $\Delta\overline{G}_f^0$, and $\log K_p$ can be found for a variety of different species as function of the temperature. From these values all other thermodynamic functions can be computed. The $\log K_p$ for the formation from the elements is useful, because it allows one to calculate the free enthalpy of formation $\Delta\overline{G}_f^0(T)$; see (4.42). A JANAF-example of data is given in Table 4.3.

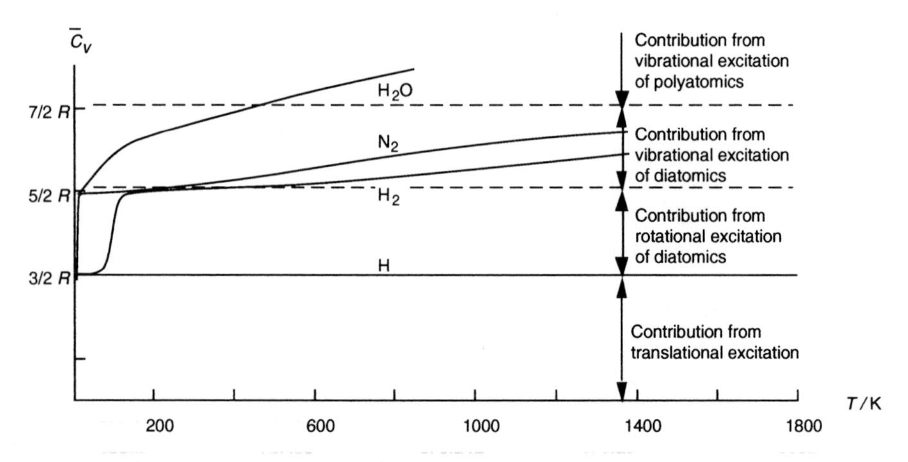

Fig. 4.3. Temperature dependence of the molar heat capacity of H, H_2, N_2, and H_2O

The temperature dependence of some heat capacities is shown in Fig. 4.3. At low temperatures, only the three translational degrees of freedom are excited and the molar heat capacity (at constant volume) is 3/2 R. At higher temperatures two rotational degrees of freedom (for diatomics; see Fig. 4.4) contribute so that the heat capacity obtains a value of 5/2 R. At even higher temperatures, the vibrational degrees of freedom are also excited, and the molar heat capacity approaches 7/2 R for diatomics and even more for polyatomics.

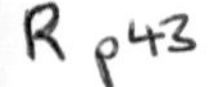

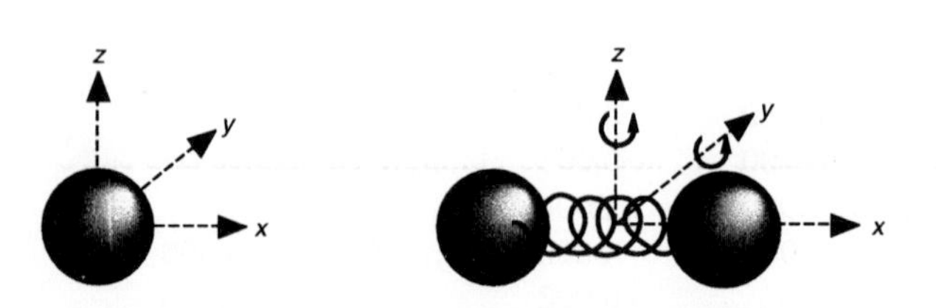

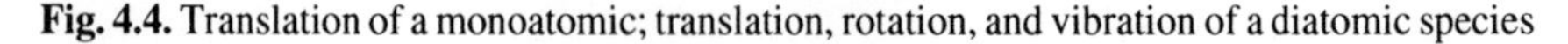
Fig. 4.4. Translation of a monoatomic; translation, rotation, and vibration of a diatomic species

Only a few of the tabulated values stem from calorimetric measurements. Most of the data is derived from theoretical calculations and spectroscopic data, leading to a higher accuracy. Unfortunately, satisfactory data is only available for a small number of compounds. Even for important species like CH_3, C_2H_3, and C_2H, which can be found in nearly any combustion system, there is still a lack of data (Baulch et al. 1991).

Tab. 4.3. Thermodynamic data of H_2O(g); $T_0 = 298.15$ K, $p^0 = 1$ bar (Stull and Prophet 1971)

Water (H_2O)	Ideal Gas	$M = 18.01528$ g·mol^{-1}	$\Delta \overline{H}^0_{f,298} = -241.826$ kJ·mol^{-1}	
T/K	$\overline{C}^0_p$/J·mol^{-1}·K^{-1}	$\overline{S}^0$/J·mol^{-1}·K^{-1}	$(\overline{H}^0 - \overline{H}^0_{298})$/kJ·mol^{-1}	log K_p
0	0.000	0.000	-9.904	INFINITE
100	33.299	152.388	-6.615	123.579
200	33.349	175.485	-3.282	60.792
298	33.590	188.834	0.000	40.047
300	33.596	189.042	0.062	39.785
400	34.262	198.788	3.452	29.238
500	35.226	206.534	6.925	22.884
600	36.325	213.052	10.501	18.631
700	37.495	218.739	14.192	15.582
800	38.721	223.825	18.002	13.287
900	39.987	228.459	21.938	11.496
1000	41.268	232.738	26.000	10.060
1100	42.536	236.731	30.191	8.881
1200	43.768	240.485	34.506	7.897
1300	44.945	244.035	38.942	7.063
1400	46.054	247.407	43.493	6.346
1500	47.090	250.620	48.151	5.724
1600	48.050	253.690	52.908	5.179
1700	48.935	256.630	57.758	4.698
1800	49.749	259.451	62.693	4.269
1900	50.496	262.161	67.706	3.885
2000	51.180	264.769	72.790	3.540
2100	51.823	267.282	77.941	3.227
2200	52.408	269.706	83.153	2.942
2300	52.947	272.048	88.421	2.682
2400	53.444	274.312	93.741	2.443
2500	53.904	276.503	99.108	2.223
3000	55.748	286.504	126.549	1.344
3500	57.058	295.201	154.768	0.713
4000	58.033	302.887	183.552	0.239
5000	59.390	315.993	242.313	-0.428

For computations, the tabulated thermodynamic data could be used with interpolation schemes. However, this requires handling of large data sets and time-consuming table-look-up algorithms. Therefore, the data is more often stored as polynomials in T. Usually the molar heat capacities $\overline{C}^0_p$ ($\overline{C}^0_p = \overline{C}^0_V + R$) are expressed as polynomials of fourth order in T,

$$\overline{C}^0_p / R \;=\; \overline{C}^0_{p,1} + \overline{C}^0_{p,2} \cdot T + \overline{C}^0_{p,3} \cdot T^2 + \overline{C}^0_{p,4} \cdot T^3 + \overline{C}^0_{p,5} \cdot T^4 \,. \qquad (4.49)$$

In addition, two integration constants are needed to compute enthalpies and entropies, where $\overline{C}^0_{p,6} \cdot R = \overline{H}^0_{298}$ and $\overline{C}^0_{p,7} \cdot R = \overline{S}^0_{298}$,

$$\overline{H}^0_T = \overline{C}^0_{p,6} \cdot R + \int_{T'=298\,\mathrm{K}}^{T} \overline{C}^0_p \,\mathrm{d}T' \quad \text{and} \quad \overline{S}^0_T = \overline{C}^0_{p,7} \cdot R + \int_{T'=298\,\mathrm{K}}^{T} \frac{\overline{C}^0_p}{T'} \,\mathrm{d}T' \,. \qquad (4.50)$$

In order to improve accuracy, usually two different polynomials are used for low ($T < 1000$ K) and high ($T > 1000$ K) temperatures. Examples of these polynomial coefficients are given in Table 4.4 for some species related to combustion.

Tab. 4.4. Examples of thermodynamic data in polynomial form (Kee et al. 1987, Burcat 1984); two sets of 7 polynomial coefficients are stored in rows 2 to 4

N2	J 3/77	G	300.000	5000.000	1
0.28532899E+01	0.16022128E-02	-0.62936893E-06	0.11441022E-09	-0.78057465E-14	2
-0.89008093E+03	0.63964897E+01	0.37044177E+01	-0.14218753E-02	0.28670392E-05	3
-0.12028885E-08	-0.13954677E-13	-0.10640795E+04	0.22336285E+01		4
CO	J 9/65	G	300.000	5000.000	1
0.29840696E+01	0.14891390E-02	-0.57899684E-06	0.10364577E-09	-0.69353550E-14	2
-0.14245228E+05	0.63479156E+01	0.37100928E+01	-0.16190964E-02	0.36923594E-05	3
-0.20319674E-08	0.23953344E-12	-0.14356310E+05	0.29555351E+01		4
CO2	J 9/65	G	300.000	5000.000	1
0.44608041E+01	0.30981719E-02	-0.12392571E-05	0.22741325E-09	-0.15525954E-13	2
-0.48961442E+05	-0.98635982E+00	0.24007797E+01	0.87350957E-02	-0.66070878E-05	3
0.20021861E-08	0.63274039E-15	-0.48377527E+05	0.96951457E+01		4
H2	J 3/77	G	300.000	5000.000	1
0.30667095E+01	0.57473755E-03	0.13938319E-07	-0.25483518E-10	0.29098574E-14	2
-0.86547412E+03	-0.17798424E+01	0.33553514E+01	0.50136144E-03	-0.23006908E-06	3
-0.47905324E-09	0.48522585E-12	-0.10191626E+04	-0.35477228E+01		4
H2O	J 3/79	G	300.000	5000.000	1
0.26110472E+01	0.31563130E-02	-0.92985438E-06	0.13331538E-09	-0.74689351E-14	2
-0.29868167E+05	0.72091268E+01	0.41677234E+01	-0.18114970E-02	0.59471288E-05	3
-0.48692021E-08	0.15291991E-11	-0.30289969E+05	-0.73135474E+00		4

4.11 Exercises

Exercise 4.1. (a) Determine the equilibrium constant K_p of the reaction $C_2H_4 + H_2 = C_2H_6$ at the temperature $T = 298$ K. (b) For the reaction in (a) determine the equilibrium composition (i. e., the partial pressures of the different species) at a temperature of 298 K and a pressure of 1 bar. The ratio between the amounts of carbon and hydrogen atoms shall be given by $c_{C,H} = 1/3$.

Exercise 4.2. Calculate the adiabatic flame temperature for the stoichiometric combustion of gaseous C_3H_8 with O_2. The occurrence of dissociation products like H, O, OH, ... shall be neglected, i. e., water and carbon dioxide shall be the only products. ($T_u = 298$ K, $p = 1$ bar, ideal gas). Assume

$$\overline{C}_p(H_2O) = \overline{C}_p(CO_2) = 71\ \mathrm{J/(mol \cdot K)} + (T - 298\mathrm{K}) \cdot 0.080\ \mathrm{J/(mol \cdot K^2)}\ .$$

5 Transport Phenomena

Molecular transport processes, i. e., *diffusion*, *heat conduction*, and *viscosity*, have in common that the corresponding physical properties are transported by the movement of the molecules in the gas. Diffusion is the mass transport caused by concentration gradients, viscosity is the momentum transport caused by velocity gradients, and heat conduction is the energy transport caused by temperature gradients. Additionally, there are other phenomena such as mass transport caused by temperature gradients (*thermal diffusion* or *Soret effect*) or energy transport caused by concentration gradients (*Dufour effect*). The influence of the latter is usually very small and is often neglected in the simulation of combustion processes. A detailed discussion of the transport processes can be found in the books of Hirschfelder et al. 1964 or of Bird et al. 1960.

5.1 A Simple Physical Model of the Transport Processes

A simple model for transport processes is obtained by considering two neighboring gas layers in a system (see Fig. 5.1). If there is a gradient $\partial q/\partial z$ of a property q in z-direction, the molecules at z have the mean property q and at $z + dz$ the mean property $q + (\partial q / \partial z)\,dz$. Their motion is in complete disorder (*molecular chaos*). The statistical velocity is given by a *Maxwell-Boltzmann distribution* stating that the particle number in a velocity interval $\Delta\upsilon$ is $N(\upsilon)\Delta\upsilon \propto \upsilon^2 \cdot \exp(-\upsilon^2/kT) \cdot \Delta\upsilon$ (for review see, e. g., Atkins 1996). The molecular motion causes some molecules to move from one layer to the other. Because the gas layers have different mean properties (momentum, internal energy, concentration), different mean amounts of momentum, energy and mass are transferred in both directions (layer 1 $\rightarrow$ layer 2 or layer 2 $\rightarrow$ layer 1). A continuous molecular exchange (called a *flux*) results. From the kinetic theory of gases it follows that the transport is faster if the mean velocity of the molecules is faster and if the mean free path of the molecules (this is the distance which a molecule travels before a collision with another molecule) is larger.

The simplified kinetic theory of gases is based on the assumption that the particles (atoms, molecules) are rigid spheres which interact in a completely elastic way. In reality, there are deviations from these assumptions. Molecules have a complicated structure, which is distinctly different from a spherical geometry. Furthermore, the model of elastic collisions assumes that the particles do not interact except during the collision, whereas in reality forces of attraction are present (e. g., *van-der-Waals forces*). The intermolecular potential describing the attractive or repulsive forces between molecules or atoms differs remarkably from the ideal potential of rigid spheres.

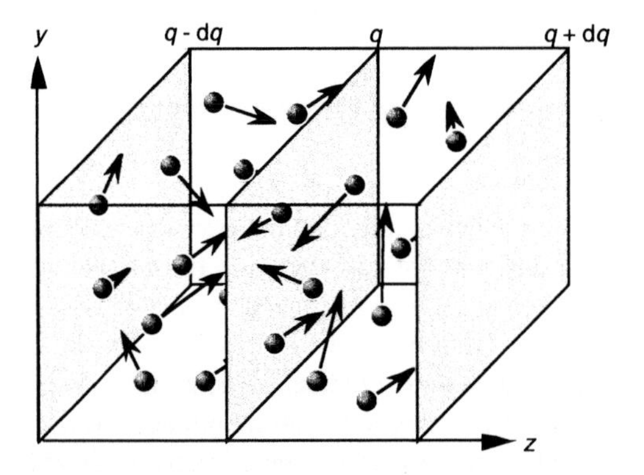

Fig. 5.1. Schematic illustration of two layers in a gas with slightly different properties with the gradient $\partial q/\partial z$ of the property q in z-direction

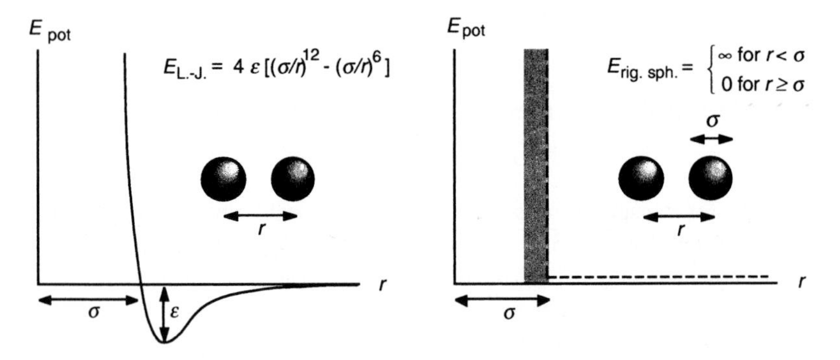

Fig. 5.2. Lennard-Jones-6-12 potential (left) and *rigid-sphere* potential (right)

In most cases, the intermolecular interactions can be described by a *Lennard-Jones-6-12-potential* (force $F = -\,\mathrm{d}E_{pot}/\mathrm{d}r$, see Fig. 5.2). This potential is characterized by the molecular diameter σ and the depth of the intermolecular potential ε (see Fig. 5.2). The parameters of some species are listed in Tab. 5.1. They are used to determine the *reduced collision integrals* that generate factors accounting for the deviation from the ideal model of rigid elastic spheres (*real gas behavior*). For example, the two constants a and b used in the *van-der-Waals real gas equation of state* $(p + a/\overline{V}^2)(\overline{V} - b) = RT$ are relatable to diameter σ ($\propto b^{-3}$) and well depth ε ($\propto a$).

Tab. 5.1. Molecular data for the calculation of transport coefficients for some species (Warnatz 1979, 1981a), $k = R/N_A$ = Boltzmann constant

Species	σ[nm]	ε/k [K]
H	0.205	145
O	0.275	80
H_2	0.292	38
O_2	0.346	107
N_2	0.362	97
H_2O	0.260	572
CO	0.365	98
CO_2	0.376	244
CH_4	0.375	140
C_2H_6	0.432	246
C_3H_8	0.498	267

A fundamental concept for the treatment of collisions of particles (atoms, molecules) is that of the *mean free path*, which is defined as the average distance of travel between two successive collisions. A specified molecule will then suffer a collision when its center approaches the distance σ from the center of any other molecule (see Fig. 5.3).

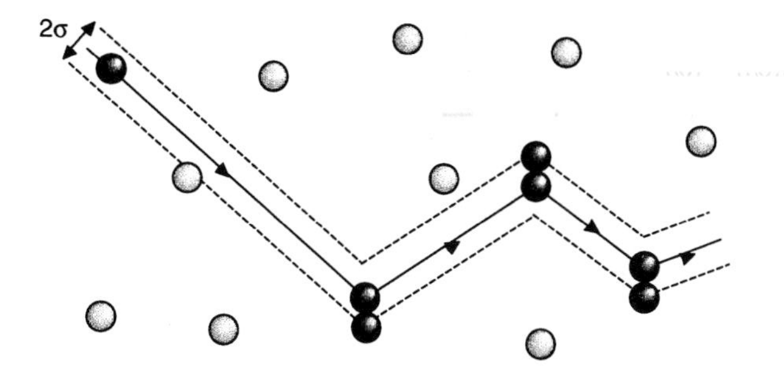

Fig. 5.3. Schematic illustration of the collision volume taken by a hard sphere particle colliding with like particles with diameter σ

For a quantitative treatment (see, e. g., Kauzmann 1966; the following arguments are not rigorous for the purpose of concise treatment) it shall be assumed that the particle A under consideration is moving at a mean velocity $\overline{v}$ and that the other particles A are standing still. The zigzag path of the particle is shown in Fig. 5.3. It is imagined now (ignoring violation of the collision laws) that this path is straightened to form a cylinder with the volume $V = \pi\sigma^2 \cdot \overline{v} \cdot \Delta t$, where Δt is the flight time and $\overline{v} \cdot \Delta t$ the cylinder length. The number of particles hit in this cylinder volume (or the *collision number*) is

$$N = \pi\sigma^2 \cdot \bar{\upsilon} \cdot \Delta t \cdot [n] \quad (= 5 \cdot 10^9\,\mathrm{s}^{-1} \text{ at } STP,\ 1\,\mathrm{atm} \text{ and } 273.15\,\mathrm{K}) \,, \tag{5.1}$$

where $[n] = c \cdot N_A$ is the particle number density (c = molar concentration, N_A = Avogadro's constant). The ratio of the distance $\bar{\upsilon} \cdot \Delta t$ travelled and of the collision number N is then the mean free path

$$l_{\mathrm{coll}} = \frac{\bar{\upsilon} \cdot \Delta t}{\pi\sigma^2 \cdot \bar{\upsilon} \cdot \Delta t \cdot [n]} = \frac{1}{\pi\sigma^2 \cdot [n]} \quad (= 60 \text{ nm at } STP) \,. \tag{5.2}$$

In the case of collisions of a single particle A with particles B, Eq. (5.1) has to be replaced by

$$N_{AB} = \pi\sigma_{AB}{}^2 \cdot \bar{\upsilon}_A \cdot \Delta t \cdot [n_B] \,, \tag{5.3}$$

where $2\sigma_{AB} = \sigma_A + \sigma_B$. The total number of collisions of all particles A with all others B at the relative velocity $\bar{\upsilon}_{AB} = (\bar{\upsilon}_A^2 + \bar{\upsilon}_B^2)^{1/2}$ is obtained by multiplication with the number $V \cdot [n_A]$ of particles A, giving the *collision rate per unit time and unit volume*

$$N_{AB} = \pi\sigma_{AB}^2 \cdot \bar{\upsilon}_{AB} \cdot [n_A] \cdot [n_B] \quad (\text{with } \pi\sigma_{AB}^2 \cdot \bar{\upsilon}_{AB} \approx 10^{14} \frac{\mathrm{cm}^3}{\mathrm{mol} \cdot \mathrm{s}} \text{ at } STP) \,. \tag{5.4}$$

5.2 Heat Conduction in Gases

For the transport of energy Q through area A, the empirically determined *Fourier law* of heat conduction states that the *heat flux density* j_q is proportional to the temperature gradient (Bird et al. 1960, Hirschfelder et al. 1964),

$$j_q = \frac{\partial Q}{\partial t \cdot A} = -\lambda \frac{\partial T}{\partial z} \quad [\mathrm{W/m^2}] \,. \tag{5.5}$$

This means that a heat flux occurs from a region of high temperature towards one with a lower temperature (see Fig. 5.4). The coefficient of proportionality λ is called *thermal conductivity*.

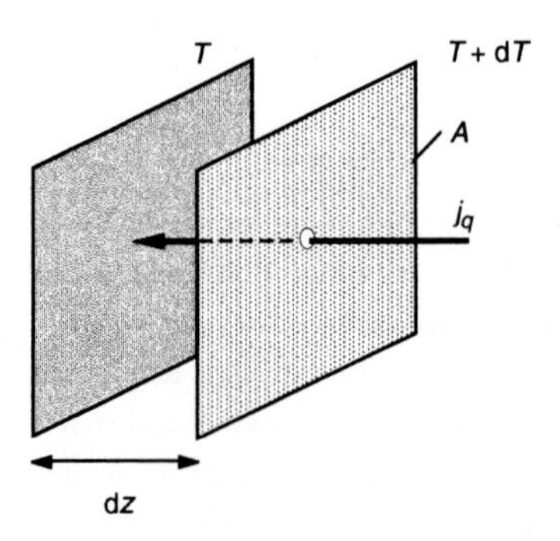

Fig. 5.4. Schematic illustration of a heat flux j_q caused by a temperature gradient

A device for the measurement of thermal conductivities usually consists of a hot wire or a heated cylinder, concentric to another cylinder at a different temperature. The gas under investigation is between these cylinders. Heat conduction in the gas changes the temperature of the two bodies, and the thermal conductivity can be calculated from the temperatures of the bodies.

The kinetic theory of rigid sphere gases (Bird et al. 1960, Hirschfelder et al. 1964) yields λ to be a product of particle density $[n]$, mean velocity $\bar{v}$, molecular heat capacity c_V, and mean free path l_{coll} (see Section 5.1), $\lambda \propto [n] \cdot \bar{v} \cdot c_V \cdot l_{\mathrm{coll}}$, leading to

$$\lambda = \frac{25}{32} \cdot \frac{\sqrt{\pi m k T}}{\pi \sigma^2} \cdot \frac{c_V}{m} , \tag{5.6}$$

where m = mass of the particle (atom, molecule) , $k = R/N_A$ = Boltzmann constant, T = absolute temperature, σ= particle diameter, and $c_V = \overline{C}_V / N_A$ = molecular heat capacity. The inclusion of real gas effects modifies this rigid sphere result by a factor $\Omega^{(2,2)*}$ (called the *reduced collision integral*), see Fig. 5.5,

$$\lambda = \frac{25}{32} \cdot \frac{\sqrt{\pi m k T}}{\pi \sigma^2 \Omega^{(2,2)*}} \cdot \frac{c_V}{m} = \frac{\lambda_{\text{rigid sphere}}}{\Omega^{(2,2)*}} , \tag{5.7}$$

If a Lennard-Jones-6-12-potential is assumed, the reduced collision integral $\Omega^{(2,2)*}$ is a unique function of the reduced temperature T^* which is the ratio of the absolute temperature T and the depth of the intermolecular potential, i. e., $T^* = kT/\varepsilon$. The temperature dependence of the collision integral $\Omega^{(2,2)*}$ is shown in Fig. 5.5; note that the correction can be as much as a factor of 2 (Hirschfelder et al. 1964).

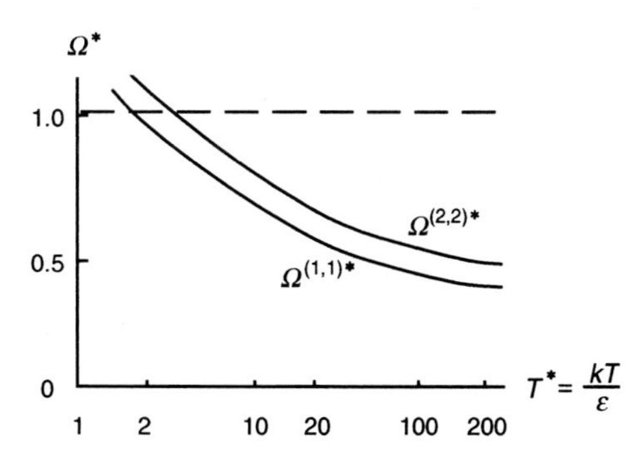

Fig. 5.5. Temperature dependence of $\Omega^{(2,2)*}$ and $\Omega^{(1,1)*}$ (see Section 5.4)

For practical purposes the more convenient formulation (for monoatomic gases with $c_V = 3/2\ k$)

$$\lambda = 8.323 \cdot 10^{-4} \frac{\sqrt{T/M}}{\sigma^2 \Omega^{(2.2)*}} \frac{\mathrm{J}}{\mathrm{m \cdot K \cdot s}} \tag{5.8}$$

is often used. In this equation, the values have to be inserted with the following units:

T in K, M in g/mol, σ in nm. One can see that the thermal conductivity is proportional to the square root of the temperature ($\lambda \sim T^{1/2}$) and does not depend on the pressure.

In combustion processes, the gas is composed of a large number of different species. In this case the thermal conductivity of the mixture has to be known. For gas mixtures the thermal conductivity can be calculated from the thermal conductivities λ_i and the mole fractions x_i of the pure species i with an accuracy of some 10-20% using the empirical law (Mathur et al. 1967)

$$\lambda = \frac{1}{2} \cdot \left[\sum_i x_i \lambda_i + \left(\sum \frac{x_i}{\lambda_i} \right)^{-1} \right]. \tag{5.9}$$

When greater accuracy is required (5-10%) the more complicated formula

$$\lambda = \sum_{i=1}^{S} \frac{\lambda_i}{1 + \sum_{k \neq i} x_k \cdot 1.065\, \Phi_{ik}} \tag{5.10}$$

can be used, where the correction factors Φ_{ik} depend on the coefficients of viscosity μ_i (see Section 5.3) and the molar masses M_i of the species i,

$$\Phi_{ik} = \frac{1}{2\sqrt{2}} \left(1 + \frac{M_i}{M_k} \right)^{-\frac{1}{2}} \cdot \left[1 + \left(\frac{\mu_i}{\mu_k} \right)^{\frac{1}{2}} \cdot \left(\frac{M_k}{M_i} \right)^{\frac{1}{4}} \right]^2. \tag{5.11}$$

5.3 Viscosity of Gases

Newton's law of viscosity states for the transport of momentum the *momentum flux density* j_{mv} is proportional to the velocity gradient (see, e. g., Bird et al. 1960, Hirschfelder et al. 1964),

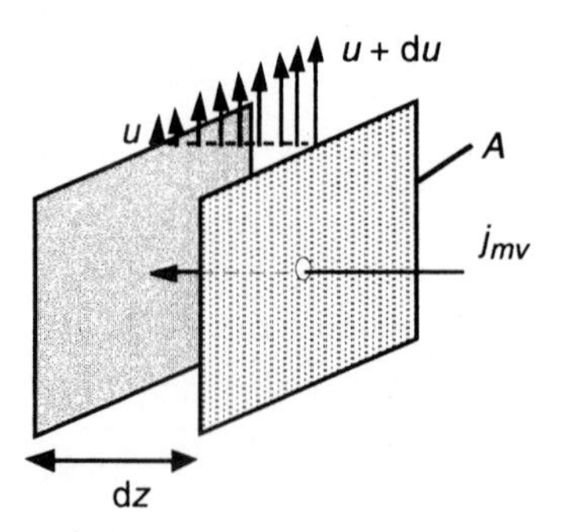

Fig. 5.6. Schematic illustration of a momentum flux j_{mv} caused by a velocity gradient

$$j_{mv} = \frac{\partial(mu)}{\partial t \cdot A} = -\mu \frac{\partial u}{\partial z} . \tag{5.12}$$

Thus, momentum is transported from regions of high velocity to regions of low velocity (see Fig. 5.6). The coefficient of proportionality μ is called *coefficient of viscosity*. (5.12) is valid for the simple example in Fig. 5.6, where only a gradient in z-direction occurs. The general case is complicated by the vector nature of velocity, and will be discussed in Chapter 11.

Coefficients of viscosity can be measured by situating the gas between two rotating, concentric cylinders (see measurement of thermal conductivities, Section 5.2). The coefficient of viscosity is calculated from the torque. Another method is based on *Hagen-Poiseulle's law* for laminar flow, which states that the volume ΔV flowing through a capillary tube per time Δt is inversely proportional to the coefficient of viscosity (with r = radius of the capillary tube, l = length, Δp = pressure difference),

$$\frac{\Delta V}{\Delta t} = \frac{\pi r^4 \Delta p}{8 \mu l} , \tag{5.13}$$

which is a result of integration of (5.12). The kinetic theory of gases, using the intermolecular potential model of rigid spheres, (see, e. g., Bird et al. 1960, Hirschfelder et al. 1964) predicts μ to be proportional to a product of particle density $[n]$, mean velocity $\bar{\upsilon}$, molecular mass m, and mean free path l_{coll} (see Section 5.1 for explanation), $\mu \propto [n] \cdot \bar{\upsilon} \cdot m \cdot l_{\text{coll}}$, leading to

$$\mu = \frac{5}{16} \frac{\sqrt{\pi m k T}}{\pi \sigma^2} \quad \text{or} \quad \mu = \frac{2}{5} \cdot \frac{m}{c_V} \cdot \lambda . \tag{5.14}$$

A refined prediction of μ, again, includes real gas effects by considering an intermolecular potential of the Lennard-Jones-6-12 type. The inclusion of real gas effects modifies the rigid sphere result by a factor $\Omega^{(2,2)*}$ (reduced collision integral),

$$\mu = \frac{5}{16} \frac{\sqrt{\pi m k T}}{\pi \sigma^2 \, \Omega^{(2,2)*}} = \frac{\mu_{\text{rigid sphere}}}{\Omega^{(2,2)*}} . \tag{5.15}$$

Analogous to the thermal conductivity, the coefficient of viscosity does not depend on the pressure and is proportional to the square root of the temperature ($\mu \sim T^{1/2}$). For practical applications the convenient formulation

$$\mu = 2.6693 \cdot 10^{-8} \frac{\sqrt{MT}}{\sigma^2 \Omega^{(2,2)*}} \frac{\text{kg}}{\text{m} \cdot \text{s}} \tag{5.16}$$

is used, where M is in g/mol, T in K, and σ in nm. For mixtures, analogous to the thermal conductivity, an empirical approximation (~10% error) can be used,

$$\mu = \frac{1}{2} \cdot \left[\sum_i x_i \mu_i + \left(\sum_i \frac{x_i}{\mu_i} \right)^{-1} \right] . \tag{5.17}$$

The formulation by Wilke (1950) has an improved accuracy (ca. 5%), where Φ_{ik} is calculated according to (5.11) from the molar masses and the coefficients of viscosity,

$$\mu = \sum_{i=1}^{S} \frac{\mu_i}{1 + \sum_{k \neq i} \frac{x_k}{x_i} \Phi_{ik}} . \tag{5.18}$$

5.4 Diffusion in Gases

For the mass transport caused by concentration gradients (see Fig. 5.7) *Fick's law* states that the *mass flux* j_m is proportional to the concentration gradient (see, e.g., Bird et al. 1960, Hirschfelder et al. 1964):

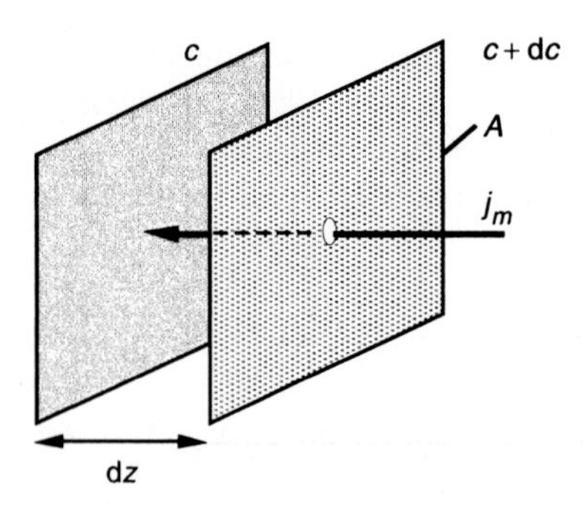

Fig. 5.7. Schematic illustration of a mass flux j_m caused by a concentration gradient

$$j_m = \frac{\partial m}{\partial t \cdot A} = -D\rho \frac{\partial c}{\partial z} \quad \left[\frac{\text{kg}}{\text{cm}^2\,\text{s}}\right] . \tag{5.19}$$

The coefficient of proportionality D is called the *diffusion coefficient*. Diffusion coefficients can be determined by measuring the movement of particles which have been labeled by isotopes. It is important to eliminate convection, which confuses the results fatally.

Predictions of D for an ideal gas, with the intermolecular model of rigid spheres, show the *coefficient of self-diffusion* to be proportional to a product of mean velocity $\bar{\upsilon}$ and mean free path l_{coll} (see Section 5.1), $D \propto \bar{\upsilon} \cdot l_{\text{coll}}$, leading to (see, e. g., Bird et al. 1960, Hirschfelder et al. 1964; $\nu = \mu/\rho$ = *kinematic viscosity*)

$$D = \frac{3}{8} \frac{\sqrt{\pi m k T}}{\pi \sigma^2} \frac{1}{\rho} = \frac{6}{5} \frac{\mu}{\rho} = \frac{6}{5} \nu , \tag{5.20}$$

leading, together with (5.14), to $D = \nu = \alpha$ for rigid spheres ($\alpha = \lambda \cdot (m/\rho c_V)$ = *thermal diffusivity*). When the molecular interaction is accounted for by the Lennard-Jones-6-12-potential, a correction factor $\Omega^{(1,1)*}$ (reduced collision integral) modifies the ideal gas result, again, to include deviations from the rigid sphere model,

$$D = \frac{3}{8} \frac{\sqrt{\pi m k T}}{\pi \sigma^2 \Omega^{(1,1)*}} \frac{1}{\rho} = \frac{D_{\text{rigid sphere}}}{\Omega^{(1,1)*}} . \quad (5.21)$$

For a Lennard-Jones-6-12-potential, the reduced collision integral $\Omega^{(1,1)*}$ is (analogous to the collision integral $\Omega^{(2,2)*}$) a function of the reduced temperature $T^* = kT/\varepsilon$ only. The temperature dependence of the collision integral $\Omega^{(1,1)*}$ is shown together with $\Omega^{(2,2)*}$ in Fig. 5.5; note that corrections can be as high as a factor of 2.

For a mixture of two compounds, the theory modifies the result (5.21) by replacing the mass m with the *reduced mass* $m_1 m_2/(m_1+m_2)$. The mean molecular parameters σ_{12} (and, thus, also the reduced temperature T_{12}^*) for this case are satisfactorily approximated from the parameters of the molecules using the combination rules

$$\sigma_{12} = \frac{\sigma_1 + \sigma_2}{2} \quad \text{and} \quad \varepsilon_{12} = \sqrt{\varepsilon_1 \cdot \varepsilon_2} . \quad (5.22)$$

Thus, for the *binary diffusion coefficient* D_{12} of a compound 1 into a compound 2, one obtains

$$D_{12} = \frac{3}{8} \frac{\sqrt{\pi k T \cdot 2 \cdot \frac{m_1 \cdot m_2}{m_1 + m_2}}}{\pi \sigma_{12}^2 \, \Omega^{(1,1)*}\left(T_{12}^*\right)} \frac{1}{\rho} . \quad (5.23)$$

In practical applications, (5.23) is rendered to the formula for the binary diffusion coefficients D_{12},

$$D_{12} = 2.662 \cdot 10^{-9} \frac{\sqrt{T^3 \cdot \frac{1}{2} \frac{M_1 + M_2}{M_1 \cdot M_2}}}{p \sigma_{12}^2 \, \Omega^{(1,1)*}\left(T_{12}^*\right)} \frac{\text{m}^2}{\text{s}} , \quad (5.24)$$

with the pressure p in bar, the temperature T in K, the molecular radius σ in nm, and the molar mass M in g.

In contrast to thermal conductivity and viscosity, $D \sim T^{3/2}$ and $D \sim 1/p$ for the diffusion coefficients. Diffusion coefficients depend on the pressure!

For mixtures, empirical laws can be used for the diffusion of a compound i into the mixture of the other compounds M (Stefan 1874),

$$D_i^{\text{M}} = \frac{1 - w_i}{\sum_{j \neq i} \frac{x_j}{D_{ij}}} , \quad (5.25)$$

where w_i denotes the mass fraction of species i, x_j the mole fraction of the species j, and D_{ij} the binary diffusion coefficients. The errors of this well-known formula are in the range of 10%. The *Chapman-Enskog theory* (see, e. g., Curtiss and Hirschfelder 1959, Bird et al. 1960, Hirschfelder et al. 1964) yields more accurate expressions for thermal conductivity, coefficients of viscosity, and diffusion coefficients, with an attendant increase in computational effort for evaluation of these expressions.

5.5 Thermal Diffusion, Dufour Effect, and Pressure Diffusion

Thermal diffusion (*Soret effect*) denotes the diffusion of mass caused by temperature gradients. This effect occurs in addition to regular mass diffusion. If thermal diffusion is taken into account, the diffusion flux density $j_{m,i}$ of the species i is given by Eq. 3.9 (Bird et al. 1960, Hirschfelder et al. 1964)

$$j_{m,i} \quad = \quad -D_i^{\mathrm{M}} \rho \frac{w_i}{x_i} \frac{\partial x_i}{\partial z} - \frac{D_i^{\mathrm{T}}}{T} \frac{\partial T}{\partial z} , \qquad (5.26)$$

where D_i^{T} is called *coefficient of thermal diffusion*. Thermal diffusion is only important for light species (H, H_2, He) and at low temperatures. Thus, it is very often omitted in the simulation of combustion processes, although this is a rough approximation (for details, see textbooks on thermodynamics of irreversible processes such as Hirschfelder et al. 1964).

Driving force ⇒ / ⇓ Flux	Velocity gradient	Temperature gradient	Concentration gradient
Momentum	Newton's law [μ]		
Energy		Fourier's law [λ]	Dufour effect [D_i^{T}]
Mass		Soret effect [D_i^{T}]	Fick's law [D]

Fig. 5.8. Fluxes and driving forces in transport processes (Onsager 1931, Hirschfelder et al. 1964)

According to the thermodynamics of irreversible processes, there is a heat transport caused by concentration gradients, a reciprocal process of thermal diffusion. This *Dufour effect* is negligibly small in combustion processes, although it may not be in other chemically reacting flows. Fig. 5.8 shows fluxes and driving forces in transport processes in a systematic scheme; the symmetry of the matrix of transport coefficients defined in this way is leading to the prediction that the coefficients for the Soret and Dufour effects are equal (Onsager 1931).

One further effect negligible in combustion processes is *pressure diffusion* which is diffusion caused by a pressure gradient (Hirschfelder et al. 1964).

5.6 Comparison with Experiments

The following figures show some examples for measured and calculated transport coefficients μ, λ and D.

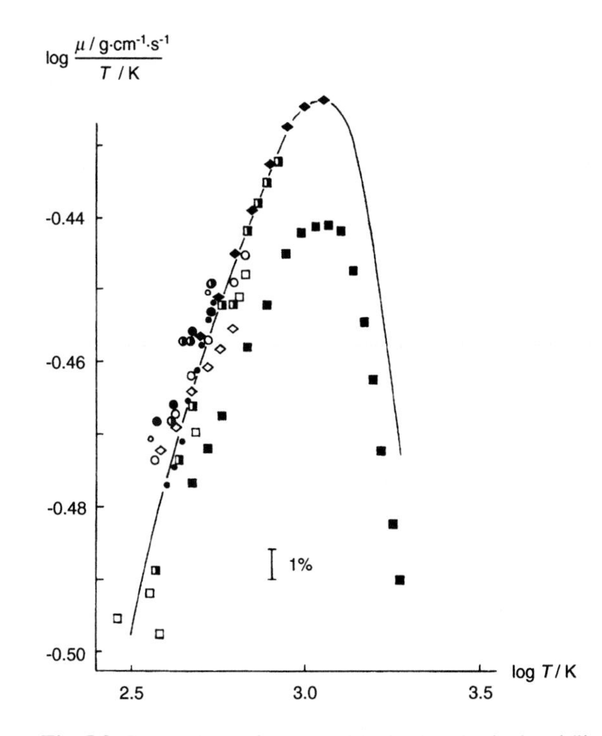

Fig. 5.9. Comparison of measured (points) and calculated (lines; (5.16) is used) coefficients of viscosity of water vapor (Warnatz 1978b).

The first example (see Fig. 5.9) is a comparison of measured and calculated viscosities (Warnatz 1978b). For a better presentation of the temperature dependence, the viscosity is divided by the temperature. It can be seen that measurements which have a large absolute error (filled squares) can nevertheless provide useful information (here on the temperature dependence of μ). The deviations between measured and calculated viscosities are usually < 1 %; for this reason, viscosities are preferentially used to determine molecular potential parameters, like those given in Table 5.1 (heat conductivities, e. g., would not be appropriate for this purpose due to the Eucken correction discussed below).

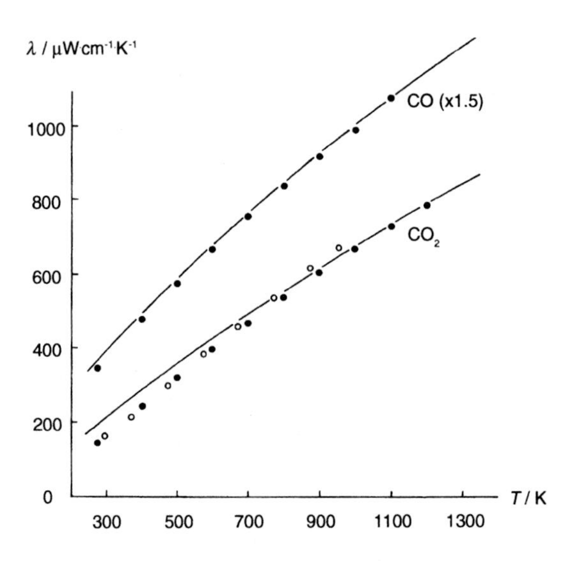

Fig. 5.10. Measured (points) and calculated (lines, with Eucken correction) thermal conductivities (Warnatz 1979); multiplication by 1.5 means that the λ plotted is enlarged by a factor of 1.5

The second example (Fig. 5.10) shows a comparison between measured and calculated thermal conductivities (Warnatz 1979). The deviations are larger in this case due to the internal degrees of freedom of polyatomic molecules. The *Eucken correction* is used to correct for this effect (Hirschfelder et al. 1964),

$$\lambda_{\text{polyatomic}} = \lambda_{\text{monoatomic}} \cdot \left(\frac{4}{15} \frac{\overline{C}_V}{R} + \frac{3}{5} \right) .$$

The Eucken correction can be derived by accounting for diffusion of vibrationally and rotationally hot molecules down the temperature gradient considered. A detailed treatment of this subject is outside the scope of the book.

Figures 5.11 and 5.12 (Warnatz 1978b, 1979) show comparisons of measured and calculated binary diffusion coefficients. It is worth mentioning that several measurements for atoms exist (example in Fig. 5.11). However, the accuracy is considerably less than for stable species due to the delicate treatment of the short-living atoms. The systematic errors in connection with the water molecule are due to its polar nature which cannot be treated with the spherical molecular potential used here.

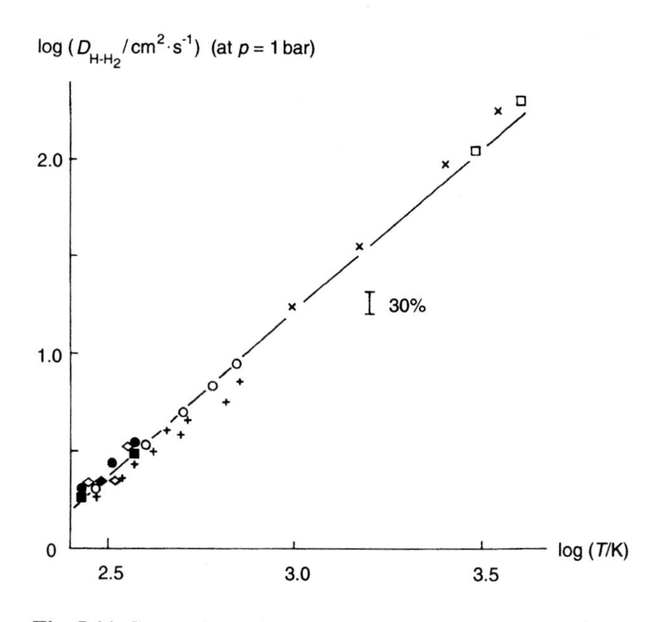

Fig. 5.11. Comparison of measured (points) and calculated (lines; (5.24) is used) binary diffusion coefficients of hydrogen atoms and hydrogen molecules (Warnatz 1978b); the double-logarithmic scale is masking the large deviation of the calculation from the experiments

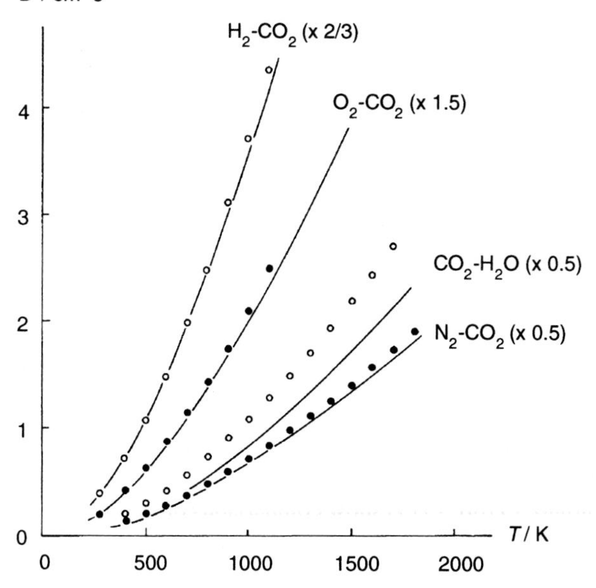

Fig. 5.12. Comparison of measured (points) and calculated (lines; (5.24) is used) binary diffusion coefficients at $p = 1$ bar; multiplication by 1.5 , e. g., means that the diffusion coefficient plotted is larger than the original one by a factor 1.5 (Warnatz 1979)

The last example (Fig. 5.13) shows comparisons of measured and calculated coefficients of thermal diffusion; see (5.26). Usually the *thermal diffusion ratio* k^{T} is used, which in a binary mixture approximately is given as ($M_1 > M_2$ is taken here without loss of generality)

$$k_1^{\mathrm{T}} = -k_2^{\mathrm{T}} = x_1 \cdot x_2 \frac{105}{118} \frac{M_1 - M_2}{M_1 + M_2} R_{\mathrm{T}} . \tag{5.27}$$

Here the reduction factor R_{T} is a universal function of the reduced temperature T^*. Tabulated values of R_{T} versus T^* can be found in the book of Hirschfelder et al. (1964).

For binary mixtures k^{T} is correlated with the coefficient of thermal diffusion D_i^{T} according to (Fristrom and Westenberg 1965)

$$D_i^{\mathrm{T}} = k_i^{\mathrm{T}} \frac{c^2 M_1 M_2}{\rho} D_{12} \quad , \quad i = 1, 2 . \tag{5.28}$$

In analogy to the binary expression (5.28), the thermal diffusion ratio and the thermal diffusion coefficient in a multicomponent mixture is given as (Paul and Warnatz 1998)

$$k_{i,\mathrm{mix}}^{\mathrm{T}} = x_i \cdot \sum_{j=1}^{N} x_j k_{ij}^{\mathrm{T}} \quad \text{and} \quad D_{i,\mathrm{mix}}^{\mathrm{T}} = k_{i,\mathrm{mix}}^{\mathrm{T}} \frac{c^2 M_i \overline{M}}{\rho} \cdot D_i^M . \tag{5.29}$$

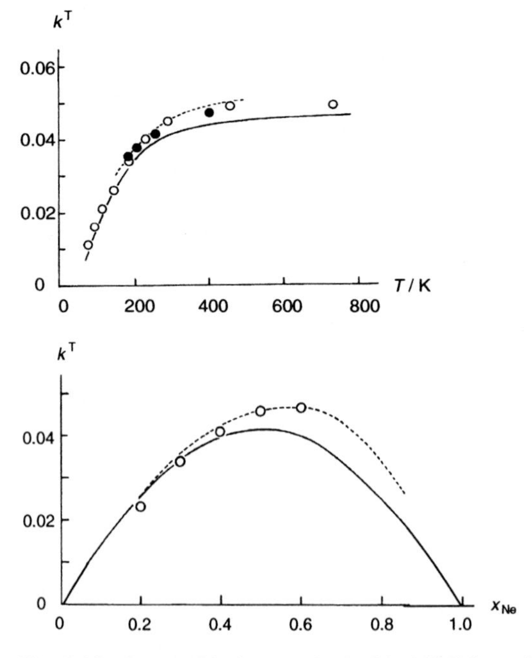

Fig. 5.13. Measured (points) and calculated (full lines, (5.28) is used) thermal diffusion ratios in the Ar-Ne system; upper: temperature dependence, lower: composition dependence. Dashed curves: calculated multicomponent thermal diffusion ratios (Hirschfelder et al. 1964)

For multicomponent mixtures this is a very simplified formula, which, however, is justified by the fact that thermal diffusion is important for species only with molar masses very different from the mean molar mass (in combustion: H, H_2, He, or large soot prcursors; see Rosner 2000); therefore, a binary approach seems to be adequate. *Multicomponent thermal diffusion coefficients* (Hirschfelder et al. 1964) can be formulated, but require a prohibitive amount of computational work.

In contrast to numerous data on μ, λ, and D, measurements of the thermal diffusion ratio k^{T} are available for only a small number of different gases (see, e. g., Bird et al. 1960, Warnatz 1982).

5.7 Exercises

Exercise 5.1. The viscosity of carbon dioxide CO_2 has been compared with the viscosity of argon by comparing the flow through a long narrow tube (1 mm diameter) using the Hagen-Poiseuille formula (5.13) $\mathrm{d}V/\mathrm{d}t = \pi r^4 \Delta p/(8\mu l)$. At the same pressure difference, equal volumes of carbon dioxide and argon needed 55 s and 83 s, respectively, to pass through the tube. Argon has a viscosity of $2.08 \cdot 10^{-5}$ kg/(m·s) at 25 °C. What is the viscosity of carbon dioxide? What is the diameter of the carbon dioxide? (Use $\Omega^{(2,2)*} = 1$ for the reduced collision integral; the mass of a proton or a neutron is $1.6605 \cdot 10^{-27}$ kg.) Compute the Reynolds number $Re = \rho \cdot 2r \cdot \bar{v}/\mu$ and insure that the flow is laminar by showing that $Re < 2100$ (for round pipes).

Exercise 5.2. In a cool wine cellar (10 °C, 1 bar) fermentation of a good vintage has lead to an enrichment of the cellar air to a carbon dioxide content of $w_{\mathrm{CO_2},0} = 5\ \%$. The carbon dioxide is diffusing from the cellar, via a 1 m by 2 m doorway and along a hallway with length $L = 10$ m (with the same cross section as the doorway), to the outside where the CO_2 mass fraction is essentially zero, $w_{\mathrm{CO_2},L} = 0$. (1) Determine the diffusion coefficient $D_{\mathrm{CO_2,Air}}$ assuming that air behaves as N_2 with $\Omega^{(1,1)*} = 1$. (2) Use Eq. 3.11 for CO_2 and air to show that the CO_2 flux $j_{\mathrm{CO_2}}$ (and hence the fermentation rate) is

$$j_{\mathrm{CO_2}} = \frac{\rho \cdot D_{\mathrm{CO_2,Air}}}{L} \cdot \ln \frac{1 - w_{\mathrm{CO_2},L}}{1 - w_{\mathrm{CO_2},0}} \approx \rho \cdot D_{\mathrm{CO_2,Air}} \cdot \frac{w_{\mathrm{CO_2},0} - w_{\mathrm{CO_2},L}}{L} .$$

(Flow through a stagnant gas film is known as the *Stefan problem*, see, e. g., Bird et al. 1960.)

Excercise 5.3. Droplet evaporation is a variant of the Stefan Problem of Excercise 5.2. It will be shown in Chapter 15 that, after an initial transient phase, it is reasonable to assume that the vapor pressure of the liquid at the droplet surface, $r = r_{\mathrm{S}}$, is constant. Conservation of mass dictates that the product $4\pi r^2 \cdot j_r$ is constant, where j_r is the mass flux at radius r.

Use this mass conservation assumption together with Eq. 5.11 to conclude that the flux is

$$\dot{m} = 4\pi r_S^2 j_S = 4\pi r_S^2 \rho \cdot D_{\text{vapor,air}} \cdot \ln \frac{1-w_\infty}{1-w_{r_S}} .$$

Furthermore, with the droplet mass $m_d = \rho_{\text{liquid}} \cdot (4/3) \cdot \pi r_S^3$ changing temporally as $dm_d/dt = -\dot{m}$, show that

$$\frac{dr_S^2}{dt} = \frac{2\rho \cdot D_{\text{vapor,air}}}{\rho_{\text{liquid}}} \cdot \ln \frac{1-w_\infty}{1-w_{r_S}} = \text{const.}$$

Thus, the square of the droplet radius recedes at a constant rate, consistent with experiments. A doubling of the droplet initial diameter will quadruple the time for complete evaporation.

Exercise 5.4. Suppose that you have a 50 l tank containing N_2 you wish to keep at atmospheric pressure. You connect the chamber to the atmosphere by a tube of length L. To avoid contamination of the chamber with O_2, you slightly overpressure the N_2 chamber so that a flow of N_2, at velocity v, counters the diffusion of O_2 from the air toward the chamber. At what velocity v is the steady state flux of O_2 into the chamber zero? Let the diffusivity of O_2 in N_2 be D and let O_2 concentration be c.

Exercise 5.5. It is often the case that reactive gases are mixed into a tube and then ignited to create high temperature and pressure. This high temperature and pressure can be used to drive a "shock wave tube" or a "gas gun" where the high pressures push a large massive piston down a tube containing hydrogen. As the massive piston nears the end of the tube, the hydrogen gas pressure rapidly increases as well as the temperature. At the right moment, the gas is allowed to push and accelerate a projectile.

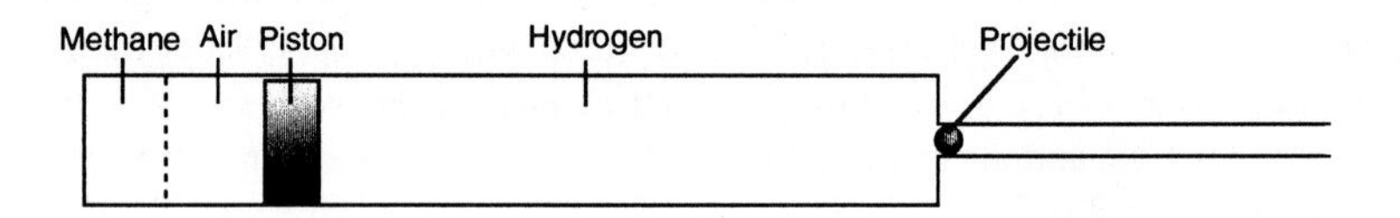

Space launch velocities are possible, or satellites could possibly be put into orbit. When filling the combustion tank, often one thinks of adding the air (or oxygen) first and the gas (say methane) second. The expectation is that the two gases will rapidly diffuse into one another. In fact, the characteristic diffusion time is surprisingly long. Show this, assuming the combustion tube is 10 m long, 1 m diameter. For simplicity, assume the tube is full of methane for 5 m and the balance is air for 5 m. Assume that the pressure is 60 bar and that the diffusivity for CH_4-air is $D = 10^{-5}$ m^2/s at $p = 1$ bar. What is the characteristic time for diffusive mixing?

6 Chemical Kinetics

The thermodynamic laws discussed in Chapter 4 allow the determination of the equilibrium state of a chemical reaction system. If one assumes that the chemical reactions are fast compared to the other processes like diffusion, heat conduction, and flow, thermodynamics alone allow the description of the system locally (see, e. g., Section 13.1). In most cases, however, chemical reactions occur on time scales comparable with that of the flow and of the molecular transport processes. Therefore, information is needed about the rate of chemical reactions, i. e., the *chemical kinetics*. Thus, the basic laws of chemical kinetics shall be discussed in the following, which are based on macroscopic observation. The chapter will show that these macroscopic rate laws are a consequence of the underlying microscopic phenomena of collisions between molecules.

6.1 Rate Laws and Reaction Orders

The so-called *rate law* shall be discussed for a chemical reaction, which in its general case (as introduced in Chapter 1) can be described by the equation

$$\mathrm{A} + \mathrm{B} + \mathrm{C} + \dots \xrightarrow{k} \mathrm{D} + \mathrm{E} + \mathrm{F} + \dots \,, \tag{6.1}$$

where A, B, C, . . . denote the different species involved in the reaction. A rate law describes an empirical formulation of the *reaction rate,* i. e., the rate of formation or consumption of a species in a chemical reaction (see, e. g., Homann 1975). Looking at the consumption of species A, the reaction rate can be expressed according to

$$\frac{\mathrm{d}[\mathrm{A}]}{\mathrm{d}t} = -k \cdot [\mathrm{A}]^a \, [\mathrm{B}]^b \, [\mathrm{C}]^c \dots . \tag{6.2}$$

Here $a, b, c, \dots$ are *reaction orders* with respect to the species A, B, C, . . . and k is the *rate coefficient* of the reaction. The sum of all exponents is the *overall reaction order*.

Frequently some species are in excess. In this case their concentrations do not change noticeably. If, e. g., [B], [C], ... remain nearly constant during the reaction, an effective rate coefficient can be generated from the rate coefficient and the nearly constant concentrations of the species in excess, and, using $k_{exp} = k \cdot [B]^b [C]^c \ldots$, one obtains as a simplified version of (6.2)

$$\frac{d[A]}{dt} = -k_{exp} [A]^a . \tag{6.3}$$

The temporal change of the concentration of species A can be calculated by integrating this differential equation, as it will be shown next for some typical cases.

For *first-order reactions* ($a = 1$) integration of (6.3) yields the first-order time behavior (provable by insertion of (6.4) into (6.3))

$$\ln \frac{[A]_t}{[A]_0} = -k_{exp} (t - t_0) , \tag{6.4}$$

where $[A]_0$ and $[A]_t$ denote the concentrations of species A at time t_0 and t, respectively.

Accordingly, one obtains for *second-order reactions* ($a = 2$) the temporal behavior

$$\frac{1}{[A]_t} - \frac{1}{[A]_0} = k_{exp} (t - t_0) \tag{6.5}$$

and for *third-order reactions* ($a = 3$) the temporal behavior

$$\frac{1}{[A]_t^2} - \frac{1}{[A]_0^2} = 2 k_{exp} (t - t_0) . \tag{6.6}$$

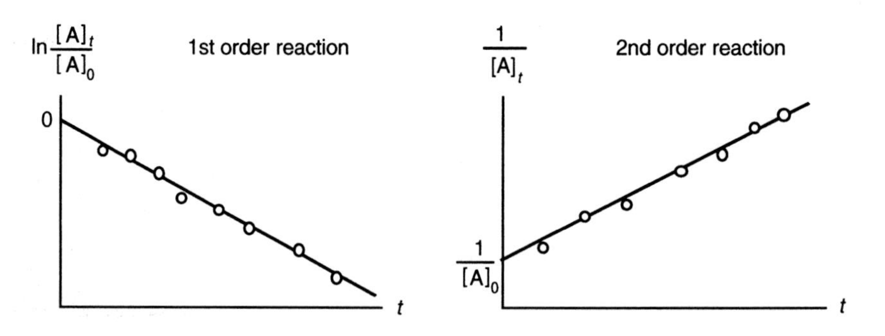

Fig. 6.1. Time behavior of the concentrations for first- and second-order reactions

If the time behavior is measured, the reaction order can be determined. Logarithmic plots of the concentrations versus time for first-order reactions lead to linear dependences with the slope $-k_{exp}$ and plots of $1/[A]_t$ versus the time for second order reactions lead to linear dependences with the slope k_{exp}; see examples given in Fig. 6.1. A similar treatment is possible for other integer reaction orders a.

6.2 Relation of Forward and Reverse Reactions

For the reverse reaction of (6.1) one obtains, analogous to (6.2), the rate law for the production of A

$$\frac{\mathrm{d}[\mathrm{A}]}{\mathrm{d}t} = k^{(\mathrm{r})}\,[\mathrm{D}]^d\,[\mathrm{E}]^e\,[\mathrm{F}]^f\,..... \tag{6.7}$$

At chemical equilibrium, forward and backward reactions have the same rate on a microscopic level (the forward reaction here is characterized by the superscript (f), the reverse reaction by the superscript (r)). On a macroscopic level, no net reaction can be observed. Therefore, at chemical equilibrium one has

$$k^{(\mathrm{f})}\,[\mathrm{A}]^a\,[\mathrm{B}]^b\,[\mathrm{C}]^c\,..... = +k^{(\mathrm{r})}\,[\mathrm{D}]^d\,[\mathrm{E}]^e\,[\mathrm{F}]^f\,.....$$

or (6.8)

$$\frac{[\mathrm{D}]^d\,[\mathrm{E}]^e\,[\mathrm{F}]^f\cdot...}{[\mathrm{A}]^a[\mathrm{B}]^b\,[\mathrm{C}]^c\cdot...} = \frac{k^{(\mathrm{f})}}{k^{(\mathrm{r})}}\,.$$

The expression on the left hand side corresponds to the equilibrium constant of the reaction, which can be calculated from thermodynamic data (discussed in Section 4.7), and, thus, an important relation between the rate coefficients of forward and reverse reaction can be obtained,

$$K_c = \frac{k^{(\mathrm{f})}}{k^{(\mathrm{r})}} = \exp(-\Delta_{\mathrm{R}}\,\overline{A}^0/RT)\,. \tag{6.9}$$

6.3 Elementary Reactions, Reaction Molecularity

An *elementary reaction* is one that occurs on a molecular level exactly in the way which is described by the reaction equation (see, e. g., Homann 1975). The reaction of hydroxy radicals (OH) with molecular hydrogen (H_2) forming water and hydrogen atoms is such an elementary reaction,

$$\mathrm{OH} + \mathrm{H_2} \rightarrow \mathrm{H_2O} + \mathrm{H}\,. \tag{6.10}$$

Due to the molecular motion in the gas, hydroxy radicals collide with hydrogen molecules. In the case of non-reactive collisions the molecules collide and bounce apart. But in the case of reactive collisions, the molecules react and the products H_2O and H emerge from the collision. On the contrary, the reaction

$$2\,\mathrm{H_2} + \mathrm{O_2} \rightarrow 2\,\mathrm{H_2O} \tag{6.11}$$

is not an elementary one. Detailed investigations show that water is not produced by a single collision between the three reacting molecules. Instead, many reactive intermediates like H, O, and OH are formed. Reactions such as (6.11) are called *net reactions* or *overall reactions*. Usually these net reactions have very complicated rate laws like (6.2) or even more complex; the reaction orders $a, b, c, \ldots$ are usually not integers, can be negative, depend on time and reaction conditions, and an extrapolation to conditions where no experiments exist is not reliable or even completely wrong. In many cases, a mechanistic interpretation of non-elementary rate laws is not possible.

Overall reactions are a consequence of a large number of elementary reactions. Resolution of these elementary reactions is a hard and time-consuming task. The formation of water (6.11), e. g., can be described by the first 38 elementary reactions shown in Tab. 6.1 (Baulch et al. 1991, Warnatz et al. 1999).

The concept of using elementary reactions has many advantages: The reaction order of elementary reactions is always constant (in particular, independent of time and of experimental conditions) and can be determined easily. One only has to look at the *molecularity of the reaction*. This is the number of species which form the reaction complex (the transition state between reactants and products during the reaction). In fact, only three possible values of the reaction molecularity are observed:

Unimolecular reactions describe the rearrangement or dissociation of a molecule,

$$\mathrm{A} \quad \rightarrow \quad \text{products}\,. \tag{6.12}$$

Unimolecular reactions have a first-order time behavior. If the initial concentration is doubled, the reaction rate is also doubled.

Bimolecular reactions are the reaction type found most (see Tab. 6.1). They proceed according to the reaction equations

$$\begin{aligned} \mathrm{A} + \mathrm{B} \quad &\rightarrow \quad \text{products} \\ \text{or} \qquad & \\ \mathrm{A} + \mathrm{A} \quad &\rightarrow \quad \text{products}\,. \end{aligned} \tag{6.13}$$

Bimolecular reactions have always a second-order rate law. Doubling of the concentration of each reaction partner quadruples the reaction rate; e. g., doubling of the pressure quadruples the rate, see (5.4) in Section 5.1 for further explanation.

Trimolecular reactions are usually recombination reactions (see Reactions 5 to 8 in Tab. 6.1). They obey a third-order rate law,

$$\begin{aligned} \mathrm{A} + \mathrm{B} + \mathrm{C} \quad &\rightarrow \quad \text{products} \\ \text{or} \qquad & \\ \mathrm{A} + \mathrm{A} + \mathrm{B} \quad &\rightarrow \quad \text{products} \\ \text{or} \qquad & \\ \mathrm{A} + \mathrm{A} + \mathrm{A} \quad &\rightarrow \quad \text{products}\,. \end{aligned} \tag{6.14}$$

In general the molecularity equals the order for elementary reactions. Thus, the rate laws are easily derived. If the equation of an elementary reaction r is given by

$$\sum_{s=1}^{S} \nu_{rs}^{(e)} \mathrm{A}_s \xrightarrow{k_r} \sum_{s=1}^{S} \nu_{rs}^{(p)} \mathrm{A}_s\,, \tag{6.15}$$

then the rate law for the formation of species i in reaction r is given by the expression

$$\left(\frac{\partial c_i}{\partial t}\right)_{\mathrm{chem},r} = k_r\left(\nu_{ri}^{(\mathrm{p})} - \nu_{ri}^{(\mathrm{e})}\right) \prod_{s=1}^{S} c_s^{\nu_{rs}^{(\mathrm{e})}} \quad . \tag{6.16}$$

Here $\nu_{rs}^{(e)}$ and $\nu_{rs}^{(p)}$ denote stoichiometric coefficients of reactants (sometimes called *educts* in the literature) or products and c_s concentrations of the S different species s.

Tab. 6.1. Elementary reactions in the H_2-CO-C_1-C_2-O_2 system at $p = 1$ bar for high temperature ($T > 1200$ K); rate coefficients are presented in the form $k = A \cdot T^b \cdot \exp(-E/RT)$ as described in Section 6.5, $[M^*] = [H_2]+6.5 \cdot [H_2O]+0.4 \cdot [O_2]+0.4 \cdot [N_2]+0.75 \cdot [CO]+1.5 \cdot [CO_2]+3.0 \cdot [CH_4]$; $\rightarrow$: only forward reaction is considered, =: reverse reaction to be calculated via (6.9)

Reaction					A [cm,mol,s]	b	E / kJ·mol⁻¹
----	01. - 04. H_2-CO Oxidation						
----	01. H_2-O_2 Reactions (HO_2, H_2O_2 not included)						
O_2	+H		=OH	+O	$2.00 \cdot 10^{14}$	0.0	70.3
H_2	+O		=OH	+H	$5.06 \cdot 10^{04}$	2.67	26.3
H_2	+OH		$=H_2O$	+H	$1.00 \cdot 10^{08}$	1.6	13.8
OH	+OH		$=H_2O$	+O	$1.50 \cdot 10^{09}$	1.14	0.42
H	+H	+M*	$=H_2$	+M*	$1.80 \cdot 10^{18}$	-1.0	0.00
O	+O	+M*	$=O_2$	+M*	$2.90 \cdot 10^{17}$	-1.0	0.00
H	+OH	+M*	$=H_2O$	+M*	$2.20 \cdot 10^{22}$	-2.0	0.00
----	02. HO_2 Formation/Consumption						
H	$+O_2$	+M*	$=HO_2$	+M*	$2.30 \cdot 10^{18}$	-0.8	0.00
HO_2	+H		=OH	+OH	$1.50 \cdot 10^{14}$	0.0	4.20
HO_2	+H		$=H_2$	$+O_2$	$2.50 \cdot 10^{13}$	0.0	2.90
HO_2	+H		$=H_2O$	+O	$3.00 \cdot 10^{13}$	0.0	7.20
HO_2	+O		=OH	$+O_2$	$1.80 \cdot 10^{13}$	0.0	-1.70
HO_2	+OH		$=H_2O$	$+O_2$	$6.00 \cdot 10^{13}$	0.0	0.00
----	03. H_2O_2 Formation/Consumption						
HO_2	$+HO_2$		$=H_2O_2$	$+O_2$	$2.50 \cdot 10^{11}$	0.0	-5.20
OH	+OH	+M*	$=H_2O_2$	+M*	$3.25 \cdot 10^{22}$	-2.0	0.00
H_2O_2	+H		$=H_2$	$+HO_2$	$1.70 \cdot 10^{12}$	0.0	15.7
H_2O_2	+H		$=H_2O$	+OH	$1.00 \cdot 10^{13}$	0.0	15.0
H_2O_2	+O		=OH	$+HO_2$	$2.80 \cdot 10^{13}$	0.0	26.8
H_2O_2	+OH		$=H_2O$	$+HO_2$	$5.40 \cdot 10^{12}$	0.0	4.20
----	04. CO Reactions						
CO	+OH		$=CO_2$	+H	$6.00 \cdot 10^{06}$	1.5	-3.10
CO	$+HO_2$		$=CO_2$	+OH	$1.50 \cdot 10^{14}$	0.0	98.7
CO	+O	+M*	$=CO_2$	+M*	$7.10 \cdot 10^{13}$	0.0	-19.0
CO	$+O_2$		$=CO_2$	+O	$2.50 \cdot 10^{12}$	0.0	200.
----	10. - 19. C_1-Hydrocarbons Oxidation						
----	10. CH Reactions						
CH	+O		=CO	+H	$4.00 \cdot 10^{13}$	0.0	0.00
CH	$+O_2$		=CHO	+O	$6.00 \cdot 10^{13}$	0.0	0.00
CH	$+CO_2$		=CHO	+CO	$3.40 \cdot 10^{12}$	0.0	2.90

CH	+H_2O	=CH_2O	+H		$3.80\cdot10^{12}$	0.0	-3.20
CH	+H_2O	=$^3CH_2$	+OH		$1.90\cdot10^{12}$	0.0	-3.20
CH	+OH	=CHO	+H		$3.00\cdot10^{13}$	0.0	0.00
----	11. CHO Reactions						
CHO	+M*	=CO	+H	+M*	$7.10\cdot10^{14}$	0.0	70.3
CHO	+H	=CO	+H_2		$9.00\cdot10^{13}$	0.0	0.00
CHO	+O	=CO	+OH		$3.00\cdot10^{13}$	0.0	0.00
CHO	+O	=CO_2	+H		$3.00\cdot10^{13}$	0.0	0.00
CHO	+OH	=CO	+H_2O		$1.00\cdot10^{14}$	0.0	0.00
CHO	+O_2	=CO	+HO_2		$3.00\cdot10^{12}$	0.0	0.00
CHO	+CHO	=CH_2O	+CO		$3.00\cdot10^{13}$	0.0	0.00
----	12. CH_2 Reactions						
$^3CH_2$	+H	=CH	+H_2		$6.00\cdot10^{12}$	0.0	-7.50
$^3CH_2$	+O	→CO	+H	+H	$8.40\cdot10^{12}$	0.0	0.00
$^3CH_2$	+$^3CH_2$	=C_2H_2	+H_2		$1.20\cdot10^{13}$	0.0	3.40
$^3CH_2$	+$^3CH_2$	=C_2H_2	+H	+H	$1.10\cdot10^{14}$	0.0	3.40
$^3CH_2$	+CH_3	=C_2H_4	+H		$4.20\cdot10^{13}$	0.0	0.00
$^3CH_2$	+O_2	=CO	+OH	+H	$1.30\cdot10^{13}$	0.0	6.20
$^3CH_2$	+O_2	=CO_2	+H_2		$1.20\cdot10^{13}$	0.0	6.20
$^1CH_2$	+M*	=$^3CH_2$	+M*		$1.20\cdot10^{13}$	0.0	0.00
$^1CH_2$	+O_2	=CO	+OH	+H	$3.10\cdot10^{13}$	0.0	0.00
$^1CH_2$	+H_2	=CH_3	+H		$7.20\cdot10^{13}$	0.0	0.00
$^1CH_2$	+CH_3	=C_2H_4	+H		$1.60\cdot10^{13}$	0.0	-2.38
----	13. CH_2O Reactions						
CH_2O	+M*	=CHO	+H	+M*	$5.00\cdot10^{16}$	0.0	320.
CH_2O	+H	=CHO	+H_2		$2.30\cdot10^{10}$	1.05	13.7
CH_2O	+O	=CHO	+OH		$4.15\cdot10^{11}$	0.57	11.6
CH_2O	+OH	=CHO	+H_2O		$3.40\cdot10^{09}$	1.2	-1.90
CH_2O	+HO_2	=CHO	+H_2O_2		$3.00\cdot10^{12}$	0.0	54.7
CH_2O	+CH_3	=CHO	+CH_4		$1.00\cdot10^{11}$	0.0	25.5
CH_2O	+O_2	=CHO	+HO_2		$6.00\cdot10^{13}$	0.0	171.
----	14. CH_3 Reactions						
CH_3	+M*	=$^3CH_2$	+H	+M*	$6.90\cdot10^{14}$	0.0	345.
CH_3	+M*	=$^3CH_2$	+H	+M*	$1.00\cdot10^{16}$	0.0	379.
CH_3	+O	=CH_2O	+H		$8.43\cdot10^{13}$	0.0	0.00
CH_3	+H	=CH_4			$1.93\cdot10^{36}$	-7.0	38.0
CH_3	+OH	→CH_3O	+H		$2.26\cdot10^{14}$	0.0	64.8
CH_3O	+H	→CH_3	+OH		$4.75\cdot10^{16}$	-.13	88.0
CH_3	+O_2	→CH_2O	+OH		$3.30\cdot10^{11}$	0.0	37.4
CH_3	+HO_2	=CH_3O	+OH		$1.80\cdot10^{13}$	0.0	0.00
CH_3	+HO_2	=CH_4	+O_2		$3.60\cdot10^{12}$	0.0	0.00
CH_3	+CH_3	→C_2H_4	+H_2		$1.00\cdot10^{16}$	0.0	134.
CH_3	+CH_3	=C_2H_6			$1.69\cdot10^{53}$	-12.	81.2
----	15a. CH_3O Reactions						
CH_3O	+M*	=CH_2O	+H	+M*	$5.00\cdot10^{13}$	0.0	105.
CH_3O	+H	=CH_2O	+H_2		$1.80\cdot10^{13}$	0.0	0.00
CH_3O	+O_2	=CH_2O	+HO_2		$4.00\cdot10^{10}$	0.0	8.90
CH_2O	+CH_3O	→CHO	+CH_3OH		$6.00\cdot10^{11}$	0.0	13.8

Reaction	A	b	E
$CH_3OH + CHO \rightarrow CH_2O + CH_3O$	$6.50 \cdot 10^{09}$	0.0	57.2
$CH_3O + O = O_2 + CH_3$	$1.10 \cdot 10^{13}$	0.0	0.00
$CH_3O + O = OH + CH_2O$	$1.40 \cdot 10^{12}$	0.0	0.00
---- 15b. CH_2OH Reactions			
$CH_2OH + M^* = CH_2O + H + M^*$	$5.00 \cdot 10^{13}$	0.0	105.
$CH_2OH + H = CH_2O + H_2$	$3.00 \cdot 10^{13}$	0.0	0.00
$CH_2OH + O_2 = CH_2O + HO_2$	$1.00 \cdot 10^{13}$	0.0	30.0
---- 16. CH_3O_2 Reactions			
$CH_3O_2 + M^* \rightarrow CH_3 + O_2 + M^*$	$7.24 \cdot 10^{16}$	0.0	111.
$CH_3 + O_2 + M^* \rightarrow CH_3O_2 + M^*$	$1.41 \cdot 10^{16}$	0.0	-4.60
$CH_3O_2 + CH_2O \rightarrow CHO + CH_3O_2H$	$1.30 \cdot 10^{11}$	0.0	37.7
$CH_3O_2H + CHO \rightarrow CH_3O_2 + CH_2O$	$2.50 \cdot 10^{10}$	0.0	42.3
$CH_3O_2 + CH_3 \rightarrow CH_3O + CH_3O$	$3.80 \cdot 10^{12}$	0.0	-5.00
$CH_3O + CH_3O \rightarrow CH_3O_2 + CH_3$	$2.00 \cdot 10^{10}$	0.0	0.00
$CH_3O_2 + HO_2 \rightarrow CH_3O_2H + O_2$	$4.60 \cdot 10^{10}$	0.0	-10.9
$CH_3O_2H + O_2 \rightarrow CH_3O_2 + HO_2$	$3.00 \cdot 10^{12}$	0.0	163.
$CH_3O_2 + CH_3O_2 \rightarrow CH_2O + O_2 + CH_3OH$	$1.80 \cdot 10^{12}$	0.0	0.00
$CH_3O_2 + CH_3O_2 \rightarrow CH_3O + CH_3O + O_2$	$3.70 \cdot 10^{12}$	0.0	9.20
---- 17. CH_4 Reactions			
$CH_4 + H = H_2 + CH_3$	$1.30 \cdot 10^{04}$	3.00	33.6
$CH_4 + O = OH + CH_3$	$6.92 \cdot 10^{08}$	1.56	35.5
$CH_4 + OH = H_2O + CH_3$	$1.60 \cdot 10^{07}$	1.83	11.6
$CH_4 + HO_2 = H_2O_2 + CH_3$	$1.10 \cdot 10^{13}$	0.0	103.
$CH_4 + CH = C_2H_4 + H$	$3.00 \cdot 10^{13}$	0.0	-1.70
$CH_4 + {}^3CH_2 = CH_3 + CH_3$	$1.30 \cdot 10^{13}$	0.0	39.9
---- 18. CH_3OH Reactions			
$CH_3OH = CH_3 + OH$	$9.51 \cdot 10^{29}$	-4.3	404.
$CH_3OH + H = CH_2OH + H_2$	$4.00 \cdot 10^{13}$	0.0	25.5
$CH_3OH + O = CH_2OH + OH$	$1.00 \cdot 10^{13}$	0.0	19.6
$CH_3OH + OH = CH_2OH + H_2O$	$1.00 \cdot 10^{13}$	0.0	7.10
$CH_3OH + HO_2 \rightarrow CH_2OH + H_2O_2$	$6.20 \cdot 10^{12}$	0.0	81.1
$CH_2OH + H_2O_2 \rightarrow HO_2 + CH_3OH$	$1.00 \cdot 10^{07}$	1.7	47.9
$CH_3OH + CH_3 = CH_4 + CH_2OH$	$9.00 \cdot 10^{12}$	0.0	41.1
$CH_3O + CH_3OH \rightarrow CH_2OH + CH_3OH$	$2.00 \cdot 10^{11}$	0.0	29.3
$CH_2OH + CH_3OH \rightarrow CH_3O + CH_3OH$	$2.20 \cdot 10^{04}$	1.7	45.4
$CH_3OH + CH_2O \rightarrow CH_3O + CH_3O$	$1.53 \cdot 10^{12}$	0.0	333.
$CH_3O + CH_3O \rightarrow CH_3OH + CH_2O$	$3.00 \cdot 10^{13}$	0.0	0.00
---- 19. CH_3O_2H Reactions			
$CH_3O_2H = CH_3O + OH$	$4.00 \cdot 10^{15}$	0.0	180.
$OH + CH_3O_2H = H_2O + CH_3O_2$	$2.60 \cdot 10^{12}$	0.0	0.00
---- 20. - 29. C_2-Hydrocarbons Oxidation			
---- 20. C_2H Reactions			
$C_2H + O = CO + CH$	$1.00 \cdot 10^{13}$	0.0	0.00
$C_2H + O_2 = HCCO + O$	$3.00 \cdot 10^{12}$	0.0	0.00
---- 21. HCCO Reactions			
$HCCO + H = {}^3CH_2 + CO$	$1.50 \cdot 10^{14}$	0.0	0.00
$HCCO + O \rightarrow CO + H + CO$	$9.60 \cdot 10^{13}$	0.0	0.00

Reaction					A	n	E
HCCO	$+^3CH_2$	$=C_2H_3$	+CO		$3.00 \cdot 10^{13}$	0.0	0.00
---- 22. C_2H_2 Reactions							
C_2H_2	+M*	$=C_2H$	+H	+M*	$3.60 \cdot 10^{16}$	0.0	446.
C_2H_2	$+O_2$	=HCCO	+OH		$2.00 \cdot 10^{08}$	1.5	126.
C_2H_2	+H	$=C_2H$	$+H_2$		$6.02 \cdot 10^{13}$	0.0	116.
C_2H_2	+O	$=^3CH_2$	+CO		$1.72 \cdot 10^{04}$	2.8	2.10
C_2H_2	+O	=HCCO	+H		$1.72 \cdot 10^{04}$	2.8	2.10
C_2H_2	+OH	$=H_2O$	$+C_2H$		$6.00 \cdot 10^{13}$	0.0	54.2
C_2H_2	$+C_2H$	$=C_4H_2$	+H		$3.00 \cdot 10^{13}$	0.0	0.00
---- 23. CH_2CO Reactions							
CH_2CO	+M*	$=^3CH_2$	+CO	+M*	$1.00 \cdot 10^{16}$	0.0	248.
CH_2CO	+H	$=CH_3$	+CO		$3.60 \cdot 10^{13}$	0.0	14.1
CH_2CO	+O	=CHO	+CHO		$2.30 \cdot 10^{12}$	0.0	5.70
CH_2CO	+OH	$=CH_2O$	+CHO		$1.00 \cdot 10^{13}$	0.0	0.00
---- 24. C_2H_3 Reactions							
C_2H_3		$=C_2H_2$	+H		$4.73 \cdot 10^{40}$	-8.8	194.
C_2H_3	+OH	$=C_2H_2$	$+H_2O$		$5.00 \cdot 10^{13}$	0.0	0.00
C_2H_3	+H	$=C_2H_2$	$+H_2$		$1.20 \cdot 10^{13}$	0.0	0.00
C_2H_3	+O	$=C_2H_2$	+OH		$1.00 \cdot 10^{13}$	0.0	0.00
C_2H_3	+O	$=CH_3$	+CO		$1.00 \cdot 10^{13}$	0.0	0.00
C_2H_3	+O	=CHO	$+^3CH_2$		$1.00 \cdot 10^{13}$	0.0	0.00
C_2H_3	$+O_2$	=CHO	$+CH_2O$		$5.40 \cdot 10^{12}$	0.0	0.00
---- 25a. CH_3CO Reactions							
CH_3CO		$=CH_3$	+CO		$2.32 \cdot 10^{26}$	-5.0	75.1
CH_3CO	+H	$=CH_2CO$	$+H_2$		$2.00 \cdot 10^{13}$	0.0	0.00
---- 25b. CH_2CHO Reactions							
CH_2CHO	+H	$=CH_2CO$	$+H_2$		$2.00 \cdot 10^{13}$	0.0	0.00
---- 26. C_2H_4 Reactions							
C_2H_4	+M*	$=C_2H_2$	$+H_2$	+M*	$7.50 \cdot 10^{17}$	0.0	320.
C_2H_4	+M*	$=C_2H_3$	+H	+M*	$8.50 \cdot 10^{17}$	0.0	404.
C_2H_4	+H	$=C_2H_3$	$+H_2$		$5.67 \cdot 10^{14}$	0.0	62.9
C_2H_4	+O	=H	$+CH_2CHO$		$1.40 \cdot 10^{06}$	2.08	0.00
C_2H_4	+O	=CHO	$+CH_3$		$2.42 \cdot 10^{06}$	2.08	0.00
C_2H_4	+OH	$=C_2H_3$	$+H_2O$		$2.11 \cdot 10^{13}$	0.0	24.9
---- 27. CH_3CHO Reactions							
CH_3CHO	+M*	$=CH_3$	+CHO	+M*	$7.00 \cdot 10^{15}$	0.0	343.
CH_3CHO	+H	$=CH_3CO$	$+H_2$		$2.10 \cdot 10^{09}$	1.16	10.1
CH_3CHO	+H	$=H_2$	$+CH_2CHO$		$2.00 \cdot 10^{09}$	1.16	10.1
CH_3CHO	+O	$=CH_3CO$	+OH		$5.00 \cdot 10^{12}$	0.0	7.60
CH_3CHO	+O	=OH	$+CH_2CHO$		$8.00 \cdot 10^{11}$	0.0	7.60
CH_3CHO	$+O_2$	$=CH_3CO$	$+HO_2$		$4.00 \cdot 10^{13}$	0.0	164.
CH_3CHO	+OH	$=CH_3CO$	$+H_2O$		$2.30 \cdot 10^{10}$	.73	-4.70
CH_3CHO	$+HO_2$	$=CH_3CO$	$+H_2O_2$		$3.00 \cdot 10^{12}$	0.0	50.0
CH_3CHO	$+^3CH_2$	$=CH_3CO$	$+CH_3$		$2.50 \cdot 10^{12}$	0.0	15.9
CH_3CHO	$+CH_3$	$=CH_3CO$	$+CH_4$		$2.00 \cdot 10^{-06}$	5.64	10.3
---- 28. C_2H_5 Reactions							
C_2H_5		$=C_2H_4$	+H		$1.02 \cdot 10^{43}$	-9.1	224.

C_2H_5	$+H$	$=CH_3$	$+CH_3$	$3.00 \cdot 10^{13}$	0.0	0.00
C_2H_5	$+O$	$=H$	$+CH_3CHO$	$5.00 \cdot 10^{13}$	0.0	0.00
C_2H_5	$+O$	$=CH_2O$	$+CH_3$	$1.00 \cdot 10^{13}$	0.0	0.00
C_2H_5	$+O_2$	$=C_2H_4$	$+HO_2$	$1.10 \cdot 10^{10}$	0.0	-6.30
C_2H_5	$+CH_3$	$=C_2H_4$	$+CH_4$	$1.14 \cdot 10^{12}$	0.0	0.00
C_2H_5	$+C_2H_5$	$=C_2H_4$	$+C_2H_6$	$1.40 \cdot 10^{12}$	0.0	0.00
---- 29. C_2H_6 Reactions						
C_2H_6	$+H$	$=C_2H_5$	$+H_2$	$1.40 \cdot 10^{09}$	1.5	31.1
C_2H_6	$+O$	$=C_2H_5$	$+OH$	$1.00 \cdot 10^{09}$	1.5	24.4
C_2H_6	$+OH$	$=C_2H_5$	$+H_2O$	$7.20 \cdot 10^{06}$	2.0	3.60
C_2H_6	$+HO_2$	$=C_2H_5$	$+H_2O_2$	$1.70 \cdot 10^{13}$	0.0	85.9
C_2H_6	$+O_2$	$=C_2H_5$	$+HO_2$	$6.00 \cdot 10^{13}$	0.0	217.
C_2H_6	$+^3CH_2$	$=C_2H_5$	$+CH_3$	$2.20 \cdot 10^{13}$	0.0	36.3
C_2H_6	$+CH_3$	$=C_2H_5$	$+CH_4$	$1.50 \cdot 10^{-07}$	6.0	25.4

If, as an example, the reaction $H + O_2 \rightarrow OH + O$ is considered, the rate laws are given by the equations

$$\begin{aligned} \mathrm{d[H]/d}t &= -k\,[\mathrm{H}]\,[\mathrm{O_2}] \\ \mathrm{d[O_2]/d}t &= -k\,[\mathrm{H}]\,[\mathrm{O_2}] \\ \mathrm{d[OH]/d}t &= k\,[\mathrm{H}]\,[\mathrm{O_2}] \\ \mathrm{d[O]/d}t &= k\,[\mathrm{H}]\,[\mathrm{O_2}]\,. \end{aligned}$$

For the elementary reaction $OH + OH \rightarrow H_2O + O$ (or $2\,OH \rightarrow H_2O + O$) one obtains the rate equations

$$\mathrm{d[OH]/d}t = -2\,k\,[\mathrm{OH}]^2 \qquad \mathrm{d[H_2O]/d}t = k\,[\mathrm{OH}]^2$$

$$\mathrm{d[O]/d}t = k\,[\mathrm{OH}]^2\,.$$

Thus, rate laws can always be specified for elementary reaction mechanisms. If the *reaction mechanism* is composed of all possible elementary reactions in the system (complete mechanism), the mechanism is valid for all conditions (i. e., temperatures and mixture compositions). Complete mechanisms are rarely available.

For an elementary mechanism composed of R reactions of S species, which are given by

$$\sum_{s=1}^{S} \nu_{rs}^{(e)} \mathrm{A}_s \xrightarrow{k_r} \sum_{s=1}^{S} \nu_{rs}^{(p)} \mathrm{A}_s \quad \text{with} \quad r = 1,\ldots,R\,, \tag{6.17}$$

the rate of formation of a species i is given by summation over the rate equations (6.16) of all elementary reactions,

$$\left(\frac{\partial c_i}{\partial t}\right)_{\text{chem}} = \sum_{r=1}^{R} k_r \left(\nu_{ri}^{(p)} - \nu_{ri}^{(e)}\right) \prod_{s=1}^{S} c_s^{\nu_{rs}^{(e)}} \quad \text{with} \quad i = 1,\ldots,S\,. \tag{6.18}$$

6.4 Experimental Investigation of Elementary Reactions

Experimental setups for the measurement of elementary reactions can be divided into three components: the kind of reactor, the method of generation of the reactive species, and the kind of analysis (see, e. g., Homann 1975).

Reactors: Usually *static reactors* (an isothermal vessel is filled with reactants and then the time behavior of the concentrations is measured) or *flow reactors* (spatial profiles of the concentrations give information about their time behavior) are used.

Generation of the Reactive Species: Usually reactive atoms (e. g., H, O, N, ...) or *radicals* (e. g., OH, CH, CH_3, C_2H_5, ...) have to be produced as reactants. This is either done by a microwave discharge (H_2, O_2, ... form H, O, ... -atoms), flash photolysis, laser photolysis (dissociation of molecules by UV light), or thermally using high temperatures (e. g., dissociation by rapid heating in a shock tube). High dilution with noble gases (He, Ar) slows down the reaction of those species with themselves.

Analysis: The concentration measurement has to be sensitive (if, e. g., the reaction rate of bi- or trimolecular reactions is slowed down by dilution) and fast. Methods used are *mass spectroscopy*, *electron spin resonance*, all kinds of *optical spectroscopy* and *gas chromatography*.

Figure 6.2 shows a setup for the measurement of rate coefficients. The generation of radicals (here H-atoms and O-atoms) is done by a microwave discharge. The chemical reaction (here, e. g., with butadiyne, C_4H_2, a stable hydrocarbon) takes place in a flow reactor, and the products are monitored using mass spectroscopy.

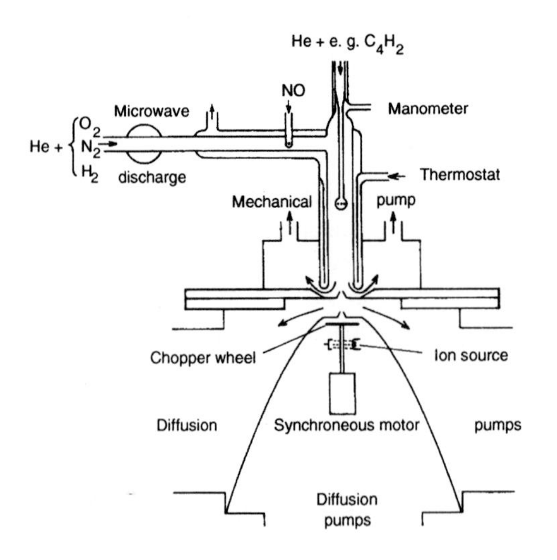

Fig. 6.2. Combination of microwave discharge and mass spectroscopy for the investigation of elementary reactions; here the reaction of H-, O- or N-atoms with a stable species (Schwanebeck and Warnatz 1972)

6.5 Temperature Dependence of Rate Coefficients

It is characteristic for chemical reactions that their rate coefficients depend strongly and in a nonlinear way on the temperature. According to Arrhenius (1889), this temperature dependence can be described by a simple formula (*Arrhenius law*)

$$k = A' \cdot \exp\left(-\frac{E'_a}{RT}\right). \tag{6.19}$$

More recently, accurate measurements showed a temperature dependence of the *preexponential factor* A', which, however, is usually small in comparison to the exponential dependence,

$$k = A\,T^b \cdot \exp\left(-\frac{E_a}{RT}\right). \tag{6.20}$$

The *activation energy* E_a corresponds to an energy barrier which has to be overcome during the reaction (see Fig. 6.3). Its maximum value corresponds to the bond energies in the molecule (in dissociation reactions, e. g., the activation energy is approximately equal to the bond energy of the bond, which is split), but it can also be much smaller (or even zero), if new bonds are formed simultaneously with the breaking of the old bonds.

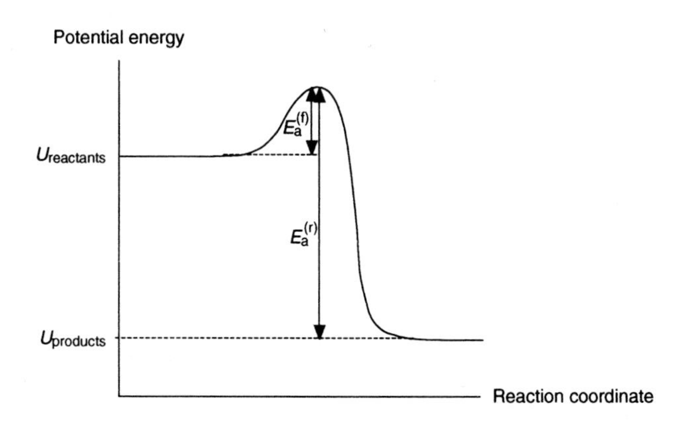

Fig. 6.3. Energy diagram for a chemical reaction. The relation $E_a^{(f)} - E_a^{(r)} = U_{products} - U_{reactants}$ is a result of (6.9). The reaction coordinate is the path of minimum potential energy from reactants to products with respect to the changing interatomic distance (e. g., Atkins 1996)

Figure 6.4 shows the temperature dependence of some elementary reactions (here: reactions of halogen atoms with molecular hydrogen). Plotted are the logarithms of the rate coefficients k versus the reciprocal temperature. According to (6.19) a linear dependence is obtained ($\log k = \log A$ - const./T); a temperature dependence of the preexponential factor $A \cdot T^b$ is often obscured by the experimental uncertainty.

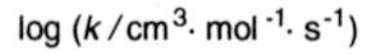

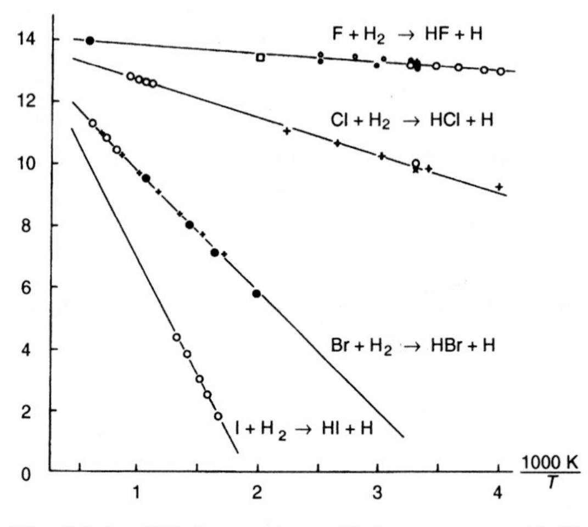

Fig. 6.4. $k = k(T)$ for reactions of halogen atoms with H_2 (data from Homann et al. 1970)

For vanishing activation energies, or for very high temperatures, the exponential term in (6.19) goes to 1. Then the reaction rate coefficient is only governed by the preexponential factor A' or $A{\cdot}T^b$, respectively. This preexponential factor has different meanings in uni-, bi- and trimolecular reactions:

For unimolecular reactions, the reciprocal value of A corresponds to a mean lifetime of an activated (reactive) molecule. In dissociation reactions, this lifetime is determined by the frequency of the vibration of the bond which is broken. Typically, the preexponential factor is given by twice the frequency of the bond vibration ($A' \approx 10^{14}$-$10^{15}\ s^{-1}$).

For bimolecular reactions the preexponential factor A' corresponds to a product of collision rate and probability of reaction. This collision rate is an upper limit for the reaction rate. The kinetic theory of gases yields values for A' in (5.4) in the range from 10^{13} to $10^{14}\ cm^3mol^{-1}s^{-1}$.

In trimolecular reactions, a third collision partner has to remove the energy of the reaction. If, for example, two hydrogen atoms collide and form a hydrogen molecule for a short time, this molecule will dissociate immediately because of the excess of energy in the molecule. On closer inspection, a three-body reaction is actually two bimolecular reactions in rapid succession. Expressing sequential bimolecular reactions as trimolecular reactions is a convenience that leads to a wide range of preexponential factors and effective E_a that are small or even negative.

6.6 Pressure Dependence of Rate Coefficients

The apparent pressure dependence of rate coefficients of dissociation (unimolecular) and recombination (trimolecular) reactions (see, e. g., Reactions 5-8 in Tab. 6.1) is an indication that these reactions are not elementary; they are in fact a sequence of reactions. In the simplest case, the pressure dependence can be understood using the *Lindemann model* (1922). According to this model, a unimolecular decomposition is only possible, if the energy in the molecule is sufficient to break the bond. Therefore, it is necessary that, prior to the decomposition reaction, energy is added to the molecule by collision with other molecules M (e. g., for the excitation of the molecular vibrations). Then the excited molecule may decompose into the products, or it can deactivate through a collision,

$$\begin{array}{llll} A + M & \xrightarrow{k_a} & A^* + M & \text{(activation)} \\ A^* + M & \xrightarrow{k_{-a}} & A + M & \text{(deactivation)} \\ A^* & \xrightarrow{k_u} & \text{P(roducts)} & \text{(unimolecular reaction)}\,. \end{array} \tag{6.21}$$

According to (6.3), the rate equations for this case are given by

$$\frac{\mathrm{d[P]}}{\mathrm{d}t} = k_\mathrm{u}[\mathrm{A}^*] \quad \text{and} \quad \frac{\mathrm{d[A^*]}}{\mathrm{d}t} = k_\mathrm{a}[\mathrm{A}][\mathrm{M}] - k_{-\mathrm{a}}[\mathrm{A}^*][\mathrm{M}] - k_\mathrm{u}[\mathrm{A}^*]\,. \tag{6.22}$$

Assuming that the concentration of the reactive intermediate A^* is in a *quasi-steady state* (details are discussed in Section 7.1),

$$\frac{\mathrm{d[A^*]}}{\mathrm{d}t} \approx 0\,, \tag{6.23}$$

one obtains for the concentration of the activated species $[A^*]$ and the formation of the product P

$$[\mathrm{A}^*] = \frac{k_\mathrm{a}[\mathrm{A}][\mathrm{M}]}{k_{-\mathrm{a}}[\mathrm{M}] + k_\mathrm{u}}$$

and

$$\frac{\mathrm{d[P]}}{\mathrm{d}t} = \frac{k_\mathrm{u} k_\mathrm{a}[\mathrm{A}][\mathrm{M}]}{k_{-\mathrm{a}}[\mathrm{M}] + k_\mathrm{u}}\,. \tag{6.24}$$

Two extremes can be distinguished, i. e., reaction at very low and very high pressures.

In the *low pressure range* the concentration of the collision partners M is very small; with $k_{-a}[M] << k_u$ one obtains an apparent second-order rate law

$$\frac{\mathrm{d[P]}}{\mathrm{d}t} = k_\mathrm{a} \cdot [\mathrm{A}][\mathrm{M}]\,. \tag{6.25}$$

Thus, the reaction rate is proportional to the concentrations of species A and the collision partner M, because the activation is slow (i. e., rate-limiting) at low pressures.

In the *high pressure range*, the collision partner M has a large concentration and, together with $k_{-a}[M] >> k_u$, one obtains the apparent first order rate law

$$\frac{\mathrm{d[P]}}{\mathrm{d}t} = \frac{k_\mathrm{u} k_\mathrm{a}}{k_{-\mathrm{a}}}[\mathrm{A}] = k_\infty \cdot [\mathrm{A}]\,. \tag{6.26}$$

Here the reaction rate does not depend on the concentrations of the collision partners, because at high pressures collisions occur very often and, thus, the decomposition of the activated molecule A^* is rate-limiting instead of the activation.

The Lindemann mechanism illustrates the fact that the reaction orders of complex (i. e., non-elementary) reactions depend on the conditions chosen. However, the Lindemann mechanism itself is a simplified model. More accurate results for the pressure dependence of unimolecular reactions can be obtained from the *theory of unimolecular reactions* (see, e. g., Robinson and Holbrook 1972, Atkins (1996), Golden (1994). This theory takes into account that not only one activated species can be defined, but a large number of activated molecules with different levels of activation.

If the rate law of a unimolecular reaction is written as $d[P]/dt = k[A]$, then the rate coefficient k depends on pressure and temperature. The theory of unimolecular reactions yields *fall-off* curves which describe the pressure dependence of k for different temperatures. Usually the logarithm of the rate coefficient is plotted versus the logarithm of the pressure.

Tab. 6.2. Fitted Arrhenius parameters for pressure-dependent reactions in Tab. 6.1 (1000 K $< T <$ 2500 K); the parameters do not have any physical meaning for complex reactions

p/bar	A[cm, mol,s]	b	E/kJ·mol^{-1}	A[cm, mol,s]	b	E/kJ·mol^{-1}
	$CH_3 + H = CH_4$			$CH_3 + CH_3 = C_2H_6$		
.0253	$3.77\cdot10^{35}$	-7.30	36.0	$3.23\cdot10^{58}$	-14.0	77.8
0.120	$1.26\cdot10^{36}$	-7.30	36.7	$2.63\cdot10^{57}$	-13.5	80.8
1.000	$1.93\cdot10^{36}$	-7.00	38.0	$1.69\cdot10^{53}$	-12.0	81.2
3.000	$4.59\cdot10^{35}$	-6.70	39.3	$1.32\cdot10^{49}$	-10.7	75.7
9.000	$8.34\cdot10^{33}$	-6.10	38.0	$8.32\cdot10^{43}$	-9.10	67.0
20.00	$2.50\cdot10^{32}$	-5.60	36.5	$1.84\cdot10^{39}$	-7.70	57.8
50.00	$1.39\cdot10^{30}$	-4.90	32.8	$3.37\cdot10^{33}$	-6.00	45.3
	$CH_3OH = CH_3 + OH$			$C_2H_3 = C_2H_2 + H$		
.0267	$2.17\cdot10^{24}$	-3.30	368.	$0.94\cdot10^{38}$	-8.50	190.
0.120	$3.67\cdot10^{26}$	-3.70	381.	$3.77\cdot10^{38}$	-8.50	190.
1.000	$9.51\cdot10^{29}$	-4.30	404.	$4.73\cdot10^{40}$	-8.80	194.
3.000	$2.33\cdot10^{29}$	-4.00	407	$1.89\cdot10^{42}$	-9.10	200.
9.000	$8.44\cdot10^{27}$	-3.50	406.	$3.63\cdot10^{43}$	-9.30	205.
20.00	$2.09\cdot10^{26}$	-3.00	403.	$4.37\cdot10^{43}$	-9.20	208.
50.00	$4.79\cdot10^{24}$	-2.50	400.	$0.95\cdot10+45$	-9.50	220.
	$CH_3CO = CH_3 + CO$			$C_2H_5 = C_2H_4 + H$		
.0253	$4.13\cdot10^{23}$	-4.70	68.5	$2.65\cdot10^{42}$	-9.50	210.
0.120	$3.81\cdot10^{24}$	-4.80	70.0	$1.76\cdot10^{43}$	-9.50	215.
1.000	$2.32\cdot10^{26}$	-5.00	75.1	$1.02\cdot10^{43}$	-9.10	224.
3.000	$4.37\cdot10^{27}$	-5.20	80.9	$6.09\cdot10^{41}$	-8.60	226.
9.000	$8.79\cdot10^{28}$	-5.40	88.3	$6.67\cdot10^{39}$	-7.90	227.
20.00	$2.40\cdot10^{29}$	-5.40	92.9	$2.07\cdot10^{37}$	-7.10	224.
50.00	$7.32\cdot10^{29}$	-5.40	98.4	$1.23\cdot10^{34}$	-6.10	219.

Typical fall-off curves are shown in Fig. 6.5. For $p \rightarrow \infty$, $k = k_u k_a[M]/(k_{-a}[M]+k_u)$ tends to the limit k_∞, i. e., the rate coefficient becomes independent of the pressure, see (6.26). At low pressures the rate coefficient k is proportional to $[M] = p/RT$ (6.25), and a linear dependence results. By the same reasoning, the reaction rate coefficient k will decrease with temperature if the effective activation energy of k_∞ is low.

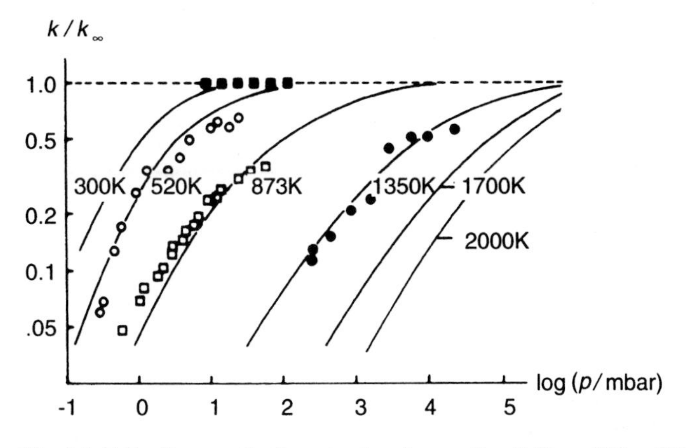

Fig. 6.5. Fall-off curves for the unimolecular reaction $C_2H_6 \rightarrow CH_3 + CH_3$ (Warnatz 1983, 1984)

As can be seen from Fig. 6.5, the fall-off curves depend very much on the temperature. Thus, the rate coefficients of unimolecular reactions show different temperature dependences at different values of the pressure. Figure 6.6 shows rate coefficients for the reverse of the "unimolecular" reaction considered in Fig. 6.5. Other reaction products of competing bimolecular steps are possible, leading to a rather complex situation.

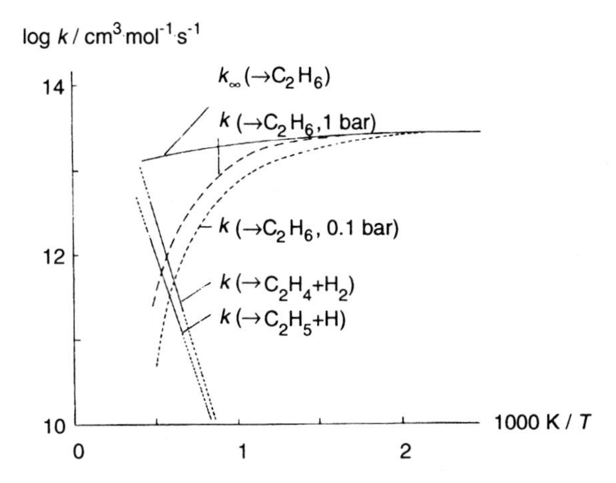

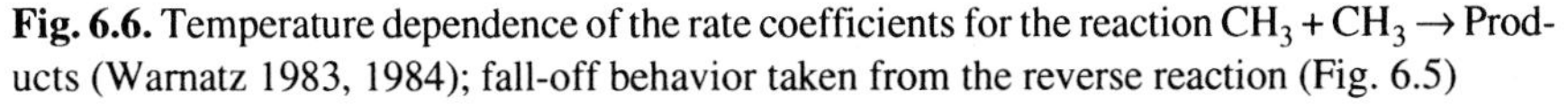
Fig. 6.6. Temperature dependence of the rate coefficients for the reaction $CH_3 + CH_3 \rightarrow$ Products (Warnatz 1983, 1984); fall-off behavior taken from the reverse reaction (Fig. 6.5)

Arrhenius parameters for the pressure-dependent reactions in Tab. 6.1 are listed in Tab. 6.2. These Arrhenius parameters are the result of a fitting procedure $(1000\ \text{K} < T < 2500\ \text{K})$ of curves like that given in Fig. 6.6. The positive slopes in this figure lead to apparent negative "activation energies" that lack physical interpretation as illustrated in Fig. 6.3. The appropriate treatment of unimolecular and trimolecular reactions is important because many experiments on reaction kinetics are done at atmospheric or lower pressure while many combustion processes run at elevated pressure.

6.7 Surface Reactions

Surface reactions play an important part in many combustion applications, e. g., in wall recombination processes during autoignition (Section 10.3), in coal combustion (Section 15.2), in soot formation and oxidation (Chapter 18), or in catalytic combustion (see below).

The main property of surface reactions (in comparison to gas-phase reactions) is the inclusion of surface sites and species adsorbed on these sites into the description of reaction rates. Surface sites and surface species (species that are attached to the surface site) have a *surface concentration* measured, e. g., in mol/cm^2; for bare platinum metal the surface-site concentration is $2.72 \cdot 10^{-9}\ \text{mol/cm}^2$. These surface concentrations in turn lead to initially unfamiliar units for reaction rates and rate coefficients. There may be more than one rate coefficient for the same material, since surface sites with different adsorption energies (e. g., on terraces and at steps; see Hsu et al. 1987) have to be treated as different species.

A review of heterogeneous reactions can be found in Atkins' textbook (1996). More details are presented by Boudart and Djega-Mariadassou (1984), Bond (1990), and Christmann (1991).

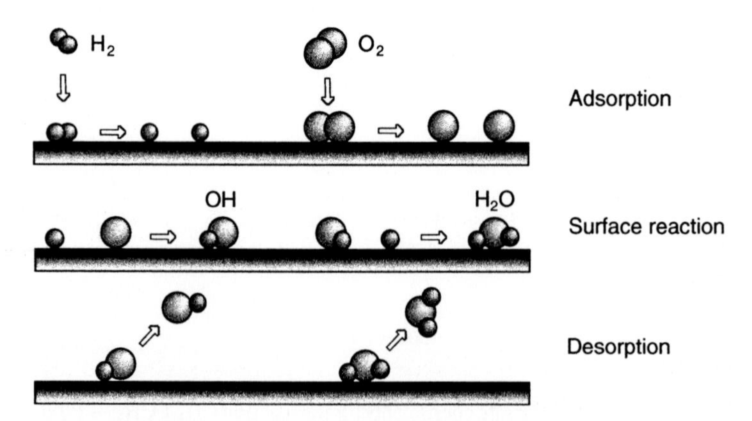

Fig. 6.7. Surface reaction mechanism of hydrogen oxidation (schematic).

For numerical calculations, a general formalism for the treatment of heterogeneous reaction and details of the chemical reaction-rate formulation can be found in the users manuals for the SURFACE CHEMKIN (Coltrin et al. 1993) software, together with a review of basic phenomena.

As an example, a H_2 oxidation reaction mechanism for Pt surfaces with associated rate coefficients following the work of Ljungström et al. (1989) is given in Tab. 6.3 and illustrated in Fig. 6.7. The mechanism consists of the steps of *dissociative adsorption* of both H_2 and O_2, which leads to H atoms and O atoms absorbed on the surface. These atoms are very mobile on the surface. Thus, the adsorbed atoms collide while attached to the surface, forming first OH and then adsorbed H_2O in *surface reactions*. Finally, the *desorption* of H_2O into the gas phase (see Fig. 6.7). This mechanism is based on LIF measurements of desorbed OH and is similar to reaction schemes postulated by Hsu et al. (1987) and by Williams et al. (1992). A similar mechanisms are existing for the oxidation of CH_4 (Deutschmann et al. 1994).

Tab. 6.3. Detailed surface reaction mechanism of hydrogen oxidation on a platinum surface in terms of elementary steps (Warnatz et al. 1994)

Reaction	*S*	*A*(cm,mol,s)	E_a(kJ/mol)
1. H_2/O_2 Adsorption/Desorption			
H_2 + Pt(s) = H_2(s)	0.10		0.0
H_2(s) + Pt(s) = H(s) + H(s)		$1.50 \cdot 10^{23}$	17.8
O_2 + Pt(s) = O_2(s)	.046		0.0
O_2(s) + Pt(s) = O(s) + O(s)		$5.00 \cdot 10^{24}$	0.0
2. Surface Reactions			
H(s) + O(s) = OH(s) + Pt(s)		$3.70 \cdot 10^{21}$	19.3
H(s) + OH(s) = H_2O(s) + Pt(s)		$3.70 \cdot 10^{21}$	0.0
OH(s) + OH(s) = H_2O(s) + O + Pt(s)		$3.70 \cdot 10^{21}$	0.0
3. Product Adsorption/Desorption			
H + Pt(s) = H(s)	1.00		0.0
O + Pt(s) = O(s)	1.00		0.0
H_2O + Pt(s) = H_2O(s)	0.75		0.0
OH + Pt(s) = OH(s)	1.00		0.0

S = sticking coefficient, see (6.27); rate coefficient $k = A \cdot \exp(-E_a/RT)$, neglecting temperature dependence of the preexponential factor *A*; (s) denotes surface species and Pt(s) free surface sites

Adsorption: Rates of adsorption cannot exceed the rate that gas-phase molecules collide with a surface. On the basis of simple hard-sphere gas kinetics the maximum adsorption rate cannot exceed (see, e. g., Atkins 1996)

$$k_{\max} = c \cdot \bar{\upsilon}/4 \ ,$$

where c = concentration, $\bar{\upsilon}$ = mean velocity. The actual rate is the product of this maximum rate and a *sticking coefficient* S, which is the probability that a molecule sticks when colliding with the surface. The adsorption rate is then given by (Coltrin et al. 1993)

$$k_{\text{ads}} = S \cdot k_{\max} = S \cdot \sqrt{\frac{RT}{2\pi M}} \quad . \tag{6.27}$$

The theoretically unresolved problem is the estimate of sticking coefficient S, which, of course, should have a maximum value of 1, but can have even extremely small values as low as 10^{-6} (Bond 1990).

Surface Reaction: In order to react, the surface species must be able to move on the surface. The mechanism for this movement is that the adsorbed species overcome a small energy barrier and "hop" to an adjacent site. Interestingly, a strongly adsorbed species will be immobile and, thus, the substrate in this case is a poor catalyst for that species. (Furthermore, a strongly adsorbed species will not leave and will *poison* the surface by permanently occupying the sites; common poisons are sulfur and lead.)

The reaction rate $\dot{s}$ of a surface reaction A(s) + A(s) is estimated analogously to bimolecular reactions in the gas phase (for review, see Fig. 5.3). The velocity υ is the product of the hopping frequency ν and the distance σ hopped, where σ is the diameter of the molecule A. The zigzag path of the hopping species has then a collision area $2\sigma \cdot \upsilon \cdot \Delta t$, where Δt is the time interval considered. Thus, the number N of collisions per unit time of the molecule considered with other ones in the collision area is

$$N = 2\sigma \cdot \upsilon \cdot [n] \ ,$$

where $[n]$ is the particle density of the surface species A(s). Consequently, the total number of collisions per unit time of all particles A(s) is $2\sigma \cdot \upsilon \cdot [n] \cdot [n]$. Finally, multiplication by the probability of sufficient energy (described by an Arrhenius term) gives the result

$$2\nu \cdot \sigma^2 \cdot \exp(-E/kT) \cdot [n][n] = \dot{s} = A_{\text{surf}} \cdot \exp(-E/kT) \cdot [n][n] \ .$$

With estimates $\sigma = 2 \cdot 10^{-8}$ cm, $\nu = 10^{14}$ s^{-1}, the preexponential A_{surf} can be calculated to be

$$A_{\text{surf}} = 8 \cdot 10^{-2} \, \frac{\text{cm}^2}{\text{s}} \approx 5 \cdot 10^{22} \, \frac{\text{cm}^2}{\text{mol} \cdot \text{s}}$$

in rough agreement with the values of the preexponentials given in Tab. 6.3.

Desorption: Desorption requires that the molecules have sufficient energy to overcome the bond strength between the surface and the adsorbed species. The desorption behaves typically Arrhenius-like with an activation energy E_{des} that is comparable to bond strength,

$$k_{\text{des}} = A_{\text{des}} \cdot \exp(-E_{\text{des}}/kT) \ .$$

It is often proposed that the preexponential factor A_{des} may be estimated from vibrational frequencies of the corresponding bond, which are related to the bond energy. Quantum mechanical calculations of surface bond-energies are increasingly available allowing improved estimation of frequencies and activation energies of adsorption/desorption processes. The preexponentials for OH and H_2O desorption ($\sim 10^{13}$ s^{-1}) are in rather good agreement with these ideas.

The limiting expressions for rate coefficients mentioned above are extremely helpful, even if they give only rough estimates, due to the fact that the understanding of surface reactions is far behind that of gas-phase reactions. This is due to

- missing experimental rate data on surface reactions, which fortunately often is compensated by the fact that the adsorption/desorption reactions are rate-limiting, not the surface processes (Behrendt et al. 1995) and
- the fact that surface species are not uniformly distributed as gas-phase particles can be assumed to be. Instead, nonuniform surface concentrations can coexist, leading to phenomena like island formation and oscillating structures (Bar et al. 1995, Kissel-Osterrieder et al. 1998).

Recent advances in the development of surface species diagnostics (Lauterbach et al. 1995, Härle et al. 1998) indicate that this unsatisfactory situation is improving.

Some Typical Results: Results can easily produced for a counterflow geometry, where the surface is presented by a (boundary) point. The governing equations in this case are one-dimensional and similar to that treated in Chapter 3 (see Section 9.1 for more details). Variation of the surface temperature at a specified mixture composition in the gas phase leads to the determination of catalytic ignition temperatures; an example is given in Fig. 6.8.

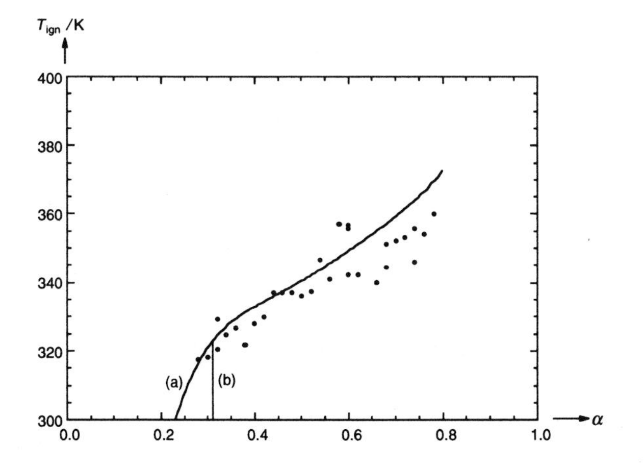

Fig. 6.8. Ignition temperature T_{ign} in H_2-O_2-N_2 mixtures (94 % N_2, p = 1 bar) on Pd as function of $\alpha = p_{H_2}/(p_{O_2} + p_{H_2})$; points: measurements (Behrendt et al. 1996), lines: computations (a) with initial H surface coverage, (b) with initial O surface coverage (Deutschmann et al. 1996)

Typical of catalytic processes is a transition from kinetically controlled behavior at low temperature (indicated by a high surface concentration and a low gas-phase gradient in the boundary layer; see Fig. 6.9) to transport control at high temperature (indicated by a low surface concentration and a high gas-phase gradient).

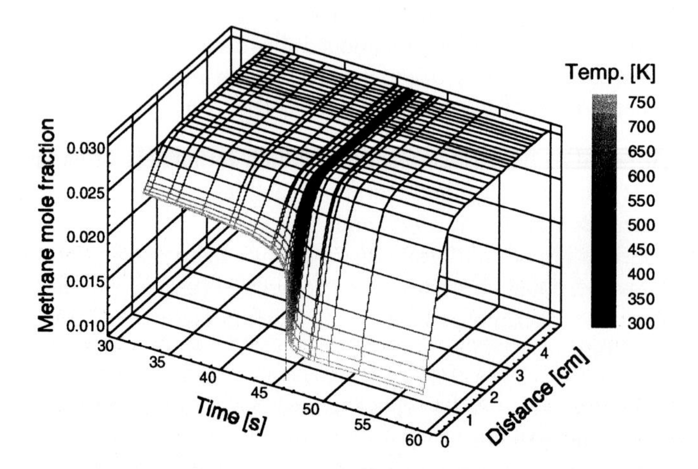

Fig. 6.9. Methane mole fraction as function of time and distance from the Pt catalyst surface during heterogeneous ignition of a CH_4/O_2 mixture for $\alpha = 0.5$ (Deutschmann et al. 1996)

6.8 Exercises

Exercise 6.1. Write down the rate law for the formation of hydrogen atoms (H) according to the reaction mechanism (all reactions are elementary reactions)

1.	$H + H + M$	$\rightarrow H_2 + M$
2.	$H_2 + M$	$\rightarrow H + H + M$
3.	$H + H + H$	$\rightarrow H_2 + H$
4.	$H + O_2$	$\rightarrow OH + O$
5.	$H + O_2 + M$	$\rightarrow HO_2 + M$.

When has the number $\nu_{rs}^{(e)}$ in (6.18) to be interpreted as a stoichiometric number, when as a reaction order?

Exercise 6.2. What is the rate law for the decomposition of octane into two butyl radicals,

$$C_8H_{18} \rightarrow C_4H_9 + C_4H_9 \ ?$$

How do you have to plot measured C_8H_{18} concentrations versus time, if a linear graph has to result? What is the decay time of the decomposition of C_8H_{18} at 1500 K if $k = 1 \cdot 10^{16} \exp(-340 \text{ kJ} \cdot \text{mol}^{-1}/RT) \text{ s}^{-1}$?

7 Reaction Mechanisms

In Chapter 6 it has been shown that the combustion of a simple fuel like hydrogen (global reaction $2\ H_2 + O_2 \rightarrow 2\ H_2O$) requires nearly 40 elementary reactions for a satisfactory chemical mechanism. For combustion of hydrocarbon fuels, as simple as methane CH_4, the number of elementary reactions in the chemical mechanism is much larger. In some cases several thousands of elementary reactions (e. g., in the case of autoignition of the Diesel fuel with the typical component *cetane* $C_{16}H_{34}$; see Chapter 16) influence the overall process.

The interaction of these elementary reactions governs the whole combustion. However, independent of the specific properties of the fuel, all reaction mechanisms show properties characteristic for all combustion processes. For example, only a few of the many elementary reactions determine the rate of the overall process (*rate-limiting reactions*).

In this chapter, characteristic properties of mechanisms, methods for the analysis of mechanisms, basic concepts for their simplification, and the consequences for the mathematical simulation shall be discussed in detail. This is of particular interest, because the use of reaction mechanisms with more than 1000 chemical species is possible in the simulation of a homogeneous reactor (see Chapter 16). However, such large mechanisms leads to a prohibitive amount of computational work in the simulation of practical systems such as found in engines or furnaces due to the spatially inhomogeneous nature of three-dimensional, turbulent flows with changing concentrations and temperatures.

7.1 Characteristics of Reaction Mechanisms

Independent of the specific problems, reaction mechanisms show several characteristic properties. A knowledge of those properties improves the understanding of the chemical reaction system and provides valuable information that leads to the simplification of the mechanisms by eliminating those steps that are irrelevant for the problem at the actual conditions. Two often used means to mechanism simplification are *quasi-steady states* and *partial equilibria*, which will be discussed below.

7.1.1 Quasi-Steady States

A simple reaction chain, consisting of two elementary steps, will be considered as an example:

$$S_1 \xrightarrow{k_{12}} S_2 \xrightarrow{k_{23}} S_3 \,. \tag{7.1}$$

The rate laws for the different species S_1, S_2, and S_3 in this case are given by the expressions

$$\frac{d[S_1]}{dt} = -k_{12}[S_1] \,, \tag{7.2}$$

$$\frac{d[S_2]}{dt} = k_{12}[S_1] - k_{23}[S_2] \,, \tag{7.3}$$

and

$$\frac{d[S_3]}{dt} = k_{23}[S_2] \,. \tag{7.4}$$

If one assumes that at time $t = 0$ only the compound S_1 is present, then a lengthy calculation (together with initial conditions $[S_1]_{t=0} = [S_1]_0$, $[S_2]_{t=0} = 0$ and $[S_3]_{t=0} = 0$; see Section 7.2.3) gives the analytic solution

$$\begin{aligned}
[S_1] &= [S_1]_0 \exp(-k_{12}t) \\
[S_2] &= [S_1]_0 \frac{k_{12}}{k_{12}-k_{23}} \left\{\exp(-k_{23}t) - \exp(-k_{12}t)\right\} \\
[S_3] &= [S_1]_0 \left\{1 - \frac{k_{12}}{k_{12}-k_{23}} \exp(-k_{23}t) + \frac{k_{23}}{k_{12}-k_{23}} \exp(-k_{12}t)\right\} .
\end{aligned} \tag{7.5}$$

This solution can be verified by insertion into (7.2 - 7.4). As an example, it is assumed that S_2 is very reactive, and thus has a short life time (i. e., $k_{23} >> k_{12}$). The solution (7.5) is illustrated in Fig. 7.1 for $k_{12}/k_{23} = 0.1$.

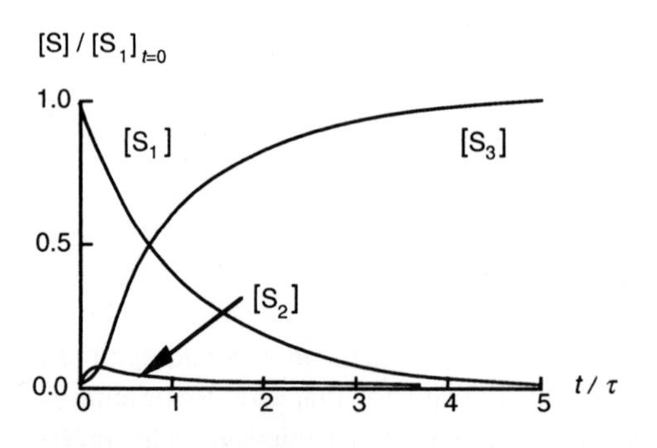

Fig. 7.1. Exact temporal behavior of the species concentrations in the reaction sequence $S_1 \rightarrow S_2 \rightarrow S_3$ (τ is the characteristic *life time* for $[S_1]_0$ to decay to $[S_1]_{t=\tau} = [S_1]_0/e$); here $\tau = 1/k_{12}$

Because S_2 is assumed to be a very reactive species, the rate of consumption of S_2 is approximately equal to the rate of formation of S_2, such that it can be written as an approximation (*quasi-steady state* assumption)

$$\frac{\mathrm{d}[S_2]}{\mathrm{d}t} = k_{12}[S_1] - k_{23}[S_2] \approx 0 . \tag{7.6}$$

The time behavior of the concentration of S_1 can be determined very easily, because (7.2) is easy to integrate. The result is (see Section 7.5)

$$[S_1] = [S_1]_0 \cdot \exp(-k_{12} \cdot t) . \tag{7.7}$$

If one is interested in the rate of formation of the product S_3, (7.4) yields only limited information, because only the concentration of the intermediate S_2 appears in the rate law for S_3. However, using the quasi-steady state assumption (7.6), one obtains the simple relationship

$$\frac{\mathrm{d}[S_3]}{\mathrm{d}t} = k_{12}[S_1] .$$

Insertion of (7.7) leads to the differential equation

$$\frac{\mathrm{d}[S_3]}{\mathrm{d}t} = k_{12}[S_1]_0 \cdot \exp(-k_{12} \cdot t) ,$$

which can be integrated yielding

$$[S_3] = [S_1]_0 \cdot \left[1 - \exp(-k_{12} \cdot t)\right] . \tag{7.8}$$

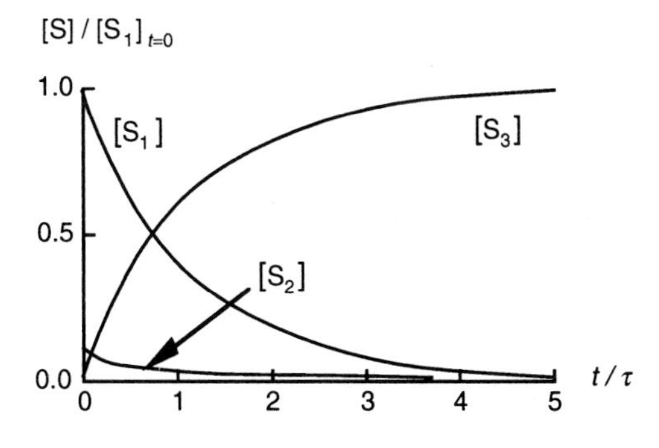

Fig. 7.2. Time behavior of the reaction $S_1 \rightarrow S_2 \rightarrow S_3$ assuming quasi-steady state for $[S_2]$

The time behavior of the concentrations, calculated according to (7.6) - (7.8), is an approximation for (7.2-7.4) using the steady-state assumption for S_2. The approximate results are shown in Fig. 7.2. Comparing Figs. 7.1 and 7.2 , one can see that the steady-state assumption is a superb approximation for the process, if $k_{23} >> k_{12}$. Only at the beginning of the reaction are there some deviations.

The combustion of hydrogen with chlorine shall be discussed now in order to provide a simple example, which is, nevertheless, of practical importance (Bodenstein and Lind 1906):

$$
\begin{array}{rlll}
Cl_2 + M & \rightarrow & Cl + Cl + M & (1) \\
Cl + H_2 & \rightarrow & HCl + H & (2) \\
H + Cl_2 & \rightarrow & HCl + Cl & (3) \\
Cl + Cl + M & \rightarrow & Cl_2 + M & (4)\ .
\end{array}
\qquad (7.9)
$$

Using the steady-state assumption, the rate laws for the intermediates H and Cl are

$$\frac{d[Cl]}{dt} = 2k_1[Cl_2][M] - k_2[Cl][H_2] + k_3[H][Cl_2] - 2k_4[H]^2[M] \approx 0 \ ,$$

$$\frac{d[H]}{dt} = k_2[Cl][H_2] - k_3[H][Cl_2] \approx 0, \quad \text{thus} \quad [H] \approx \frac{k_2[Cl][H_2]}{k_3[Cl_2]} \ .$$

Summation of these two rate laws yields as an expression for [Cl],

$$k_4[Cl]^2 = k_1[Cl_2] \ , \quad \text{thus} \quad [Cl] = \sqrt{\frac{k_1}{k_4}[Cl_2]} \ .$$

Then the rate law for the formation of HCl reads

$$
\begin{aligned}
\frac{d[HCl]}{dt} &= k_2[Cl][H_2] + k_3[H][Cl_2] = 2k_2[Cl][H_2] \\
&= 2k_2\sqrt{\frac{k_1}{k_4}}[Cl_2]^{\frac{1}{2}}[H_2] = k_{\text{total}}\,[Cl_2]^{\frac{1}{2}}[H_2] \ .
\end{aligned}
\qquad (7.10)
$$

Thus, the formation of hydrogen chloride can be expressed in terms of the concentrations of the reactants (H_2 and Cl_2). The concept of quasi-steady states allows one to obtain results despite the fact that the original equation system is a coupled set of ordinary differential equations, which cannot be solved analytically. Furthermore, the reaction of hydrogen with chlorine is an example which shows that the rate law of the overall reaction

$$H_2 + Cl_2 \rightarrow 2\,HCl \qquad (7.11)$$

does not have the reaction order 2 (as would be naively expected), but the order 1.5 (see (7.10)), because Reaction 7.11 is not an elementary one.

7.1.2 Partial Equilibrium

The concept of partial equilibrium will be illustrated using the mechanism for hydrogen combustion shown in Chapter 6 (Tab. 6.1). An analysis of experiments or simulations shows that at high temperatures ($T > 1800$ K at $p = 1$ bar) the reaction rates of forward and backward reactions are so fast that one obtains *partial equilibria* for the reactions

$$\begin{aligned} H + O_2 &= OH + O && (1,2) \\ O + H_2 &= OH + H && (3,4) \\ OH + H_2 &= H_2O + H \,. && (5,6) \end{aligned}$$

In this case, where each reaction is in equilibrium, forward and backward reaction rates are equal; the consequence is (Warnatz 1981b)

$$\begin{aligned} k_1 [H] \; [O_2] &= k_2 [OH] \; [O] \\ k_3 [O] \; [H_2] &= k_4 [OH] \; [H] \\ k_5 [OH] [H_2] &= k_6 [H_2O] [H] \,. \end{aligned}$$

This equation system can be solved for the intermediates [O], [H], and [OH]; one obtains the three relations

$$[H] = \left(\frac{k_1 k_3 k_5^2 [O_2][H_2]^3}{k_2 k_4 k_6^2 [H_2O]^2} \right)^{\frac{1}{2}} , \qquad (7.12)$$

$$[O] = \frac{k_1 k_5 [O_2][H_2]}{k_2 k_6 [H_2O]} , \qquad (7.13)$$

$$[OH] = \left(\frac{k_1 k_3}{k_2 k_4} [O_2][H_2] \right)^{\frac{1}{2}} . \qquad (7.14)$$

Thus, the concentrations of these reactive species, which are difficult to measure, can be expressed in terms of the concentrations of the stable (and, thus, easier to measure) species H_2, O_2 and H_2O. Figure 7.3 shows mole fractions of H, O, and OH in premixed stoichiometric H_2-air flames (Warnatz 1981b) at $p = 1$ bar, $T_u = 298$ K (unburnt gas temperature) as a function of the local temperature, which have been calculated using the detailed reaction mechanism and then recalculated using the partial equilibrium assumptions. It can be seen that the partial equilibrium assumption provides satisfactory results only at high temperatures. At temperatures below approximately 1600 K, partial equilibrium is not established, because the reaction times are slower than the characteristic time of combustion evaluated as the ratio of the flame thickness and the mean gas velocity $\tau = d/\bar{v}$ ($\approx 1\,\text{mm}/1\,\text{m}\cdot\text{s}^{-1} = 1$ ms typically).

Figure 7.4 shows spatial profiles of the mole fractions of oxygen atoms in a premixed stoichiometric C_3H_8-air flame at $p = 1$ bar, $T_u = 298$ K, calculated using a detailed mechanism, the (local) partial equilibrium assumption, and the (local) assumption of a complete equilibrium. While the assumption of a complete equilibrium provides unsatisfactory results at all temperatures, the partial equilibrium assumption at least works at sufficiently high temperatures. It should be noted here that the amount of oxygen atoms in a reaction system has a profound influence on the formation of nitric oxides; a subject taken up in Chapter 17.

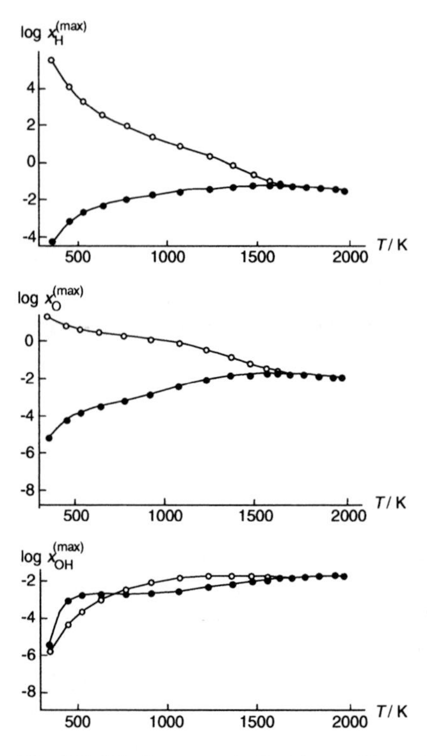

Fig. 7.3. Maximum mole fractions of the radicals H, O, and OH in premixed, stoichiometric H_2-air flames (Warnatz 1981b) at p = 1 bar, T_u = 298 K, calculated using a detailed mechanism (dark circles) and using the partial equilibrium assumption (light circles), where unphysical values ($x_i > 1$) can appear in some cases

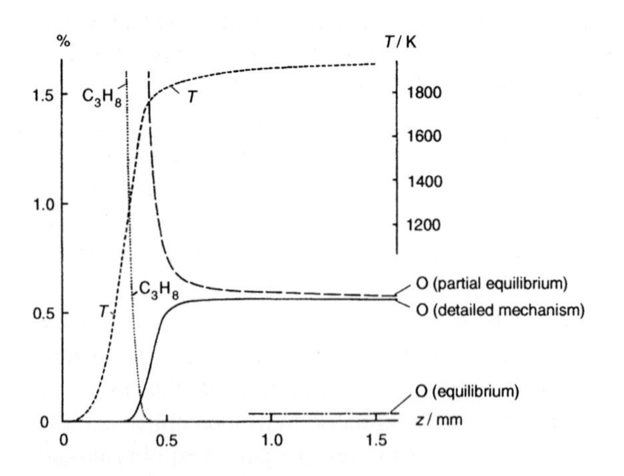

Fig. 7.4. Mole fractions of O in a premixed stoichiometric C_3H_8-air flame at p = 1 bar, T_u = 298 K, calculated using a full mechanism, the partial equilibrium assumption, and the assumption of a complete equilibrium (Warnatz 1987)

7.2 Analysis of Reaction Mechanisms

As mentioned above, complete reaction mechanisms for hydrocarbons may consist of several thousand elementary reactions. Depending on the question posed, many of these reactions can be neglected. Thus, analysis methods, which eliminate negligible reactions, are of interest. Several methods can be used:

Sensitivity analyses identify the rate limiting reaction steps. *Reaction flow analyses* determine the characteristic reaction paths. *Eigenvector analyses* determine the characteristic time scales and directions of the chemical reactions. The information obtained by these methods can be used to eliminate unimportant information and thus generate a simplified, or *reduced*, reaction mechanism.

7.2.1 Sensitivity Analysis

The rate laws for a reaction mechanism consisting of R reactions among S species can be written as a system of first order ordinary differential equations (see Chapter 6),

$$\begin{aligned} \frac{\mathrm{d}c_i}{\mathrm{d}t} &= F_i\left(c_1,\ldots,c_S\,;\,k_1,\ldots,k_R\right) \\ c_i\left(t=t_0\right) &= c_i^0 \end{aligned} \qquad i = 1, 2, \ldots, S. \tag{7.15}$$

The time t is the *independent variable*, the concentrations c_i of species i are the *dependent variables*, and k_r the *parameters* of the system; c_i^0 denote the *initial conditions* at t_0.

Here only the rate coefficients of the chemical reactions taken into account shall be considered as parameters of the system; nevertheless initial concentrations, pressure, etc. can be treated as system parameters, too, if it is desired. The solution of the differential equation system (7.15) depends on the initial conditions as well as on the parameters.

The following question is of high interest: Suppose that the system parameters are changed, i. e., the rate coefficients of the elementary reactions. How does the solution, the values of the concentrations at time t, change? For many of the elementary reactions, a change in the rate coefficients has nearly no effect on the time-dependent solution (showing that, e. g., quasi-steady states or partial equilibria are active). Even if one decides to include the reaction explicitly in the mechanism, one does not need a highly accurate rate coefficient. On the other hand, for a few of the elementary reactions, changes in the rate coefficients have really large effects on the outcome of the system. Accordingly, accurate rate coefficients are demanded. This points to where experimental resources should be expended. These few important reaction steps are *rate-determining steps* or *rate-limiting steps.*

The dependence of the solution c_i on the parameters k_r is called *sensitivity*. Absolute and relative sensitivities can be defined as

$$E_{i,r} = \frac{\partial c_i}{\partial k_r} \quad \text{and} \quad E_{i,r}^{\text{rel}} = \frac{k_r}{c_i}\frac{\partial c_i}{\partial k_r} = \frac{\partial \ln c_i}{\partial \ln k_r} . \tag{7.16}$$

The power of sensitivity analysis shall be illustrated by returning to the two-step reaction mechanism examined in Section 7.1.1 where an exact analytical solution as well as an approximate solution is given. It will be investigated how a change in the rate coefficients k_{12} and k_{23} influences the rate of formation of the concentration of the product S_3.

The sensitivity coefficients are computed by forming the partial derivative of the concentration $[S_3]$ in (7.5) with respect to the rate coefficients, holding time constant. The result is

$$E_{S_3,k_{12}}(t) = \frac{\partial[S_3]}{\partial k_{12}} = [S_1]_0 \frac{k_{23}}{(k_{12}-k_{23})^2}\left\{(k_{23}t - k_{12}t - 1)\exp(-k_{12}t) + \exp(-k_{23}t)\right\}$$

$$E_{S_3,k_{23}}(t) = \frac{\partial[S_3]}{\partial k_{23}} = [S_1]_0 \frac{k_{12}}{(k_{12}-k_{23})^2}\left\{\exp(-k_{12}t) + (k_{12}t - k_{23}t - 1)\exp(-k_{23}t)\right\}.$$

The relative sensitivities are computed, according to (7.16), as

$$E_{S_3,k_{12}}^{\text{rel}}(t) = \frac{k_{12}}{[S_3]} E_{S_3,k_{12}}(t) \quad \text{and} \quad E_{S_3,k_{23}}^{\text{rel}}(t) = \frac{k_{23}}{[S_3]} E_{S_3,k_{23}}(t) . \tag{7.17}$$

The time behaviors of the relative sensitivity coefficients are plotted together with the concentration of the product $[S_3]$ in dimensionless form in Fig. 7.5 for $k_{12} = \tau^{-1}$, $k_{23} = 100\tau^{-1}$ and $[S_1]_0 = 1$ (τ = life time; see Fig. 7.1). It can be seen that the relative sensitivity with respect to the fast reaction 23 tends to zero after a very short time, whereas the sensitivity with respect to the slow reaction 12 has a high value during the whole reaction process. Thus, sensitivity analysis distinctly shows that the product concentration $[S_3]$ has high relative sensitivity with respect to the slow (i. e., rate-limiting) reaction 12, and there is low relative sensitivity with respect to the fast (and, thus, not rate-limiting) reaction 23.

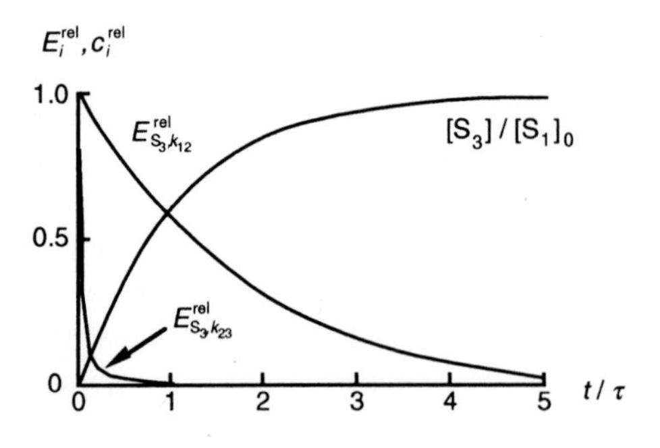

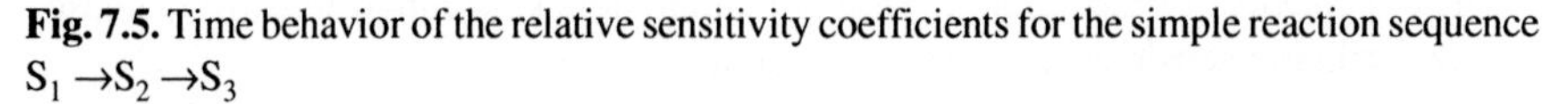

Fig. 7.5. Time behavior of the relative sensitivity coefficients for the simple reaction sequence $S_1 \rightarrow S_2 \rightarrow S_3$

Thus sensitivity analysis, which identifies rate limiting reaction steps, is a valuable tool for deeper understanding of complex reaction mechanisms (see, e. g., Nowak and Warnatz 1988).

In most applications of sensitivity analysis, an analytical solution of the differential equation system (analogous to (7.5)) and subsequent partial differentiation of the analytical solution is, of course, not possible. However, the sensitivity analysis is done numerically by generating a differential equation system for the sensitivity coefficients by formally differentiating (7.15),

$$\frac{\partial}{\partial k_r}\left(\frac{\partial c_i}{\partial t}\right) = \frac{\partial}{\partial k_r} F_i(c_1,\ldots,c_S\,;k_1,\ldots,k_R)$$

$$\frac{\partial}{\partial t}\left(\frac{\partial c_i}{\partial k_r}\right) = \left(\frac{\partial F_i}{\partial k_r}\right)_{c_l,k_{l\neq r}} + \sum_{n=1}^{S}\left\{\left(\frac{\partial F_i}{\partial c_n}\right)_{c_{l\neq n},k_l}\left(\frac{\partial c_n}{\partial k_r}\right)_{k_{l\neq j}}\right\}$$

$$\frac{\partial}{\partial t}E_{i,r} = \left(\frac{\partial F_i}{\partial k_r}\right)_{c_l,k_{l\neq r}} + \sum_{n=1}^{S}\left\{\left(\frac{\partial F_i}{\partial c_n}\right)_{c_{l\neq n},k_l} E_{n,r}\right\} \quad . \tag{7.18}$$

In these equations, c_l after the partial derivatives means that all c_l are held constant during the differentiation, and $c_{l\neq n}$ after the partial derivatives means that all c_l are held constant, except c_n. These equations (7.18) form a system of linear differential equations, which can be solved numerically together with (7.15). Software packages are available that automatically perform this sensitivity analysis (see Kramer et al. 1982, Lutz et al. 1987, Nowak and Warnatz 1988).

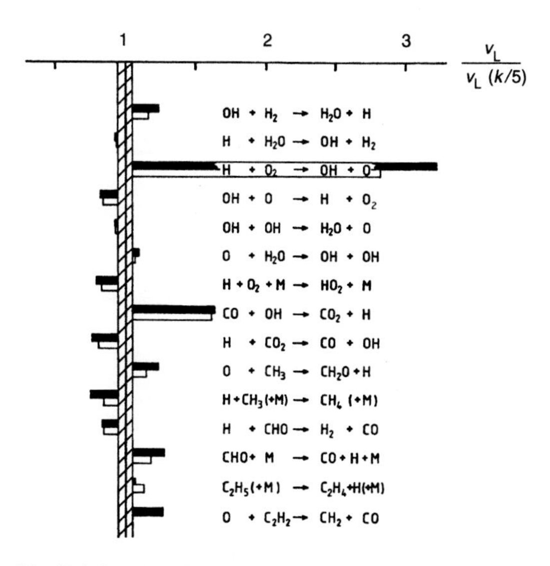

Fig. 7.6. Sensitivity analysis for the flame velocity v_L in premixed stoichiometric CH_4- (black) and C_2H_6-air flames (white) at $p = 1$ bar, $T_u = 298$ K (Warnatz 1984).

The rates of the elementary reactions in combustion processes differ greatly. Sensitivity analyses show that only a few elementary reactions are rate-limiting. Other reactions are so fast, that the accuracy of the rate coefficients has a minor influence on the simulation of the overall combustion process.

This new knowledge has important consequences for applications: The rate coefficients of elementary reactions with high sensitivities have to be well known, because they have a great influence on the results of the mathematical modelling. If reactions have low sensitivities, only approximate values for the rate coefficients have to be known. Thus, sensitivity analysis reveals a few of the many elementary reaction rates, which require accurate determination, usually through experimental measurement.

Examples of sensitivities in combustion processes are shown in Figs. 7.6 and 7.7. Shown are the maximum sensitivities over the whole combustion process. In Fig. 7.6 a sensitivity analysis is shown for the flame velocity v_L in premixed stoichiometric CH_4-air and C_2H_6-air flames. The elementary reactions which do not appear have a negligible sensitivity. Only a few reactions are sensitive. Furthermore, it can be seen that for the different systems (CH_4 and C_2H_6) the same qualitative results are obtained. This points to the important finding that some elementary reactions are always rate-limiting, independent of the fuel considered.

Figure 7.7 shows a sensitivity analysis for the OH-radical concentration in an igniting stoichiometric dodecane ($C_{10}H_{22}$)-air mixture at relatively low temperature (Nehse et al. 1996). Ignition processes generally show more sensitive reactions than stationary flames, especially at low initial temperatures, because more reactions are rate-limiting (compare Figs. 7.6 and 7.7).

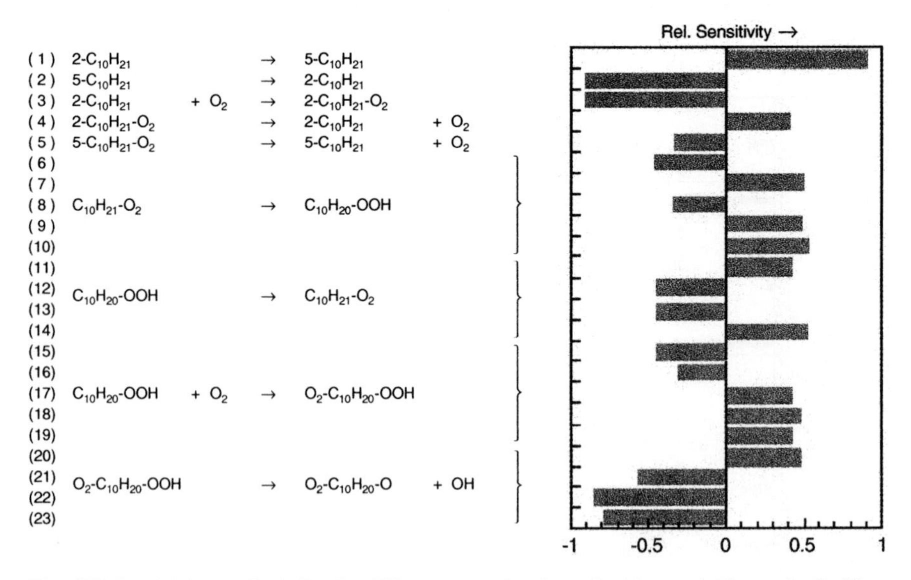

Fig. 7.7. Sensitivity analysis for the OH-concentration in an igniting stoichiometric $C_{10}H_{22}$-air mixture at $p = 13$ bar, $T_u = 800$ K (Nehse et al. 1996); for clearness, no isomeric structures are given in Reactions (6)-(23)

7.2.2 Reaction Flow Analysis

In numerical simulations of combustion processes, *reaction flow analyses* can be done very easily. Software packages are available that automatically perform this analysis. One simply considers the percentage of the contributions of different reactions r ($r=1,...,R$) to the formation (or consumption) of the chemical species s ($s=1,...,S$). A scheme of the form given in Tab. 7.1 is obtained:

Tab. 7.1. Schematic illustration of the output of a reaction flow analysis

Reaction ⇓	Species ⇒ 1	2	3	.....	S-1	S
1	20%	3%	0	.....	0	0
2	0	0	0	.....	0	0
3	2%	5%	0	.....	100%	90%
.	.	.	.		.	.
.	.	.	.		.	.
.	.	.	.		.	.
R-1	78%	90%	100%	.....	0	5%
R	0	2%	0	.....	0	0

In this example, e. g., 20% of the formation of species 1 can be attributed to reaction 1, 2% to reaction 3, and 78% to reaction R-1. The percentages in the columns have to add to 100%. Such tables allow the construction of instructive *reaction flow diagrams* examples of which are given in Figs. 7.8 and 7.9 below. *Integral reaction flow analysis* and *local reaction flow analysis* can be performed in each case.

Integral reaction flow analysis considers the overall formation or consumption during the combustion process. The results for homogeneous time-dependent systems are, e. g., integrated over the whole reaction time, and the results for stationary flames are integrated over the reaction zone. A reaction can be regarded as unimportant (here, e. g., Reaction 2), if all entries in a row (for the formation as well as for the consumption) are below a certain limit, e. g., 1% (Warnatz 1981a).

Local reaction flow analysis considers the formation and consumption of species locally, i. e., at specific times in time-dependent problems (e. g., a homogeneous ignition process) or at specific locations in steady processes (e. g., a flat flame). According to the local analysis, a reaction r is unimportant, if, for all times t or locations x, the relation

$$\left|\mathfrak{R}_{t,r,s}\right| \;<\; \varepsilon \left|\operatorname*{Max}_{r=1}^{R} \mathfrak{R}_{t,r,s}\right| \quad , \quad s = 1,..,S \quad , \quad t = 0,..,T\,. \tag{7.19}$$

holds for the reaction rate $\mathfrak{R}_{t,r,s}$. This is a much stricter requirement than in the case of integral analysis. ε is a limit, which has to be specified arbitrarily, e. g., $\varepsilon = 1\%$.

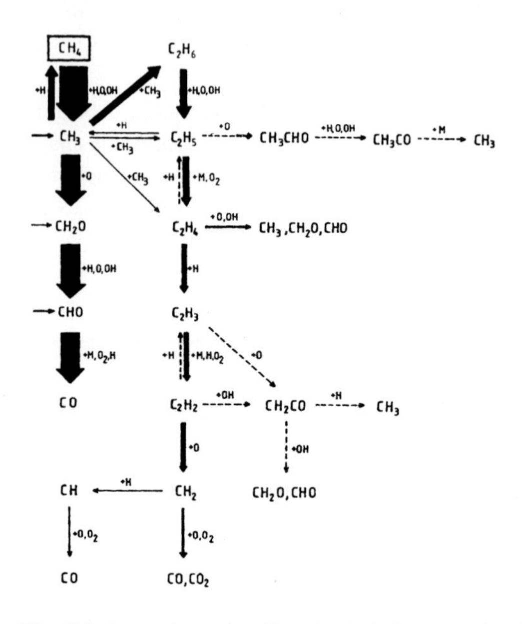

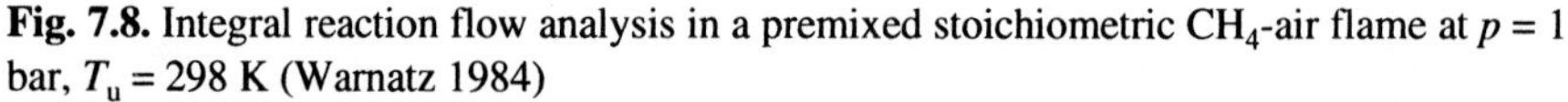

Fig. 7.8. Integral reaction flow analysis in a premixed stoichiometric CH_4-air flame at p = 1 bar, T_u = 298 K (Warnatz 1984)

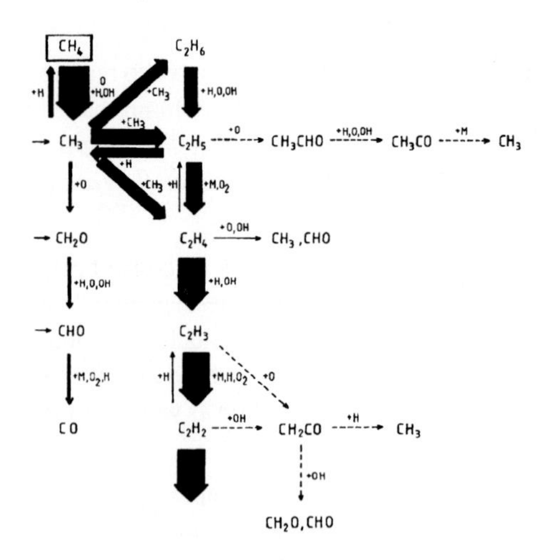

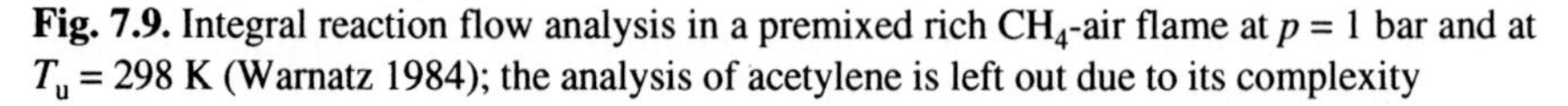

Fig. 7.9. Integral reaction flow analysis in a premixed rich CH_4-air flame at p = 1 bar and at T_u = 298 K (Warnatz 1984); the analysis of acetylene is left out due to its complexity

Figures 7.8 and 7.9 show integral reaction flow analysis in premixed stoichiometric and rich methane-air flames (Warnatz 1984). It can be clearly seen that different reaction paths are followed, depending on the stoichiometry even though the chemical mechanism is the same! In the stoichiometric flame, methane is mainly oxidized directly, whereas methyl radicals formed in the rich flame recombine to ethane (C_2H_6), which is then oxidized. Thus, one can find the rather surprising result that a satisfactory mechanism for methane oxidation also demands an additional mechanism for ethane oxidation.

7.2.3 Eigenvalue Analyses of Chemical Reaction Systems

Again the simple reaction sequence (7.1) shall be considered,

$$S_1 \xrightarrow{k_{12}} S_2 \xrightarrow{k_{23}} S_3 ,$$

where the rate laws are given by (7.2) - (7.4). These equations can be rewritten in vector notation as

$$\begin{pmatrix} d[S_1]/dt \\ d[S_2]/dt \\ d[S_3]/dt \end{pmatrix} = \begin{pmatrix} -k_{12} & 0 & 0 \\ k_{12} & -k_{23} & 0 \\ 0 & k_{23} & 0 \end{pmatrix} \begin{pmatrix} [S_1] \\ [S_2] \\ [S_3] \end{pmatrix} . \tag{7.20}$$

If one uses the definitions for $\vec{Y}$ and $\vec{Y}'$, and introduces the matrix $\boldsymbol{J}$,

$$\vec{Y} = \begin{pmatrix} [S_1] \\ [S_2] \\ [S_3] \end{pmatrix} \qquad \vec{Y}' = \begin{pmatrix} d[S_1]/dt \\ d[S_2]/dt \\ d[S_3]/dt \end{pmatrix} \qquad \boldsymbol{J} = \begin{pmatrix} -k_{12} & 0 & 0 \\ k_{12} & -k_{23} & 0 \\ 0 & k_{23} & 0 \end{pmatrix} ,$$

the result is the simple linear ordinary differential equation system

$$\vec{Y}' = \boldsymbol{J}\,\vec{Y} . \tag{7.21}$$

Now the *eigenvalues* and *eigenvectors* of the matrix $\boldsymbol{J}$ will be calculated. They have to fulfill the *eigenvalue equation*

$$\boldsymbol{J}\,\vec{v}_i = \vec{v}_i\,\lambda_i \qquad \text{or equivalently} \qquad \boldsymbol{J}\,\boldsymbol{V} = \boldsymbol{V}\,\boldsymbol{\Lambda} \tag{7.22}$$

(see textbooks on linear algebra; details cannot be given here for lack of space). Because the matrix $\boldsymbol{J}$ has a (3x3) form, there are three eigenvalues and three corresponding eigenvectors,

$$\lambda_1 = 0 \qquad \lambda_2 = -k_{23} \qquad \lambda_3 = -k_{12}$$

$$\vec{v}_1 = \begin{pmatrix} 0 \\ 0 \\ 1 \end{pmatrix} \qquad \vec{v}_2 = \begin{pmatrix} 0 \\ 1 \\ -1 \end{pmatrix} \qquad \vec{v}_3 = \begin{pmatrix} k_{12} - k_{23} \\ -k_{12} \\ k_{23} \end{pmatrix} .$$

If the matrix $\boldsymbol{V}$ of the eigenvectors and the matrix $\boldsymbol{\Lambda}$ of the eigenvalues are formed,

$$V = \begin{pmatrix} 0 & 0 & k_{12}-k_{23} \\ 0 & 1 & -k_{12} \\ 1 & -1 & k_{23} \end{pmatrix} \quad \text{and} \quad \Lambda = \begin{pmatrix} \lambda_1 & 0 & 0 \\ 0 & \lambda_2 & 0 \\ 0 & 0 & \lambda_3 \end{pmatrix} = \begin{pmatrix} 0 & 0 & 0 \\ 0 & -k_{23} & 0 \\ 0 & 0 & -k_{12} \end{pmatrix}$$

one can prove by insertion very quickly that the eigenvalue equation (7.22) is identically fulfilled.

Multiplication of the eigenvalue equations from the right with the inverse $\boldsymbol{V}^{-1}$ of the eigenvector matrix leads to a method for the decomposition of the matrix $\boldsymbol{J}$,

$$\boldsymbol{J} = \boldsymbol{V\Lambda V}^{-1}, \tag{7.23}$$

where

$$\boldsymbol{V}^{-1} = \begin{pmatrix} 1 & 1 & 1 \\ \dfrac{k_{12}}{k_{12}-k_{23}} & 1 & 0 \\ \dfrac{1}{k_{12}-k_{23}} & 0 & 0 \end{pmatrix}.$$

Insertion into the differential equation system (7.21) then yields the simple equation

$$\vec{Y}' = \boldsymbol{V\Lambda V}^{-1}\,\vec{Y} \tag{7.24}$$

and, after multiplication from the left with the inverse $\boldsymbol{V}^{-1}$,

$$\boldsymbol{V}^{-1}\vec{Y}' = \boldsymbol{\Lambda V}^{-1}\,\vec{Y}\,. \tag{7.25}$$

The rows of the matrix $\boldsymbol{V}^{-1}$ are the eigenvectors $\vec{v}_j^{-1}$. Insertion of $\boldsymbol{\Lambda}$and the $\vec{v}_j^{-1}$ yields the equation system

$$\begin{aligned} \frac{\mathrm{d}}{\mathrm{d}t}([\mathrm{S}_1]+[\mathrm{S}_2]+[\mathrm{S}_3]) &= 0 \\ \frac{\mathrm{d}}{\mathrm{d}t}\left(\frac{k_{12}}{k_{12}-k_{23}}[\mathrm{S}_1]+[\mathrm{S}_2]\right) &= -k_{23}\left(\frac{k_{12}}{k_{12}-k_{23}}[\mathrm{S}_1]+[\mathrm{S}_2]\right) \\ \frac{\mathrm{d}}{\mathrm{d}t}\left(\frac{1}{k_{12}-k_{23}}[\mathrm{S}_1]\right) &= -k_{12}\left(\frac{1}{k_{12}-k_{23}}[\mathrm{S}_1]\right), \end{aligned} \tag{7.26}$$

which shall be considered now in detail. It can be seen at once that this equation system is completely decoupled, i. e., all three differential equations can be solved separately. All three equation have the form

$$\frac{\mathrm{d}y}{\mathrm{d}t} = \mathrm{const}\cdot y$$

with the solution

$$y = y_0 \cdot \exp(\text{const} \cdot t) \ .$$

Thus, one obtains (the analytical solution (7.5) in Section 7.1.1 was actually obtained in this way)

$$[S_1]+[S_2]+[S_3] = \left([S_1]_0 + [S_2]_0 + [S_3]_0\right) \cdot \exp(0) \tag{7.27}$$

$$\left(\frac{k_{12}}{k_{12}-k_{23}}[S_1]+[S_2]\right) = \left(\frac{k_{12}}{k_{12}-k_{23}}[S_1]_0 + [S_2]_0\right) \cdot \exp(-k_{23} \cdot t) \tag{7.28}$$

$$[S_1] = [S_1]_0 \cdot \exp(-k_{12} \cdot t) \ . \tag{7.29}$$

According to this, the chemical reaction system can be grouped into three different processes with three different time scales:

The first process has (corresponding to the eigenvalue $\lambda_1 = 0$) a time scale ∞, and thus describes a value which is constant. Such values, which are constant in time, are called *conserved quantities*. Here it is the sum of the concentrations in (7.27), which in this case reflects the fact that mass is conserved in chemical reactions.

The second process proceeds (according to the eigenvalue $\lambda_2 = -k_{23}$) with the time scale $\tau_{23} = k_{23}^{-1}$. It describes the temporal change of a quantity. The corresponding eigenvector $\vec{\upsilon}_2$ is given by

$$\vec{\upsilon}_2 = \begin{pmatrix} 0 \\ 1 \\ -1 \end{pmatrix} ,$$

and it can be seen that this vector reflects the stoichiometric coefficients of the reaction $S_2 \rightarrow S_3$ ($0\ S_1 + 1\ S_2 - 1\ S_3 = 0$).

The third process proceeds (according to the eigenvalue $\lambda_3 = -k_{12}$) with the time scale $\tau_{12} = k_{12}^{-1}$. The corresponding eigenvector $\vec{\upsilon}_3$ corresponds to a linear combination of the eigenvectors of the Reactions 12 and 23.

Now the case shall be treated where one reaction is much faster than the other (remember Section 7.1.1). The situation $k_{12} >> k_{23}$ will be considered first. In this case the third process (with the time scale $\tau_{12} = k_{12}^{-1}$) is much faster than the second process. After a short time (see exponential decay according to (7.29)) the concentration $[S_1]$ of species S_1 is almost zero. In chemical terms this means that species S_1 reacts very rapidly to species S_2, which then reacts to species S_3 by a relatively slow subsequent reaction.

More interesting is the case $k_{23} >> k_{12}$. Here the exponential term in (7.28) tends to zero very rapidly. Thus, after a short time one can assume that

$$\left(\frac{k_{12}}{k_{12}-k_{23}}[S]_1 + [S]_2\right) \approx 0 \ . \tag{7.30}$$

If one compares with Section 7.1.1, one can see that (7.30) corresponds exactly to the quasi-steady state approximation (7.6) for $k_{23} >> k_{12}$. Thus, the quasi-steady-state conditions can not only be obtained by chemical considerations, but also by means of a simple eigenvalue analysis. The negative eigenvalue λ_i, which is largest in magnitude, describes the rate at which partial equilibrium or steady state is achieved. The quasi-steady-state condition, or the partial-equilibrium condition, is obtained by assuming that the scalar product of the left eigenvector $\vec{v}_j^{-1}$ and the rates of formation (right hand side of (7.20) vanishes.

Of course the differential equation systems are usually nonlinear and in their general form given by

$$\frac{\mathrm{d}Y_i}{\mathrm{d}t} = f_i(Y_1, Y_2, \ldots, Y_S) \quad ; \quad i = 1, 2, \ldots, S \tag{7.31}$$

or in vector formulation

$$\frac{\mathrm{d}\vec{Y}}{\mathrm{d}t} = \vec{F}(\vec{Y}) \,. \tag{7.32}$$

However, local eigenvalue analyses for specific conditions $\vec{Y}_0$ can be performed by approximating the function $\vec{F}$ in the neighborhood of $\vec{Y}_0$ by a Taylor series,

$$f_i(Y_{1,0} + \mathrm{d}Y_1, Y_{2,0} + \mathrm{d}Y_2, \ldots, Y_{S,0} + \mathrm{d}Y_S) = f_i(Y_{1,0}, Y_{2,0}, \ldots, Y_{S,0}) + \sum_{j=1}^{S} \left(\frac{\partial f_i}{\partial Y_j} \right)_{Y_{k \neq j}} \mathrm{d}Y_j + \ldots$$

or

$$\vec{F}(\vec{Y}_0 + \mathrm{d}\vec{Y}) = \vec{F}(\vec{Y}_0) + \boldsymbol{J}\,\mathrm{d}\vec{Y} + \ldots$$

with

$$\boldsymbol{J} = \begin{pmatrix} \frac{\partial f_1}{\partial Y_1} & \frac{\partial f_1}{\partial Y_2} & \cdots & \frac{\partial f_1}{\partial Y_S} \\ \frac{\partial f_2}{\partial Y_1} & \frac{\partial f_2}{\partial Y_2} & \cdots & \frac{\partial f_2}{\partial Y_S} \\ \vdots & \vdots & \ddots & \vdots \\ \frac{\partial f_S}{\partial Y_1} & \frac{\partial f_S}{\partial Y_2} & \cdots & \frac{\partial f_S}{\partial Y_S} \end{pmatrix} .$$

$\boldsymbol{J}$ is called *Jacobian matrix* of the system under consideration. This linearization leads to the linear differential equation system

$$\frac{\mathrm{d}\vec{Y}}{\mathrm{d}t} = \vec{F}(\vec{Y}_0) + \boldsymbol{J}(\vec{Y} - \vec{Y}_0) \,, \tag{7.33}$$

and a comparison with (7.21) yields that the eigenvalues and eigenvectors of the Jacobian reveal information about the time scales of the chemical reaction and about species in steady state or reactions in partial equilibrium (see Lam and Goussis 1989, Maas and Pope 1992).

7.3 Stiffness of Ordinary Differential Equation Systems

As noted previously, the many elementary reactions in combustion processes have greatly different reaction rates (time scales). The wide variation in times scales has severe consequences for the numerical solution of the differential equation systems, which govern the reaction system.

The eigenvalues of the *Jacobians* of these ODE-systems reflect the time scales. The ratio between the largest and smallest (in magnitude) negative eigenvalue of the Jacobian can be used as a *degree of stiffness* (though mathematicians are not pleased with this simple definition).

The stiffness characterizes the maximal differences in the time scales. Usually the smallest time scales have to be resolved in the numerical solution, even if one is only interested in the slow (and thus rate-limiting) processes. Otherwise the numerical solution tends to become unstable.

This instability problem can be avoided, if so-called *implicit solution methods* are used (Hirschfelder 1963, see Section 8.2). Another possibility to avoid this problem is the elimination of the fast processes by *decoupling the time scales*; see next section (Lam and Goussis 1989, Maas and Pope 1992,1993).

7.4 Simplification of Reaction Mechanisms

The main problem in the use of detailed reaction mechanisms is given by the fact that one species conservation equation has to be solved for each species (see Chapter 3). Therefore it is desirable to use simplified chemical kinetics schemes, which describe the reaction system in terms of only a small number of species. This can be attempted by using *reduced mechanisms*, based on quasi-steady state and partial-equilibrium assumptions (a survey of current developments can be found in Smooke 1991). However, such reduced mechanisms are usually devised for certain conditions, i. e., they provide reliable results only for a certain range of temperature and mixture composition. A reduced mechanism of this kind, e. g., which provides good results in the simulation of nonpremixed flames, will likely provide unsatisfactory results in the simulation of premixed flames of the same fuel.

Chemical reaction corresponds to a movement along a trajectory in the $(2+n_S)$-dimensional state space spanned by the enthalpy, the pressure, and the n_S species mass fractions. Starting from different initial conditions, the reactive system evolves until it reaches the equilibrium point. Chemical equilibrium is only a function of the enthalpy h, the pressure p, and the element mass fractions Z_i (see Chapter 4).

The time scales of the different processes span several orders of magnitude (see Fig. 7.10). Chemical reactions typically cover a range from 10^{-10} s to more than 1 s. On the other hand, the time scales of physical processes like molecular transport cover a much smaller range. The fast chemical processes correspond to equilibration processes, e. g., reactions in partial equilibrium and species in quasi-steady state, and they can be decoupled (see Section 7.1 for very simple examples).

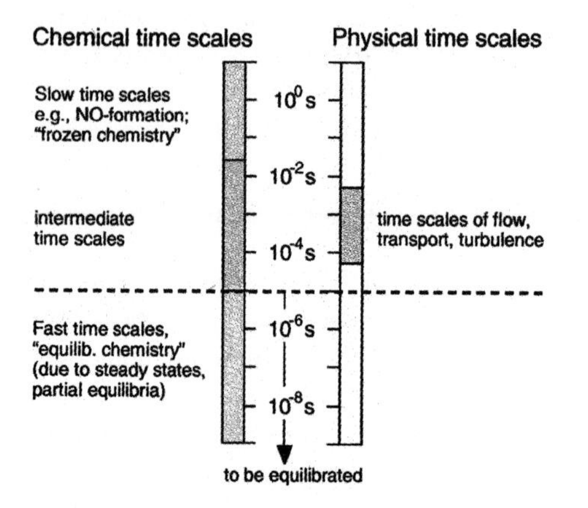

Fig. 7.10. Classification of time scales in a chemically reacting flow

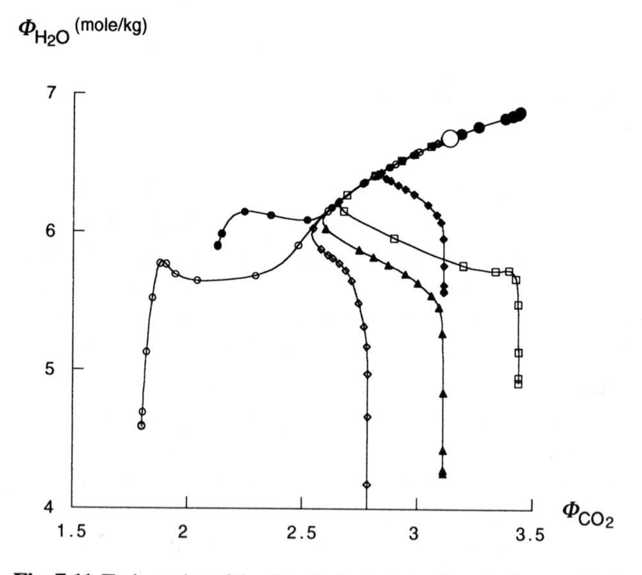

Fig. 7.11. Trajectories of the chemical reactions for a methane oxidation, ○ denotes the chemical equilibrium; projection into the CO_2-H_2O plane (Riedel et al. 1994); $\Phi_i = w_i/M_i$

In Fig. 7.11 some trajectories for a stoichiometric CH_4-air system are plotted, projected into the H_2O-CO_2 plane. The equilibrium point is marked by a circle. Realistic flows usually need some time to reach chemical equilibrium, with a time scale of the order of the physical time scales, and therefore the chemistry couples with the physical processes. For the example shown in Fig. 7.11, the chemical equilibration process takes about 5 ms. This can be seen in Fig. 7.12, which shows the system after 5 ms (i. e., the first 5 ms have simply been omitted), having nearly relaxed to the equilibrium value. Thus, if one is only interested in processes lasting longer than 5 ms, the system can be described reasonably well by the equilibrium point. In this case all the chemical dynamics is neglected, corresponding to the often applied maxim "mixed is burnt".

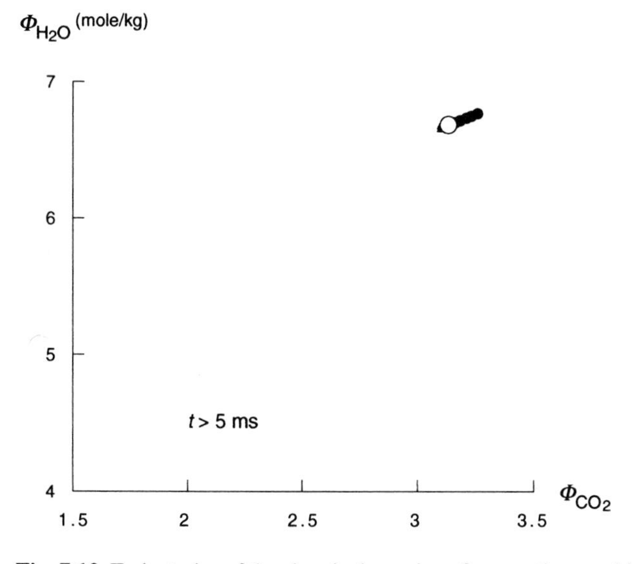

Fig. 7.12. Trajectories of the chemical reactions for a methane oxidation, where the first 5 ms have been omitted, O denotes the equilibrium; projection into the CO_2-H_2O plane (Riedel et al. 1994); $\Phi_i = w_i/M_i$

However, this is a rather crude assumption in most applications, because the physical processes themselves usually occur with time scales of the order of ms and the chemistry then couples directly with molecular transport. A next level of approximation would be to assume that only the chemical processes faster than 50 μs are in equilibrium. This corresponds to neglecting the first 50 μs of the trajectories in Fig. 7.11 and is shown in Fig. 7.13.

Instead of taking into account all the chemical dynamics, only the slower (and, thus, rate-limiting) processes (slower than 50 μs) are considered. The picture obtained is a simple line in the state space instead of the complicated curves in Fig. 7.11. This line corresponds to a *one-dimensional manifold* in the state space. All processes slower than 50 μs are described by the movement on this line, and the dynamics of the chemical system is restricted to the dynamics on this one-dimensional manifold, i. e., only one reaction progress variable is needed, describing the movement along this line.

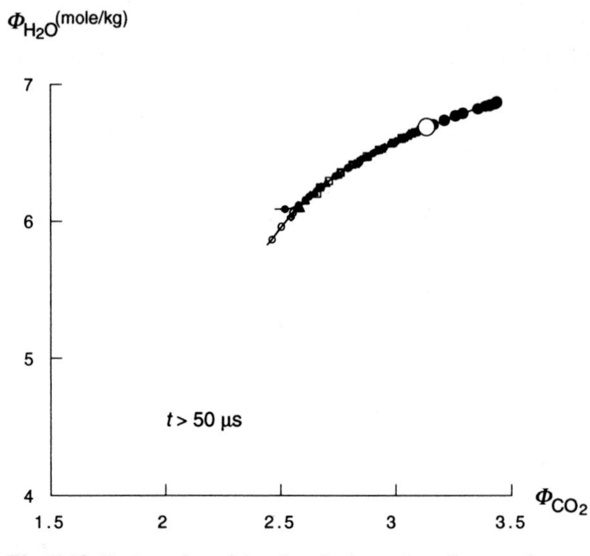

Fig. 7.13. Trajectories of the chemical reactions for a methane oxidation, where the first 50 µs have been omitted, ❍ denotes the equilibrium; projection into the projection into the CO_2-H_2O plane (Riedel et al. 1994); $\Phi_i = w_i/M_i$

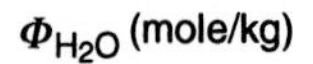

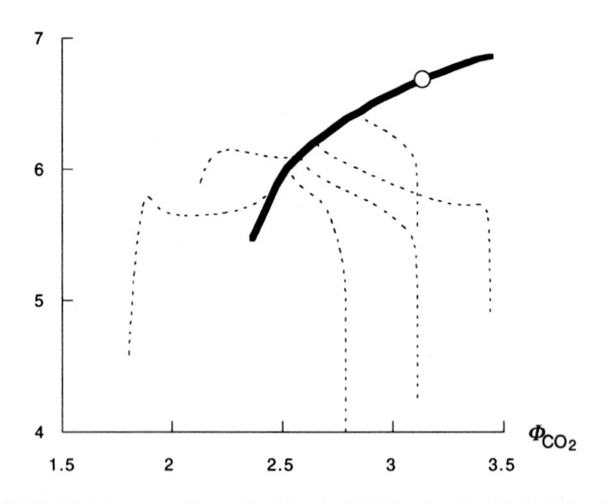

Fig. 7.14. Line: one-dimensional manifold for the stoichiometric CH_4-oxidation, broken lines: trajectories, circle: equilibrium point; projection into the CO_2-H_2O plane (Riedel et al. 1994); $\Phi_i = w_i/M_i$

Based on an eigenvector analysis (see previous section), the one-dimensional manifold can be calculated if the detailed reaction mechanism is known. Such a calculated one-dimensional manifold is plotted in Fig. 7.14 which is almost the same as Fig. 7.11. However, if one is interested in processes which are even faster than 50 μs, a plane in the state space can be obtained, corresponding to a two-dimensional manifold, and so on. This means that according to the time scales to be considered there exist m-dimensional manifolds which approximate the system, and the dynamics can be reduced to the dynamics on these manifolds.

The simplification procedure is based on a local eigenvector analysis of the Jacobian (see previous section). The eigenvalues characterize the time scales and the eigenvectors describe the characteristic directions of chemical reactions in the state space associated with these time scales. Usually there are a large number of negative eigenvalues which are large in magnitude. These large negative eigenvalues correspond to the fast relaxation processes, namely reactions in partial equilibrium and species in steady state. Of course these eigenvalues depend on the regime, i. e., they are defined locally and may change from point to point in the state space.

The attracting manifold (e.g., the line in Fig. 7.14) is composed of the points in the state space where the fast processes are in local equilibrium. This means that a low-dimensional attracting manifold can be defined by the points where the rate in direction of the n_f eigenvectors, corresponding to the n_f largest (in magnitude) negative eigenvalues, vanish. In this way a *low-dimensional manifold* in the state space is defined and can be computed numerically (for details see Maas and Pope 1992, 1993, Maas 1998).

A very simple example of a low-dimensional manifold can be derived for the simple reaction mechanism $S_1 \rightarrow S_2 \rightarrow S_3$ treated before, if the second reaction is relatively fast (see Section 7.2.3). The quasi-steady state reached after a very short time k_{23}^{-1} (see Figs. 7.1 and 7.2) is described by (7.30)

$$[S]_2 = \frac{k_{12}}{k_{23} - k_{12}} [S]_1$$

in the state space. At the same time, this equation is that of the low-dimensional manifold, leading for $t \rightarrow \infty$ to the point $[S_1] = [S_2] = 0$ due to complete conversion into the final product S_3.

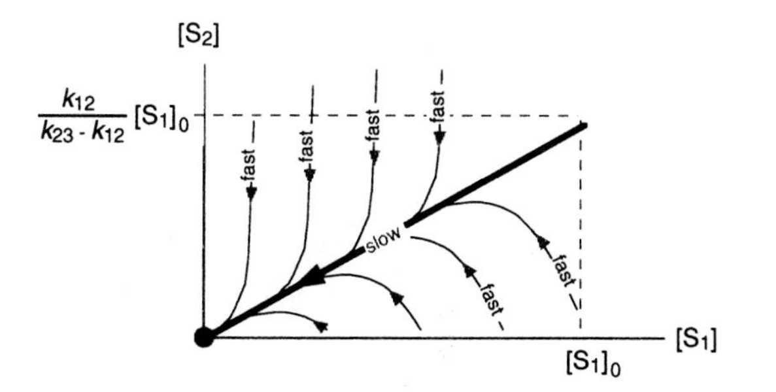

Fig. 7.15. Trajectories in the state space and low-dimensional manifold for the reaction mechanism $S_1 \rightarrow S_2 \rightarrow S_3$

The remaining task is to implement the results of the mechanism reduction procedure in calculations of laminar and turbulent flames. It is evident that there is always a coupling of the chemical kinetics with physical processes like diffusion or turbulent transport.

In the concept of intrinsic low-dimensional manifolds the physical processes perturb the chemical kinetics, i. e., they try to pull the system off the manifold. Fast chemical processes, however, relax the system back to the manifold provided that the time scale of the physical perturbation is longer that the time scales of the relaxation processes (i.e. the fast decoupled chemical time scales).

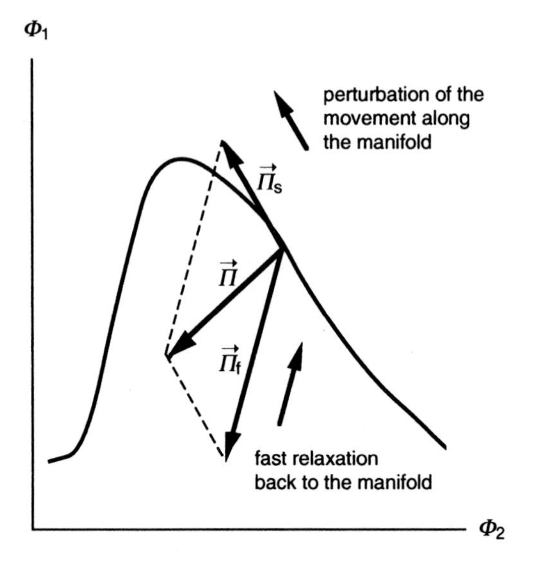

Fig. 7.16. Illustration of the coupling of chemical kinetics with molecular transport

This behavior is shown in Fig. 7.16. A physical perturbation $\vec{\Pi}$ tries to pull the system off the manifold. This perturbation can be decomposed into two components: One ($\vec{\Pi}_f$) which is relaxed back to the manifold very fast, and one ($\vec{\Pi}_s$) which leads to a net perturbation along the manifold. In laminar and turbulent flame calculations this behavior has to be taken into account by projecting the governing species conservation equations onto the manifold. Details about this procedure can be found in Maas & Pope 1993.

Laminar flame calculations using detailed and reduced reaction mechanisms allow a direct comparison and yield useful information on the quality of the reduced schemes. As an example Fig. 7.17 shows the structure of a laminar premixed flat methane air flame. The detailed mechanism for this flame comprised 34 species and 288 elementary reactions, whereas the mechanism reduction is using 3 reaction progress variables. It can be seen that quite accurate results are obtained despite of the high degree of simplification.

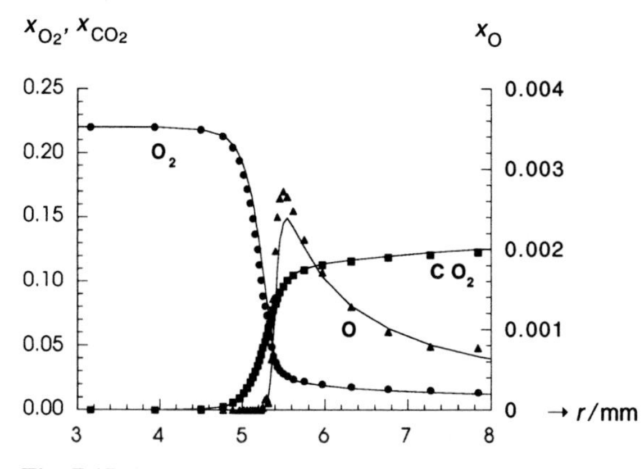

Fig. 7.17. Structure of a premixed stoichiometric free flat methane air flame (p = 1 bar, T_u = 300 K); curves denote mass fractions obtained using a detailed reaction mechanism, symbols denote results obtained using a reduced mechanism with 3 reaction progress variables (Schmidt 1996).

7.5 Radical Chain Reactions

Radical chain reactions form the basis of combustion processes. The general principle of these mechanisms will here be described by means of the hydrogen-oxygen system. The most important reactions of this system are shown in Tab. 7.2.

Tab. 7.2. Most important reactions with respect to ignition in the hydrogen-oxygen system

(0)	H_2 + O_2	=	2 OH•			*chain initiation* (reaction rate *I*)
(1)	OH• + H_2	=	H_2O	+	H•	*chain propagation*
(2)	H• + O_2	=	OH•	+	O•	*chain branching*
(3)	O• + H_2	=	OH•	+	H•	*chain branching*
(4)	H•	=	1/2 H_2			*chain termination* (*heterogeneous*)
(5)	H• + O_2 +M	=	HO_2	+	M	*chain termination (homogeneous*)
(1+1+2+3)	3 H_2 + O_2	=	2 H•	+	2 H_2O	

The reaction mechanism consists of *chain initiation steps*, where reactive species (radicals, characterized by the dot) are formed from stable species (Reaction 0), *chain propagation steps*, where reactive intermediate species react with stable species forming another reactive species (Reaction 1), *chain branching steps*, where a reactive species reacts with a stable species forming two reactive species (Reactions 2 and 3), and *chain termination steps*, where reactive species react to stable species (e. g., at the vessel surface (Reaction 4) or in the gas phase (Reaction 5). If one sums up the chain propagation and chain branching steps (1+1+2+3), one can see that in this mechanism radicals are formed from the reactants.

According to the reaction scheme given in Tab. 7.2, one obtains the three rate laws

$$\frac{d[H]}{dt} = k_1[H_2][OH] + k_3[H_2][O] - k_2[H][O_2] - k_4[H] - k_5[H][O_2][M]$$

$$\frac{d[OH]}{dt} = k_2[H][O_2] + k_3[O][H_2] - k_1[OH][H_2] + I$$

$$\frac{d[O]}{dt} = k_2[H][O_2] - k_3[O][H_2]$$

for the formation of the reactive species H, OH and O. These reactive species are *chain carriers* (I is the rate of the chain initiation). The rate of formation of the free valences (H•, •OH, O•) is obtained by summation of the three equations, where oxygen atoms have to be counted twice (two free valences per oxygen atom),

$$\frac{d([H]+[OH]+2[O])}{dt} = I + (2k_2[O_2] - k_4 - k_5[O_2][M])[H] . \tag{7.34}$$

If, as a very crude approximation, the concentration of the hydrogen atoms [H] is replaced by the concentration $[n]$ of the free valences (Homann 1975), one obtains

$$\frac{d[n]}{dt} = I + (2k_2[O_2] - k_4 - k_5[O_2][M]) \cdot [n] . \tag{7.35}$$

This differential equation can be integrated very easily. For the initial condition $[n]_{t=0} = 0$ the result is

$$[n] = I \cdot t \qquad \text{for } g = f, \tag{7.36}$$

$$[n] = \frac{I}{g-f}\left(1 - \exp[(f-g)\,t]\right) \qquad \text{for } g \neq f, \tag{7.37}$$

where $f = 2k_2[O_2]$ and $g = k_4 + k_5\,[O_2][M]$ have been defined for convenience. Three cases result, which are shown schematically in Fig. 7.18.

For $g > f$ the exponential term tends to zero with increasing time. A time-independent stationary solution is obtained for the chain carriers, and an explosion does not take place,

$$[n] = \frac{I}{g-f} . \tag{7.38}$$

For the limiting case $g = f$ a linear increase of the chain carriers $[n]$ with time t is obtained. For $g < f$ the "1" can be neglected in comparison to exponential term after a short time, and the result is an exponential growth of the radical concentration and thus an explosion,

$$[n] = \frac{I}{f-g}\exp[(f-g)\cdot t] . \tag{7.39}$$

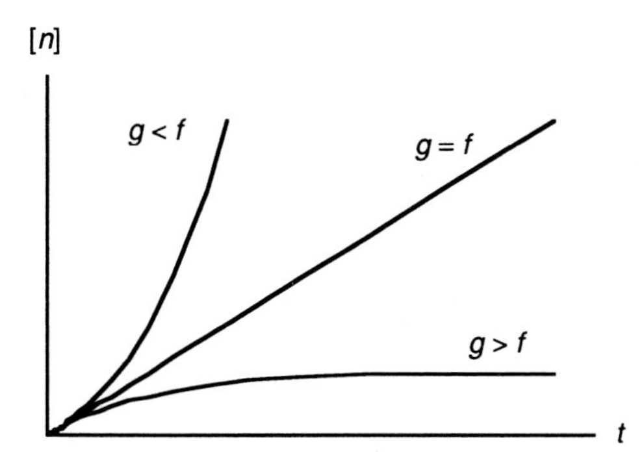

Fig. 7.18. Schematic illustration of the time behavior of the chain carriers

This simple description shows the major importance of chain branching steps in combustion processes, especially in ignition processes. Typical chain branching reactions in hydrocarbon oxidation at atmospheric pressure are, for example (the reactive species are characterized by •)

$T > 1100$ K:	H•	+ O_2	→	OH•	+ O•	
$900\ K < T < 1100$ K:	HO_2•	+ RH	→	H_2O_2	+ R•	
	H_2O_2	+ M	→	2 OH•	+ M	.

For $T > 1100$ K, these chain branching mechanisms are quite simple and relatively independent of the fuel. At temperatures below 1100 K these reactions become much more complex and fuel-specific (see Chapter 16).

7.6 Exercises

Exercise 7.1. The decomposition of ethane occurs via the initiation reaction $C_2H_6 \rightarrow$ 2 CH_3. It shall be described by the following reaction scheme: $C_2H_6 + C_2H_6 = C_2H_6 + C_2H_6{}^*$ (1); $C_2H_6{}^* \rightarrow 2\ CH_3$ (2). For the formation of the excited molecule $C_2H_6{}^*$ (Reaction 1) the reverse reaction has to be considered, too; the dissociation of $C_2H_6{}^*$ (Reaction 2) shall have a negligible reverse reaction.

(a) Formulate the steady state condition for $C_2H_6{}^*$ and determine the concentration of $C_2H_6{}^*$ as a function of the concentration of C_2H_6.
(b) Determine the rate of formation of CH_3 as a function of the C_2H_6 concentration. Suppose that the steady state condition for $C_2H_6{}^*$ is fulfilled. Show that the rate of formation of CH_3 obeys a second-order rate law for small C_2H_6 concentration and a first-order rate law for high concentrations.

(c) For certain conditions, experiments show, that the rate of formation of CH_3 obeys a first-order rate law with a rate coefficient $k_{CH_3} = 5.24 \cdot 10^{-5}\ s^{-1}$. Calculate the rate coefficient k_2 for the dissociation of the excited molecule C_2H_6*. (The equilibrium constant of Reaction 1 shall be $K_{c,1} = 1.1 \cdot 10^{-4}$.)

Exercise 7.2. The reaction mechanism in Tab. 6.1 is listing rate coefficients (using Arrhenius parameters). For the second reaction in this list, estimate how many collisions an O atom must have with H_2 in order to react, if $T = 1200$ K. You may assume that the diameter of both O and H_2 is 0.3 nm. Also recall the gas constant to be $R = 8.314\ J \cdot mol^{-1} \cdot K^{-1}$.

Exercise 7.3. Consider that the Arrhenius formula for a rate coefficient can be written as $k = A \cdot T^n \cdot \exp(-\theta_a/T)$ with $\theta_a = E_a/R$.

a) For the purpose of this exercise assume $A = 1$ (dimensionless) and $n = 0$, hand-sketch a plot of k as function of T/θ_a. Show the rate coefficient k for $T/\theta_a = 0.1$, 0.5, 1, 2, and 5.
b) How much does the rate coefficient k change when T/θ_a increases from 0.01 to a value of 0.1 ?
c) How much does the rate coefficient k change when T/θ_a increases from 1.0 to a value of 10.0 ?

It is easy to see how a reaction can rapidly transition from fast reaction (“equilibrium”) to slow reaction (“frozen”) by a drop in the temperature.

8 Laminar Premixed Flames

Measurements of laminar flame velocities and the experimental determination of concentration- and temperature profiles in laminar flame fronts were introduced in Chapter 2. A challenge to the combustion scientist is to construct a model that will match the observed concentration- and temperature profiles and allows prediction of events for which there are no measurements. Chapter 3 introduced a simple model, consisting of a system of partial differential equations derived from the conservation of mass, enthalpy, and species mass. In these equations, the terms dealing with thermodynamics were the subject of Chapter 4. The terms dealing with transport properties were the subject of Chapter 5. In the equation system, the source or sink of species and energy are the chemical reactions discussed in Chapters 6 and 7. The model now is complete; one only needs to solve the system of partial differential equations with appropriate boundary conditions.

Analytical solutions are possible and instructive, when several terms can be eliminated; examples of this were presented in Chapter 3. In general, the full system of equations was not solved until the advent of suitable digital computers (ca. 1960s). As numerical solutions began to emerge, it became apparent that the system of equations could more readily be solved if it was recast into a form that was more amenable to solution on digital computers. The solution of the equations is the large subject of this Chapter 8.

8.1 Zeldovich's Analysis of Flame Propagation

The conservation equations (3.11) and (3.12) form a system of differential equations, which are usually solved by numerical methods. Nonetheless, insight can be gained from analytical solutions obtained after much judicious elimination of terms from the complete system of conservations equations. An example is the model of thermal flame propagation by Zeldovich and Frank-Kamenetskii (1938). They assumed a time-independent solution, i. e., steady state, and further made the sweeping assumption that all of chemical kinetics could be simulated by a one-step global reaction

Fuel (F) $\rightarrow$ Products (P)

with the first-order reaction rate $r = -\rho w_F k = -\rho w_F \cdot A \cdot \exp(-E/RT)$. Furthermore, it is assumed that the thermal conductivity λ, the specific heat capacity c_p, and the product of density and diffusion coefficient ρD are constant, i. e., independent of spatial location. In addition, the term $\Sigma j_j c_{p,j}$, which describes the temperature change by different diffusion velocities of species with different specific heat capacities, is assumed negligible. When these simplifications are applied to (3.11, 3.12), one obtains the simple differential equation system

$$D\frac{\partial^2 w_F}{\partial z^2} - \upsilon\frac{\partial w_F}{\partial z} - w_F \cdot A \cdot \exp\left(-\frac{E}{RT}\right) = 0 \tag{8.1}$$

$$\frac{\lambda}{\rho c_p}\frac{\partial^2 T}{\partial z^2} - \upsilon\frac{\partial T}{\partial z} + w_F\frac{h_P - h_F}{c_p} \cdot A \cdot \exp\left(-\frac{E}{RT}\right) = 0 \tag{8.2}$$

for the variables w_F and T. Taking into consideration experimental data, it is not unreasonable to assume that mass diffusivity D and *thermal diffusivity* $\alpha\,(=\lambda/\rho c_p)$ are roughly equal. Thus, the so-called *Lewis number* $Le = D/\alpha = D\rho c_p/\lambda$ is approximately unity. This simple assumption has the great advantage that now equations (8.1) and (8.2) are *similar*. After a replacement of enthalpy with temperature (T_b = temperature of the burnt gas) via the relation

$$\delta = T_b - T = [(h_P - h_F)/c_p]\, w_F \,, \tag{8.3}$$

identical equations for fuel mass fraction and temperature result (i. e., the conservation equations for mass and energy are identical),

$$\alpha\frac{d^2\delta}{dz^2} - \upsilon\frac{d\delta}{dz} - \delta \cdot A \cdot \exp\left[-\frac{E}{R(T_b - \delta)}\right] = 0\,. \tag{8.4}$$

The solution is quite complicated and possible only for certain regions in the flame, where some terms can be neglected. Furthermore, a series expansion of the exponential term for large E is necessary for the solution. Nevertheless, it can be shown easily that solutions only exist if υ has the eigenvalue (called the *flame velocity*)

$$\upsilon_L = \sqrt{\frac{\alpha}{\tau}}\,, \tag{8.5}$$

where $\tau = 1/k = [A \cdot \exp(-E/RT)]^{-1}$ is a characteristic time of reaction at a temperature $T < T_b$ to be specified. This model states that the laminar flame velocity υ_L depends on the diffusivity α (mass and energy diffusion are the same in this model) and on the characteristic time of reaction τ. This analysis by Zeldovich illuminates the basic phenomenon that flame propagation is caused by diffusive processes and that the necessary gradients are sustained by the chemical reaction. Such insight is an advantage of analytical solutions. Unfortunately, analytical solutions are very rare.

8.2 Numerical Solution of the Conservation Equations

The details of the numerical solution of the conservation equations shall now be explored. Without loss of generality, one can revert to the mathematical model of one-dimensional laminar flames developed in Chapter 3. These conservation equations have the general form

$$\underset{\text{accumulation}}{\frac{\partial f}{\partial t}} = \underset{\text{diffusion}}{A\frac{\partial^2 f}{\partial z^2}} + \underset{\text{convection}}{B\frac{\partial f}{\partial z}} + \underset{\text{reaction}}{C}, \tag{8.6}$$

where f is a dependent variable (e. g., a mass fraction w_i or the temperature T). Time t and the spatial coordinate z are independent variables. The numerical solution is obtained by approximating the continuous problem (8.6) by a *discrete* problem, which means that the solution of the equation system is performed for distinct (discrete) points in the reaction system. Derivatives are replaced by finite differences, as will be discussed in the following.

8.2.1 Spatial Discretization

For the profiles of a dependent variable f (e. g., mass fractions w_i, temperature T, velocity υ) as a function of the independent spatial variable z, a simple *spatial discretization* can be obtained (Marsal 1976) by describing the function $z,t \rightarrow f(z,t)$ by its values only at certain *grid points* l ($l = 1, 2, \ldots, L$), where L denotes the total number of grid points (see Fig. 8.1). As an example, $f_l(t) = f(z_l,t)$ denotes the value of f at grid point l at time t.

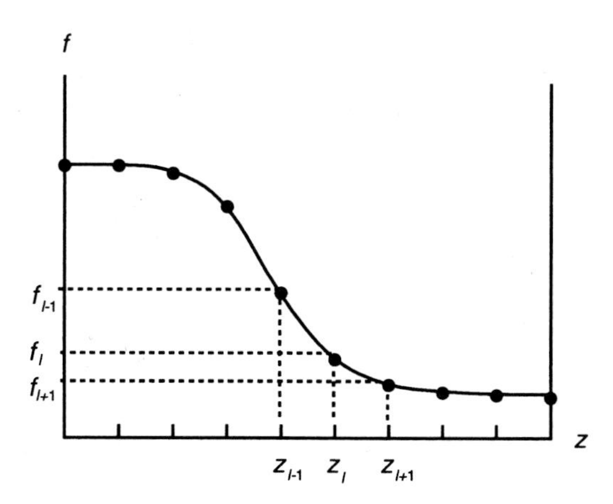

Fig. 8.1. Schematic illustration of the spatial discretization

At a point z_l the first and second derivatives $(\partial f / \partial z)_l$ and $(\partial^2 f / \partial z^2)_l$ are required. These derivatives are approximated by using the corresponding derivatives of an interpolating parabola that is determined by the points z_{l-1}, z_l, and z_{l+1}; see Fig. 8.2. (It can be shown by means of a Taylor expansion of the function f around the point z_l, that this parabola approximation for the first derivatives is second-order accurate and that the approximation for the second derivative is first-order accurate; see, e. g., Marsal 1976.) The parabolic fit to $f = az^2 + bz + c$ from three points leads to the equations

$$\begin{aligned} f_{l+1} &= a_l\, z_{l+1}^2 + b_l\, z_{l+1} + c_l \\ f_l &= a_l\, z_l^2 + b_l\, z_l + c_l \\ f_{l-1} &= a_l\, z_{l-1}^2 + b_l\, z_{l-1} + c_l \;. \end{aligned}$$

This is a linear equation system for a_l, b_l, and c_l. (The initial values of f have to be specified; see Section 8.2.2 below.)

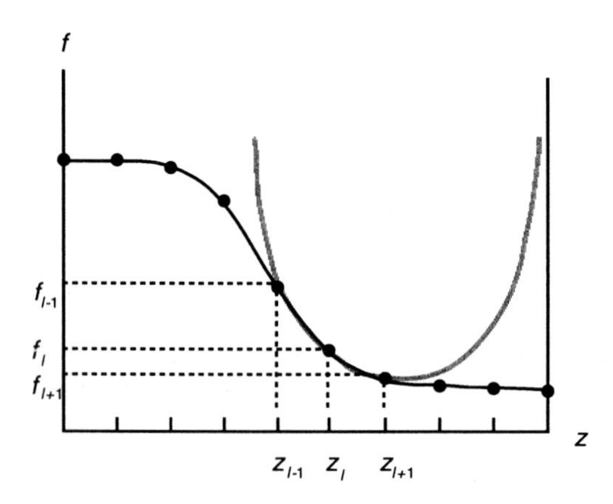

Fig. 8.2. Parabolic Approximation of a function by pieces of a parabola; Δz is not required to be uniform

By introducing the abbreviations

$$\Delta z_l = z_{l+1} - z_l \qquad \text{and} \qquad \alpha_l = \frac{\Delta z_l}{\Delta z_{l-1}} = \frac{z_{l+1} - z_l}{z_l - z_{l-1}} \;,$$

one obtains

$$a_l = \frac{f_{l+1} - (1+\alpha_l)\, f_l + \alpha_l\, f_{l-1}}{(\Delta z_l)^2 (1 + 1/\alpha_l)} \;, \quad b_l = \alpha_l \frac{f_l - f_{l-1} - a_l \dfrac{\Delta z_l}{\alpha_l}(z_l + z_{l-1})}{\Delta z_l} \;.$$

For the first derivative ($\partial f/\partial z = 2az + b$) and the second derivative ($\partial^2 f/\partial z^2 = 2a$) the result are the expressions

$$\left(\frac{\partial f}{\partial z}\right)_l = \frac{\frac{1}{\alpha_l} f_{l+1} + \left(\alpha_l - \frac{1}{\alpha_l}\right) f_l - \alpha_l f_{l-1}}{\left(1 + \frac{1}{\alpha_l}\right) \Delta z_l} \tag{8.7a}$$

and

$$\left(\frac{\partial^2 f}{\partial z^2}\right)_l = 2 \frac{f_{l+1} - (1+\alpha_l) f_l + \alpha_l f_{l-1}}{\left(1 + \frac{1}{\alpha_l}\right) (\Delta z_l)^2} , \tag{8.8a}$$

or (after some simple manipulations)

$$\left(\frac{\partial f}{\partial z}\right)_l = \frac{\Delta z_{l-1} \frac{f_{l+1} - f_l}{\Delta z_l} + \Delta z_l \frac{f_l - f_{l-1}}{\Delta z_{l-1}}}{\Delta z_l - \Delta z_{l-1}} \tag{8.7b}$$

$$\left(\frac{\partial^2 f}{\partial z^2}\right)_l = \frac{\frac{f_{l+1} - f_l}{\Delta z_l} - \frac{f_l - f_{l-1}}{\Delta z_{l-1}}}{\frac{1}{2} \cdot (\Delta z_l - \Delta z_{l-1})} . \tag{8.8b}$$

These equations clearly reflect that the derivatives are averages of first and second order difference quotients, respectively, weighted with the grid point distances.

For equidistant grids ($\Delta z_l = \Delta z_{l-1}$ or $\alpha_l = 1$, respectively, for all l), the expressions (8.7)/(8.8) reduce to the well-known *central difference* approximation (see, e. g., Forsythe and Wasow 1969). In the following examples, equidistant grids will be assumed (the generalization to non-equidistant grids is possible without difficulties),

$$\left(\frac{\partial f}{\partial z}\right)_l = \frac{f_{l+1} - f_{l-1}}{2 \Delta z} \tag{8.9}$$

$$\left(\frac{\partial^2 f}{\partial z^2}\right)_l = \frac{f_{l+1} - 2 f_l + f_{l-1}}{(\Delta z)^2} . \tag{8.10}$$

8.2.2 Initial Values, Boundary Conditions, Stationary Solution

The conservation equation for energy shall be considered now, when velocity is zero and there are no chemical reactions,

$$\frac{\partial T}{\partial t} = \lambda \frac{\partial^2 T}{\partial z^2} . \tag{8.11}$$

This equation is known as *Fourier's equation* for heat conduction. The problem is an *initial value problem* with respect to time t. The initial value of the temperature profile $T = T(z)$ must be specified at $t = t_0$, in order to begin the numerical solution (see Fig. 8.3). Furthermore, the problem is a *boundary value problem* with respect to variable z because $T(z)$ must be specified at the boundary for all times t, e. g., $T_A = T(z_A)$ and $T_E = T(z_E)$; see Forsythe and Wasow (1969).

The differential equation (8.11) describes the time behavior of the temperature profile $T = T(z)$. For sufficiently large t a *stationary* (i. e., time-independent) solution is obtained. The profile no longer changes with time t. For the Fourier equation a linear profile is obtained, since the temporal change, and thus the curvature (second spatial derivative) of the temperature, have to vanish according to (8.11).

In a similar way, the numerical simulation of a laminar flat premixed flame demands initial values for all spatial locations, and two boundary conditions (two integration constants for the second derivatives in space) are needed.

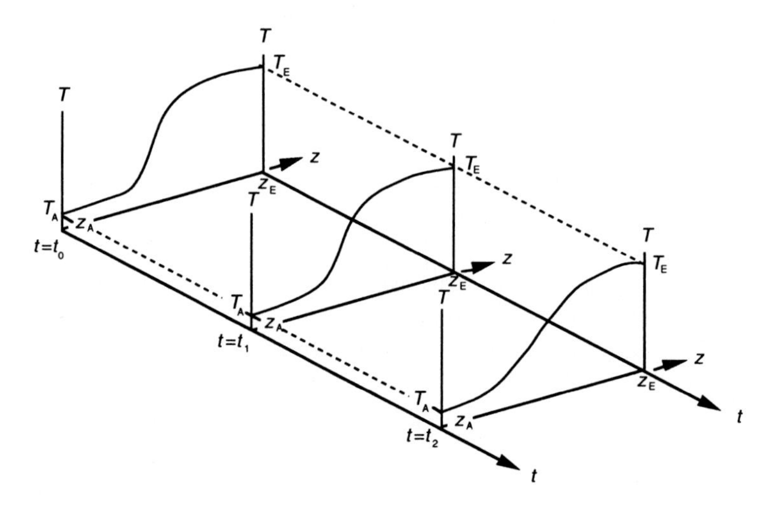

Fig. 8.3. Time behavior of the temperature profile according to the heat conduction equation

8.2.3 Explicit Solution Methods

There are principally two methods of solving the system of equations under consideration. One method is the *explicit solution method*, discussed here; the other is the *implicit solution method*, which is discussed in the next section.

The simple, but illustrative, example $\partial f/\partial t = \partial f/\partial z$ shall be considered in the following. (For the conservation equations, the second derivatives result in more complicated formulae.)

A first-order (linear) approximation is used for the time derivative and a second-order (parabolic) approximation for the spatial derivative. One obtains (see Fig. 8.4 and Section 8.2.1) the discretized equation

$$\frac{f_l^{(t+\Delta t)} - f_l^t}{\Delta t} = \frac{f_{l+1}^{(t)} - f_{l-1}^{(t)}}{2\Delta z} . \tag{8.12}$$

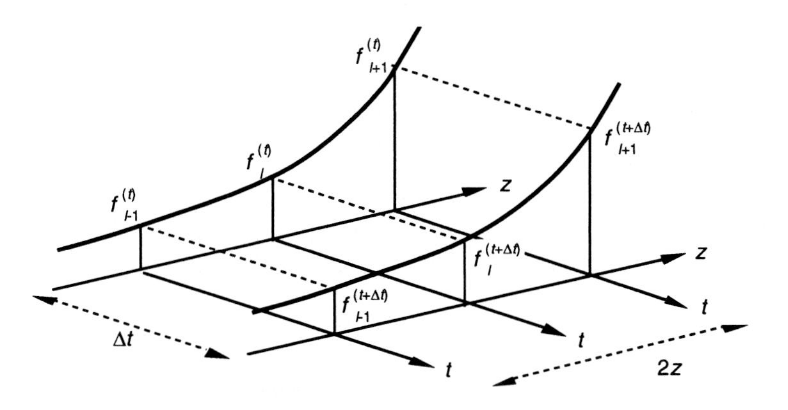

Fig. 8.4. Numerical solution of the differential equation $\partial f/\partial t = \partial f/\partial z$

Solving for $f_l^{(t+\Delta t)}$ one obtains an *explicit solution*

$$f_l^{(t+\Delta t)} \quad = \quad f_l^{(t)} \;+\; \Delta t\, \frac{f_{l+1}^{(t)} - f_{l-1}^{(t)}}{2\Delta z}\,. \tag{8.13}$$

Thus, the explicit solution can be described as a "forward shooting" from $f_l^{(t)}$ to $f_l^{(t+\Delta t)}$.

8.2.4 Implicit Solution Methods

The *implicit* solution of the above example results from evaluating the difference approximation at $t + \Delta t$, instead of at t,

$$\frac{f_l^{(t+\Delta t)} \;-\; f_l^{(t)}}{\Delta t} \quad = \quad \frac{f_{l+1}^{(t+\Delta t)} \;-\; f_{l-1}^{(t+\Delta t)}}{2\Delta z}\,. \tag{8.14}$$

Solving for the unknowns at $t + \Delta t$ one obtains (L is the number of grid points)

$$h\, f_{l-1}^{(t+\Delta t)} + f_l^{(t+\Delta t)} - h\, f_{l+1}^{(t+\Delta t)} \quad = \quad f_l^{(t)} \quad \text{with} \quad l = 2, ..., L\text{-}1\,, \tag{8.15}$$

where $h = \Delta t/2\Delta z$. This is a so-called *tridiagonal* linear equation system for $f_l^{(t+\Delta t)}$, $l = 2, ..., L$-1; f_1 and f_L result from the boundary conditions.

The *implicit* solution is an example of "backward coupling". This implicit solution can be shown to be (much) more stable than the explicit solution.

Implicit methods are important for the solution of stiff differential equation systems that are common to combustion problems, largely due to the chemical kinetics (see Section 7.3). Although a single implicit step is more expensive (due to the need of solution of a linear equation system) than an explicit step, the greater stability of the implicit steps allows considerably larger steps, and hence fewer implicit steps are needed. The net result is a substantial improvement in computational time.

8.2.5 Semi-implicit Solution of Partial Differential Equations

The general one-dimensional conservation equation shall be considered once more,

$$\frac{\partial f}{\partial t} = A\frac{\partial^2 f}{\partial z^2} + B\frac{\partial f}{\partial z} + C\,. \tag{8.16}$$

In the *semi-implicit* solution the difference approximations are formulated at $t + \Delta t$, and the coefficients $A(t)$, $B(t)$, and $C(t)$ at time t. This kind of solution can be used, if the coefficients do not change much with time,

$$\frac{f_l^{(t+\Delta t)} - f_l^{(t)}}{\Delta t} = A_l^{(t)}\frac{f_{l+1}^{(t+\Delta t)} - 2f_l^{(t+\Delta t)} + f_{l-1}^{(t+\Delta t)}}{(\Delta z)^2} + B_l^{(t)}\frac{f_{l+1}^{(t+\Delta t)} - f_{l-1}^{(t+\Delta t)}}{2\Delta z} + C_l^{(t)}. \tag{8.17}$$

Separation of the variables at time t and time $t + \Delta t$ yields a tridiagonal linear equation system for $f_l^{(t+\Delta t)}$ ($l = 2, ..., L\text{-}1$),

$$f_{l-1}^{(t+\Delta t)}\left[A_l^{(t)}\frac{\Delta t}{(\Delta z)^2} - B_l^{(t)}\frac{\Delta t}{(\Delta z)^2}\right] + f_l^{(t+\Delta t)}\left[1 - A_l^{(t)}\frac{\Delta t}{(\Delta z)^2}\right]$$

$$+ f_{l+1}^{(t+\Delta t)}\left[A_l^{(t)}\frac{\Delta t}{(\Delta z)^2} + B_l^{(t)}\frac{\Delta t}{(\Delta z)^2}\right] = f_l^{(t)} - \Delta t\cdot C_l^{(t)}\,. \tag{8.18}$$

8.2.6 Implicit Solution of Partial Differential Equations

In implicit methods, the difference approximations and the coefficients A, B, and C are evaluated at time $t + \Delta t$. If the coefficients A, B, and C are linear in the dependent variables f, this leads to *block-tridiagonal* linear equation systems for $f_l^{(t+\Delta t)}$. Block-tridiagonal linear systems are routinely solved. However, if the coefficients A, B, and C depend non-linearly on the variables f, a *linearization* is necessary to take advantage of the ease solving a block-tridiagonal linear system. An example of nonlinear dependence can illustrate the problem as follows: If $C_l^{(t+\Delta t)}$ is a reaction rate that contains second or third powers of a species concentrations or an exponential of the reciprocal temperature, then $C_l^{(t+\Delta t)}$ is nonlinear in that concentration or in temperature, or both. The linearization is done by writing the total differential of C; letting f_s = concentration of chemical species s, one obtains

$$\mathrm{d}C = \sum_{s=1}^{S}\frac{\partial C}{\partial f_s}\mathrm{d}f_s \tag{8.19}$$

or written in terms of differences ($s = 1, ..., S$ is the numbering of the species)

$$C_l^{(t+\Delta t)} = C_l^{(t)} + \sum_{s=1}^{S}\left(\frac{\partial C^{(t)}}{\partial f_s}\right)_l\left[f_{s,l}^{(t+\Delta t)} - f_{s,l}^{(t)}\right]. \tag{8.20}$$

Now $C_l^{(t+\Delta t)}$ is linear in $f_l^{(t+\Delta t)}$, and the block-tridiagonal solution procedures described above can be applied.

In reacting flows, C is often a nonlinear reaction term r. For example, using the concentrations c_i as variables, one obtains for the simple nonlinear reaction chain

$$\begin{array}{ccccc} A_1 & + & A_1 & \rightarrow & A_2 \\ A_2 & + & A_2 & \rightarrow & A_3 \end{array}$$

the results

$$r_1^{(t+\Delta t)} = \frac{dc_1^{(t+\Delta t)}}{dt} = -2k_1\left[c_1^{(t+\Delta t)}\right]^2$$

and (8.21)

$$r_2^{(t+\Delta t)} = \frac{dc_2^{(t+\Delta t)}}{dt} = k_1\left[c_1^{(t+\Delta t)}\right]^2 - 2k_2\left[c_2^{(t+\Delta t)}\right]^2 .$$

After linearization one obtains

$$\begin{aligned} r_1^{(t+\Delta t)} &= -4\,k_1\,c_1^{(t)}\left[c_1^{(t+\Delta t)} - c_1^{(t)}\right] + r_1^{(t)} \\ r_2^{(t+\Delta t)} &= 2\,k_1\,c_1^{(t)}\left[c_1^{(t+\Delta t)} - c_1^{(t)}\right] - 4\,k_2\,c_2^{(t)}\left[c_2^{(t+\Delta t)} - c_2^{(t)}\right] + r_2^{(t)} . \end{aligned} \qquad (8.22)$$

It can be seen that the reaction rates $r_i^{(t+\Delta t)}$ are now linear in $c_i^{(t+\Delta t)}$, and thus form a block-tridiagonal system of equations.

8.3 Flame Structures

The solution of the conservation equations discussed in Section 3 is now possible for concrete problems. Comparisons of experimental data, when available, and calculated profiles of flame structures will be discussed in the following examples. The numerical simulations presented below use a detailed chemical mechanism, which consists of 231 elementary reactions. In order to give an impression of the complexity of this mechanism (see also Chapters 6 and 7), a newer version of this mechanism is listed in Table 6.1. Equal signs in the reactions denote that the reverse reactions have to be considered; their rate coefficients can be calculated via thermodynamics with (6.9).

A very simple, but important, case of a premixed flame is formed by combustion of hydrogen-air mixtures. The mechanism describing hydrogen oxidation consists of the initial nineteen (reversible) reactions in Table 6.1. Resulting concentration/temperature profiles are presented in Fig. 8.5. Due to the large diffusion coefficient and heat conductivity of hydrogen, the mole fraction profile of H_2 is broadened, leading to low values at the onset of the flame front and, hence, to a maximum in the O_2 profile at the same location. A further characteristic feature is the very narrow HO_2 peak resulting from the fact that HO_2 cannot coexist with H, O, and OH.

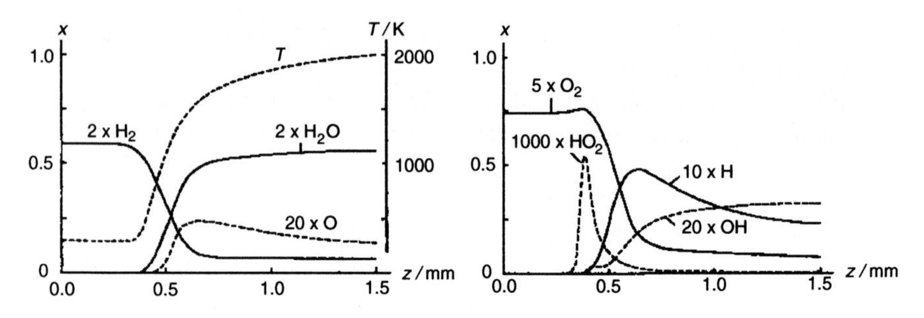

Fig. 8.5. Computed profiles of the mole fractions in a stoichiometric hydrogen-air flame, p = 1 bar, T_u = 298 K (Warnatz 1981b)

With respect to the more important combustion of hydrocarbons, it turns out that at flame conditions (T > 1100 K) the oxidation of a large aliphatic hydrocarbon R-H (like octane C_8H_{18}, see Fig. 8.6) is started by attack of H, O, or OH on a C-H bond leading to a radical R•,

$$\mathrm{H, O, OH + RH} \rightarrow \mathrm{H_2, OH, H_2O + R\bullet} \quad (\textit{H-atom abstraction}),$$

which is then thermally decomposing to an alkene and a smaller radical R',

$$\mathrm{R'{-}CH_2{-}\dot{C}H{-}R''} \rightarrow \mathrm{\bullet R' + CH_2{=}CH{-}R''} \quad (\beta\textit{-decomposition}),$$

until the relatively stable radicals methyl (CH_3) and ethyl (C_2H_5) are formed which are then oxidized. Thus, the problem of alkane oxidation can be reduced to the relatively well known oxidation of methyl and ethyl radicals (see Fig. 8.7).

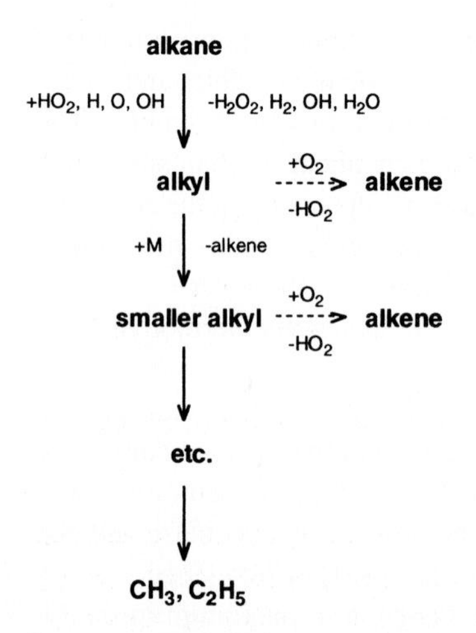

Fig. 8.6. Schematic mechanism of the radical pyrolysis of large aliphatic hydrocarbons to form CH_3 and C_2H_5 (Warnatz 1981a)

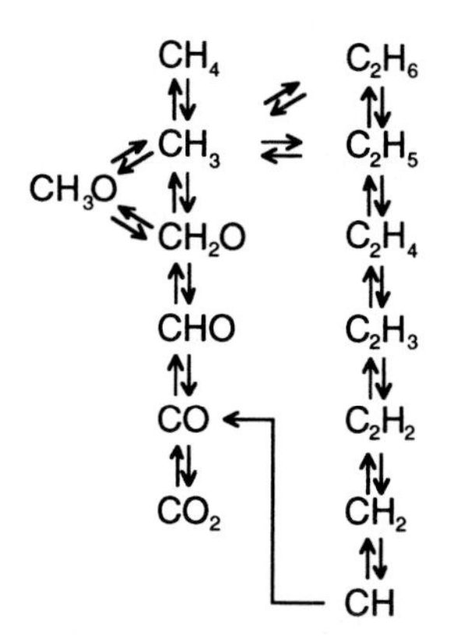

Fig. 8.7. Schematic mechanism of C_1- and C_2-hydrocarbon oxidation (Warnatz 1981a, 1993)

CH_3 radicals react mainly with O-atoms to give formaldehyde (the role of the oxidation of CH_3 by OH-radicals is not yet really established). The CHO-radical is then formed by H-atom abstraction. CHO can decompose thermally yielding CO and H-atoms, or the H-atom can be abstracted by H or O_2. This simple scheme, unfortunately, is complicated by the recombination of CH_3. In stoichiometric CH_4-air flames, this recombination path consumes about 30% of the CH_3 (if the recombination with H atoms is neglected). In rich flames, this recombination ratio increases to about 80% (see Figs. 7.8 and 7.9). CH_3/C_2H_5 oxidation is the rate-limiting (i. e., slow) part of the oxidation mechanism (see Figs. 8.13 and 8.14 below) and is therefore the reason for the similarity of all alkane and alkene flames. This is connected with the fact that hydrocarbon combustion reaction mechanisms show a hierarchical structure, as outlined in Fig. 8.8 (Westbrook and Dryer 1981).

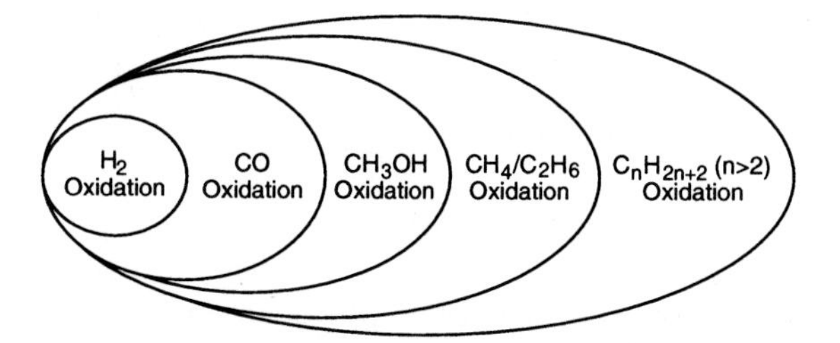

Fig. 8.8. Hierarchy in the reaction mechanism describing aliphatic hydrocarbon combustion

Figure 8.9 shows the structure of a propane-oxygen flame, which is diluted with argon in order to reduce the temperature, at $p = 100$ mbar (Bockhorn et al. 1990). Similar results are obtained for other hydrocarbons. The concentration profiles have been determined by mass spectroscopy (except for OH, which is measured by UV-light absorption), the temperature is measured by Na-D-line reversal (see Chapter 2).

Another example is an acetylene-oxygen flame (Warnatz et al. 1983) at sooting conditions (fuel-rich), see Fig. 3.5. Here the appearance of CO and H_2 as stable products and the formation of higher hydrocarbons, which are connected with the formation of precursors of soot (e. g., C_4H_2, see Chapter 18), are typical of fuel-rich combustion products.

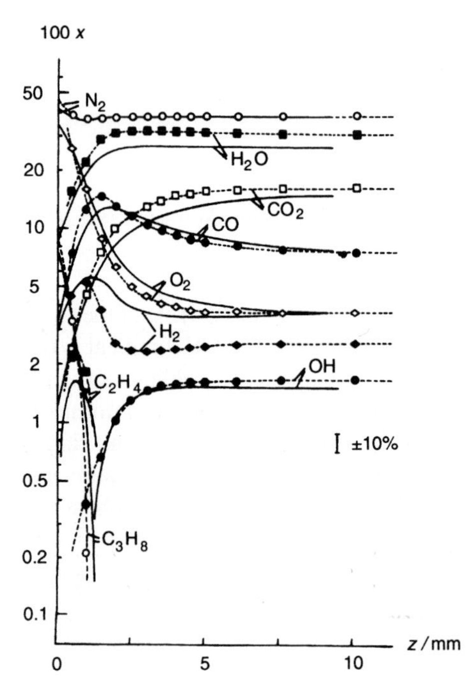

Fig. 8.9. Structure of a laminar premixed C_3H_8-O_2 flame (diluted with Ar) at p = 100 mbar

8.4 Flame Velocities

For pressure dependence and temperature dependence, in case of a single step reaction, Zeldovich's analysis (Section 8.1) delivers (n denotes the reaction order, E the activation energy, and T_b the burnt gas temperature)

$$v_L \approx p^{\frac{n}{2}-1} \exp(-\frac{E}{2RT_b}) .$$

Figure 8.10 shows the dependence of the flame velocity on the initial mixture composition (at fixed p and T_u) for H_2-air mixtures (Paul and Warnatz 1998). Furthermore, Fig. 8.11 shows the dependence of the flame velocity on pressure (at fixed initial composition and T_u) and temperature (at fixed initial composition and p) for methane-air mixtures (Warnatz 1988). In addition, Fig. 8.12 shows the dependence of flame velocity on the initial mixture composition for different fuels (Warnatz 1988).

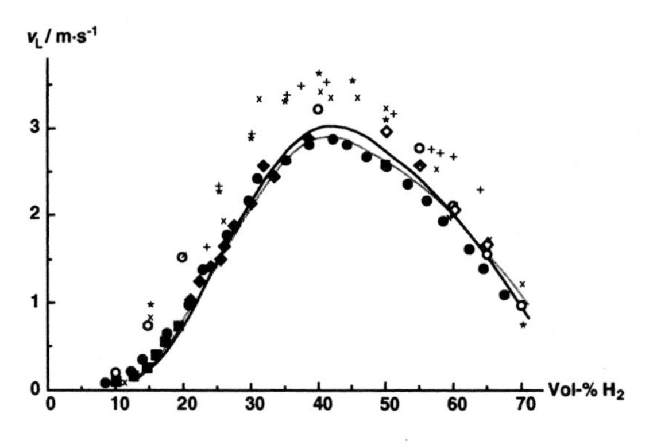

Fig. 8.10. Flame velocity in H_2-air mixtures as function of the unburnt gas composition (Paul and Warnatz 1998) calculated with two different transport models (grey and black lines) in comparison to experiments (points), p = 1 bar, T_u = 298 K

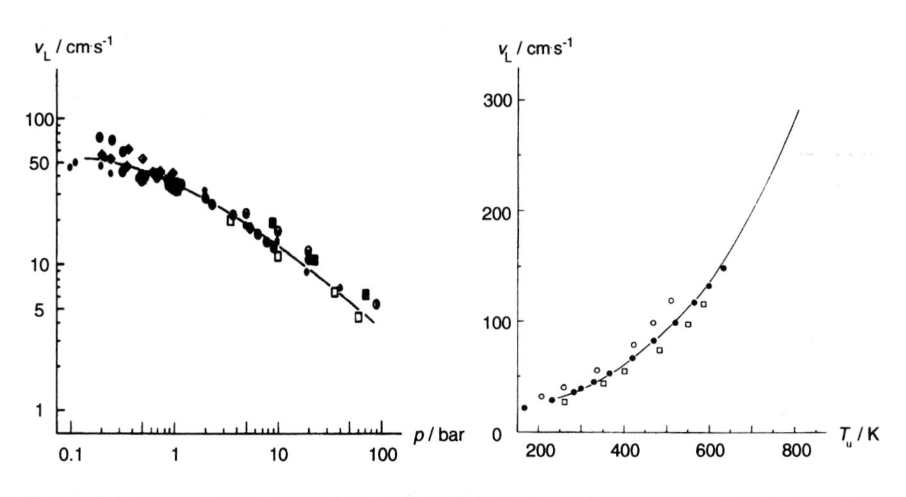

Fig. 8.11. Pressure dependence of v_L for T_u = 298 K (left) and temperature dependence of v_L for p = 1 bar (right) in stoichiometric CH_4-air mixtures (Warnatz 1988)

The numerical simulations in the figures (T_u denotes the unburnt-gas temperature) have been performed using a mechanism similar to that in Table 6.1. Figure 8.11 clearly shows the weakness of the single-step model: The reaction order of the rate-limiting steps is 2 or 3, and the simplified model predicts no pressure dependence or even a positive pressure dependence. The numerical results, however, show that the flame velocity has a negative pressure dependence. As a practical matter, extrapolation of 1 bar data to 150 bar found in Diesel engines is not credible with a single-step model.

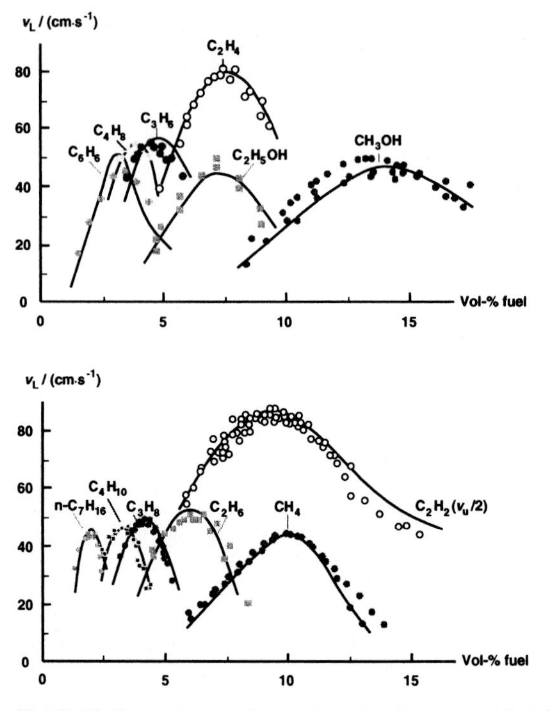

Fig. 8. 12. Dependence of v_L on the mixture composition (at p = 1 bar, T_u = 298 K) for different fuel-air mixtures (Warnatz 1993)

8.5 Sensitivity Analysis

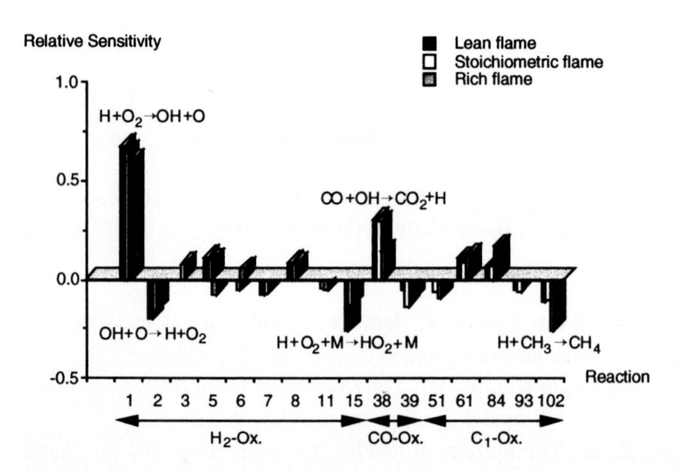

Fig. 8.13. Sensitivity analysis for the laminar flame velocity of a methane-air flame; the reaction numbers refer to an older version of the mechanism in Table 6.1 (Nowak and Warnatz 1988)

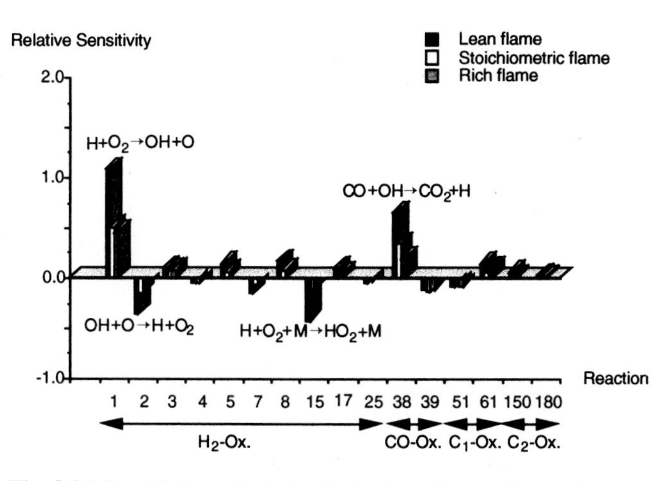

Fig. 8.14. Sensitivity analysis for the laminar flame velocity of a propane-air flame; the reaction numbers refer to a version of the mechanism in Table 6.1 (Nowak and Warnatz 1988)

Sensitivity analyses (discussed in Section 7.2) show similar results for all hydrocarbon-air mixtures, as can be seen in Figs. 8.13 and 8.14. The results are quite independent of the equivalence ratio. Note the small number of sensitive reactions. In all cases, the elementary reaction $H + O_2 \rightarrow OH + O$ (Reaction 1 in Table 6.1) is strongly rate-limiting, because it is a chain-branching step, while $H + O_2 + M \rightarrow HO_2 + M$ (Reaction 15) has a negative sensitivity because of its chain-terminating character. $CO + OH \rightarrow CO_2 + H$ governs a large part of the heat release and is thus rate-limiting, too.

8.6 Exercises

Exercise 8.1. (a) The characteristic time of reaction of a combustion process shall be given by $1/\tau = 1 \cdot 10^{10} \exp(-160 \text{ kJ} \cdot \text{mol}^{-1}/RT) \text{ s}^{-1}$. The mean diffusion coefficient shall have the value $D = 0.1\ (T/298 \text{ K})^{1.7} \text{ cm}^2/\text{s}$, and the Lewis number $Le = 1$. What are the values of the laminar flame velocity at 1000 K and 2000 K ? (b) The flame thickness is given approximately by $d = \text{const.}/(\rho_u \cdot v_L)$. In which way does the flame thickness depend on the pressure?

Exercise 8.2. A hydrocarbon-air mixture in a soap bubble with a diameter of 2 cm is ignited in the center. The unburnt gas temperature is $T_u = 300$ K, that of the burnt gas is $T_b = 1500$ K. Heat conduction between both layers shall be neglected. The propagation velocity v_b (referring to the density ρ_b) of the expanding spherical flame front is 150 cm/s. What is the laminar flame velocity (take into account that the observed v_b is caused by a combination of flame propagation and expansion)? How long does it take until the flame has reached the boundary of the soap bubble, and what is then its diameter? Sketch the time behavior of the flame radius and the soap bubble radius.

9 Laminar Nonpremixed Flames

In the previous chapter, premixed flames were discussed. In these flames, the fuel and oxidizer are mixed first with combustion occurring well after mixing. *Nonpremixed* flames were introduced as a basic flame type in Chapter 1. In nonpremixed flames, fuel and oxidizer react as they mix; examples of nonpremixed flames are given in Table 1.2. In this chapter, the standard model of laminar nonpremixed flames is developed. The extension of this model to a quantitative description of turbulent nonpremixed flames is the subject of Chapter 14.

In Section 8.1, the analysis by Zeldovich and Frank-Kamenetskii (1938) illuminated the basic phenomenon that premixed flame propagation is caused by diffusive processes and that the necessary gradients are sustained by the chemical reaction. In nonpremixed flames, fuel and oxidizer diffuse to the flame front due to the gradients sustained by the chemical reaction. The flame cannot propagate into the fuel without oxidizer or into the oxidizer without fuel and, thus, is fixed to the interface. The underlying physics is very simple: Fuel and oxidizer diffuse to the flame zone where chemical kinetics convert them to products, with attendant liberation of energy. The product species and energy diffuse away from the flame zone, both into the fuel and into the oxidizer.

As with premixed flames, even though the underlying physics is simple, the inclusion into the conservation equations of all of the relevant terms for thermodynamics, diffusive transport, and chemical reaction produces a system of partial differential equations that can rarely be solved analytically. Thus, the numerical solution of these equations for nonpremixed flames is the topic of this chapter. On a historical note, flames used to be distinguished as premixed or diffusion flames. In this book the attitude is adopted that all flames require diffusion and thus the term *diffusion* flame is not unique. Therefore, flames are distinguished as *premixed* or *nonpremixed*.

9.1 Counterflow Nonpremixed Flames

In practical devices, fuel and air are brought together by convection where they mix as a result of diffusion. In general, this is a three-dimensional problem. From a research point of view, the convection in three dimensions obfuscates the underlying physics.

Deeper understanding of nonpremixed flames has been achieved by contriving experiments in which the convection is reduced to one spatial dimension. Examples of such burners include the *Tsuji* burner (Tsuji and Yamaoka 1971), which consists of a cylinder in a cross flow (see Fig. 9.1a), and the opposed-jet flow burner (see Du et al. 1989), where a laminar flow of fuel leaves one duct and stagnates against the laminar flow of oxidizer emerging from an opposed duct; the gap between the ducts is typically the diameter of the duct (see Fig. 9.1b).

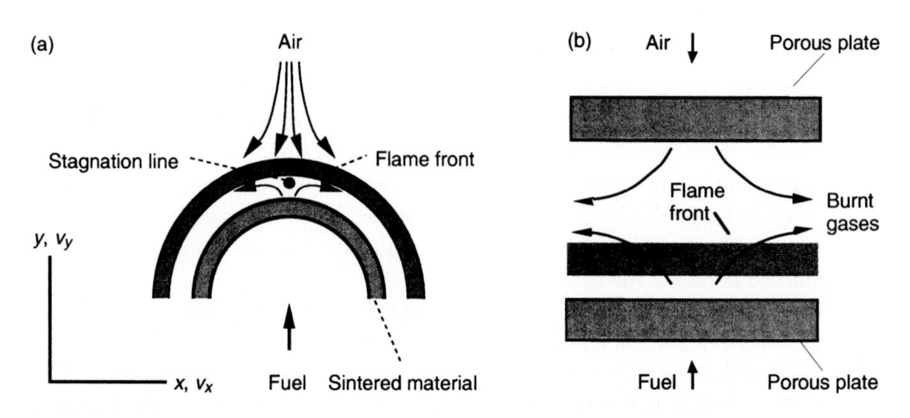

Fig. 9.1. Schematic illustration of counterflow nonpremixed flames; (a) Tsuji burner ; (b) opposed-jet burner (the cylinder diameter/burner diameter is typically 5 cm)

In either flow, the mathematical treatment can be simplified considerably if the description is restricted to the flow properties along the stagnation stream plane (Fig. 9.1a) or the stagnation stream line, respectively (Fig. 9.1b). Using the boundary layer approximation of Prandtl (~1904) (i. e., neglect of diffusion in the direction orthogonal to the stream line, in Fig. 9.1 the x direction), the problem reduces to one spatial coordinate, namely the distance from the stagnation line or point, respectively. In this way, the tangential gradients of temperature and mass fractions, and the velocity component v_x can be eliminated. Using the assumptions that

- the temperature and mass fractions of all species are functions solely of the coordinate y normal to the flame,
- the normal velocity component v_y is a function of y only,
- the tangential velocity v_x is proportional to the coordinate tangential to the flame x (which is a result of the boundary layer assumption),
- the solution is considered along the y axis (stagnation stream plane),

one obtains an equation system which has only time t and the spatial coordinate y as independent variables. For an axisymmetric opposed jet flow configuration illustrated in Fig. 9.1b (Stahl and Warnatz 1991) the equation system is

$$\frac{\partial \rho}{\partial t} + 2\rho G + \frac{\partial(\rho v_y)}{\partial y} = 0 \tag{9.1}$$

$$\frac{\partial G}{\partial t} + \frac{J}{\rho} + G^2 - \frac{1}{\rho}\frac{\partial}{\partial y}\left(\mu \frac{\partial G}{\partial y}\right) + v_y \frac{\partial G}{\partial y} = 0 \tag{9.2}$$

$$\frac{\partial v_y}{\partial t} + \frac{1}{\rho}\frac{\partial p}{\partial y} + \frac{4}{3\rho}\frac{\partial}{\partial y}(\mu G) - \frac{2\mu}{\rho}\frac{\partial G}{\partial y} - \frac{4}{3\rho}\frac{\partial}{\partial y}\left(\mu \frac{\partial v_y}{\partial y}\right) + v_y \frac{\partial v_y}{\partial y} = 0 \tag{9.3}$$

$$\frac{\partial T}{\partial t} - \frac{1}{\rho}\frac{\partial p}{\partial t} + v_y\left(\frac{\partial T}{\partial y}\right) - \frac{1}{\rho c_p}\frac{\partial}{\partial y}\left(\lambda \frac{\partial T}{\partial y}\right) + \frac{1}{\rho c_p}\sum_i c_{p,i}\, j_{i,y}\frac{\partial T}{\partial y} + \frac{1}{\rho c_p}\sum_i h_i r_i = 0 \tag{9.4}$$

$$\frac{\partial w_i}{\partial t} + v_y \frac{\partial w_i}{\partial y} - \frac{1}{\rho}\frac{\partial}{\partial y} j_{i,y} = \frac{r_i}{\rho}\ , \tag{9.5}$$

where ρ denotes the density, w_i the mass fractions, T the temperature, p the pressure, t the time, μ the viscosity, c_{pi} the specific heat capacity of species i at constant pressure, c_p the specific heat capacity of the mixture at constant pressure, λ the heat conductivity of the mixture, h_i the specific enthalpy of species i, r_i the mass rate of formation of species i (in kg/m^{3}s), and $j_{i,y}$ the diffusion flux density in y-direction. G is the tangential velocity gradient $\partial v_x/\partial x$, and J the tangential pressure gradient, $J = \partial p/\partial x$. J is constant throughout the flow field, and thus an eigenvalue of the system.

The system of equations is completed by specifying appropriate boundary conditions, which depend on the specific system that is considered. Although (9.1) is a first order equation, boundary conditions are specified for all dependent variables at both boundaries. As a result, the pressure gradient term J becomes an eigenvalue of the system, i. e., for given boundary conditions J has to adopt a certain value such that a solution to the problem exists (see Stahl and Warnatz 1991, Kee et al. 1989b).

Earlier solutions to the Tsuji problem not quite correctly assumed the pressure gradient to result from a potential flow solution (Dixon-Lewis et al. 1985). These solutions delivered nearly correct profiles of scalars and velocity, but the predictions of the position of these profiles were in error.

The equations are similar to those of premixed flames (see Chapters 3 and 8). The conservation equations for species mass (9.5) and enthalpy (9.4) are the same. (9.1)-(9.3) are conservation equations for momentum and total mass which are needed for the description of the flow field. Here the mass flux ρv_y is not constant due to the mass loss in x-direction. The solution of the above equation system (9.1)-(9.5) leads to the desired calculation of profiles of temperature, concentration, and velocity in laminar counterflow nonpremixed flames and, thus, a comparison with experimental results, which are obtained by spectroscopic methods (see Chapter 2).

Fig. 9.2 shows calculated and measured temperature profiles in a methane-air counterflow nonpremixed flame at a pressure of p = 1 bar. In these experiments the temperature has been measured by CARS spectroscopy (Sick et al. 1991). The temperature of the air (right hand side of the figure) is about 300 K. The high temperature in the combustion zone (some 1950 K) can be seen clearly. It is significant to note that the adiabatic flame temperature of 2220 K in the corresponding premixed system (see Table 4.2) is nowhere achieved; this is typical of flame fronts in the nonpremixed case.

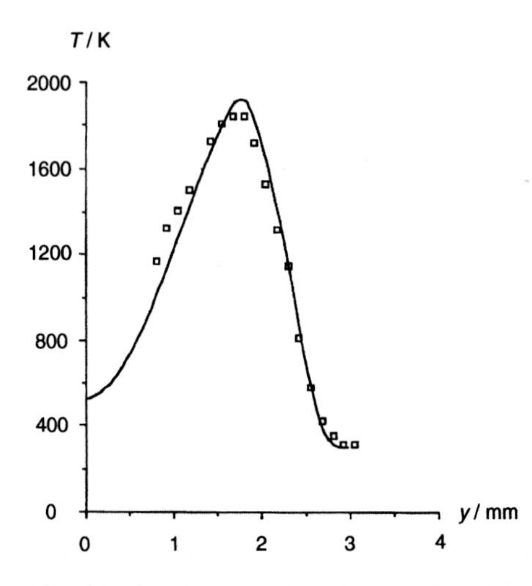

Fig. 9.2. Calculated and experimental temperature profiles in a methane-air counterflow nonpremixed flame at p = 1 bar, y denotes the distance from the burner (Sick et al. 1991)

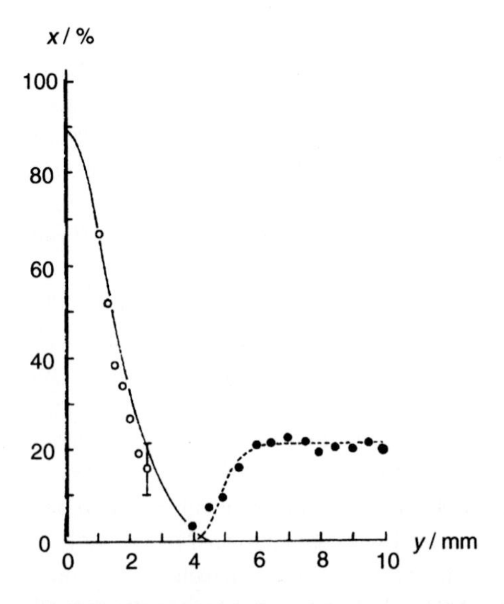

Fig. 9.3. Calculated and experimental profiles of methane and oxygen concentrations in a methane-air counterflow nonpremixed flame at p = 1 bar, y denotes the distance from the burner (Dreier et al. 1987)

Figure 9.3 shows calculated and measured concentration profiles of methane and air in a methane-air counterflow nonpremixed flame. In the experiments, the concentrations have been measured by CARS spectroscopy (Dreier et al. 1987). The fuel, as well as the oxygen, decreases towards the combustion zone. Note also that the fuel mole fraction at the cylinder surface $y = 0$ is not 100%, but is reduced due to diffusion of products from the flame zone towards the cylinder surface.

A comparison of calculated and measured (Dixon-Lewis et al. 1985) velocity profiles is shown in Fig. 9.4. The velocities have been measured from particle tracking of MgO particles (Tsuji and Yamaoka 1971).

The shape of the velocity profile can be explained very easily: A nonreactive flow is characterized by a monotonic transition between the velocities at the boundaries. In the combustion zone, however, there is a profound density change (caused by the high temperature of the burnt gas), which causes (because of monotonicity of the mass flux ρv) a deviation from the monotonic behavior in the region of the flame (here at about $y = 3$ mm) .

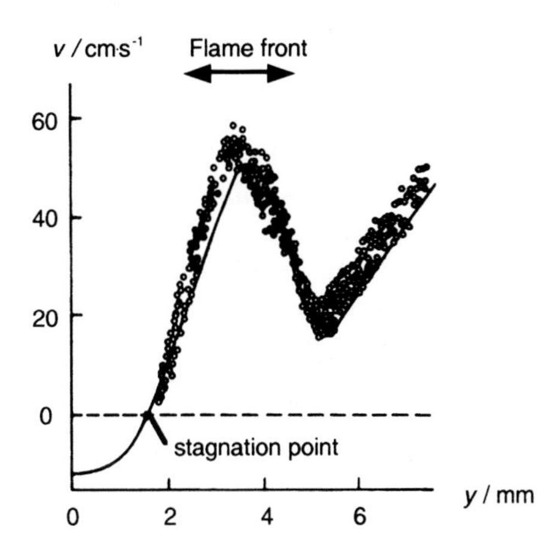

Fig. 9.4. Calculated and measured velocity profiles in a methane-air counterflow nonpremixed flame

9.2 Laminar Jet Nonpremixed Flames

The laminar jet nonpremixed flame requires at least a two-dimensional treatment (see Chapter 11). However, the laminar jet appears quite often (e. g., Bunsen burner); thus, some results shall be presented.

Figure 1.1 in Chapter 1 shows a simple Bunsen flame. Fuel is flowing into air from a round tube. Due to molecular transport (diffusion), fuel and air mix and burn in the reaction zone.

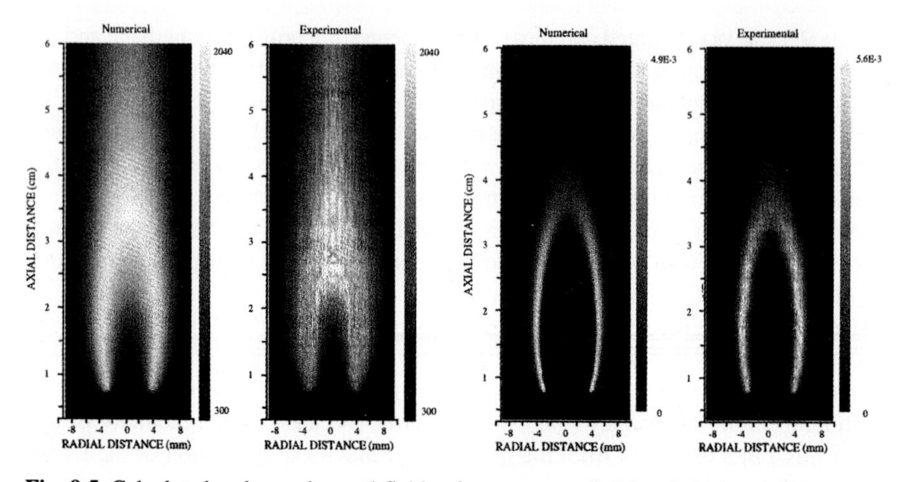

Fig. 9.5. Calculated and experimental fields of temperature (left hand side) and OH-radical mole fraction (right hand side) in a laminar jet nonpremixed flame (Smooke et al. 1989). The results can be directly compared with corresponding results from LIF-experiments on the same flame (Long et al. 1993)

The structure of such a nonpremixed Bunsen flame is shown in Fig. 9.5. The results have been obtained by a complete numerical solution of the two-dimensional conservation equations (see Smooke et al. 1989). The diameter of the fuel nozzle is 1.26 cm, the height of the flame shown in the figure is 30 cm. Temperature and concentration scales start with the lowest of the gray scales shown right; the maximum temperature is about 2000 K, the maximum OH mass fraction ~0.5 %.

The height of a jet nonpremixed flame can be calculated approximately by a simple, but crude method (Burke and Schumann 1928). The radius of the jet shall be denoted by r, the flame height by h, and the velocity in direction of the jet by υ. In the center of the cylinder, the time which is needed for the fuel to reach the tip of the flame can be calculated from the height of the nonpremixed flame and the inflow velocity ($t = h/\upsilon$). This time corresponds to the time needed for fuel and air to mix. The time of mixing can be calculated from Einstein's equation for the depth of intrusion by diffusion ($r^2 = 2Dt$, D = mean diffusion coefficient in the mixture considered, see Chapter 3). Setting equal those times t then leads to the relation

$$h = r^2 \upsilon / 2D \ . \tag{9.6}$$

Replacing the velocity υ by the volume flux $\Phi = \pi r^2 \upsilon$, one obtains $h = \Phi / 2\pi D$ or, more general (taking into account the cylindrical geometry by a correction factor θ)

$$h = \theta \Phi / \pi D \ . \tag{9.7}$$

This means hat the flame height h of a laminar jet nonpremixed flame depends on the volume flux Φ, but not on the nozzle radius r. Furthermore, the flame height is inversely proportional to the diffusion coefficient, which is reflected in the fact that the height of a hydrogen flame is smaller than a carbon monoxide flame by a factor of about 2.5. For a given mass flux, the volume flux is inversely proportional to the pressure. The diffusion coefficient is at the same time inversely proportional to the pressure (see Chapter 5). Thus, the flame height does not depend on the pressure for fixed mass flux (compensation of the pressure dependence in the nominator and denominator of (9.7).

9.3 Nonpremixed Flames With Fast Chemistry

In the case of infinitely fast (in reality very fast) chemistry, the reaction can be approximated by a fast one-step reaction of fuel and oxidizer to the products,

$$\mathrm{F} + \mathrm{Ox} \rightarrow \mathrm{P} .$$

This corresponds to the maxim "mixed = burnt", which was proposed in the 1930's by H. Rummel (see, e. g., Günther 1987).

In analogy to the species mass fractions w_i (see Chapter 1), an *element mass fraction* Z_i can be defined, which denotes the ratio between the mass of an element i and the total mass,

$$Z_i = \sum_{j=1}^{S} \mu_{ij} w_j \qquad ; \qquad i = 1, \ldots, M . \tag{9.8}$$

Here S denotes the number of different species and M the number of different elements in the mixture. The coefficients μ_{ij} denote the mass proportion of the element i in the species j (Shvab 1948, Zeldovich 1949).

As an example methane (CH_4) shall be considered here. Its molar mass can be calculated from the contributions of the elements C and H as $M_{CH_4} = 4 \cdot 1$ g/mol + $1 \cdot 12$ g/mol = 16 g/mol. The mass proportion of hydrogen is 4/16 = 1/4, and that of carbon is 12/16 = 3/4. Therefore μ_{H,CH_4} = 1/4, and μ_{C,CH_4} = 3/4 (the indices i, j have been replaced by the corresponding symbols for species and elements).

The element mass fractions have a special meaning, because they cannot be changed by reactive processes; they are changed by mixing. For simple nonpremixed flames, which can be treated as a coflow of fuel (F) and oxidizer (Ox), a *mixture fraction* ξ (independent of i; see below) can be defined based on the element mass fractions (the indices 1 and 2 denote the two streams) as

$$\xi = \frac{Z_i - Z_{i2}}{Z_{i1} - Z_{i2}} . \tag{9.9}$$

The advantage of this formulation is that ξ has a linear relation to the mass fractions because of (9.8) and (9.9). If the diffusivities are equal (which is approximately the case in many applications), the mixture fraction is independent of the choice of the element i ($i = 1, \ldots, M$) used for its definition.

As an example, a simple nonpremixed flame shall be considered, where one flow (index 1) consists of methane (CH_4), and the other (index 2) of oxygen (O_2). Furthermore an ideal, infinitely fast reaction to carbon dioxide (CO_2) and water (H_2O) shall be assumed,

$$CH_4 + 2\,O_2 \rightarrow CO_2 + 2\,H_2O\,.$$

Mixing of fuel and oxidizer is caused by diffusion. The element mass fractions can be calculated according to (9.8) to be

$$\begin{aligned}
Z_C &= \mu_{C,O_2} w_{O_2} + \mu_{C,CH_4} w_{CH_4} + \mu_{C,CO_2} w_{CO_2} + \mu_{C,H_2O} w_{H_2O} \\
Z_H &= \mu_{H,O_2} w_{O_2} + \mu_{H,CH_4} w_{CH_4} + \mu_{H,CO_2} w_{CO_2} + \mu_{H,H_2O} w_{H_2O} \\
Z_O &= \mu_{O,O_2} w_{O_2} + \mu_{O,CH_4} w_{CH_4} + \mu_{O,CO_2} w_{CO_2} + \mu_{O,H_2O} w_{H_2O}\,.
\end{aligned}$$

Based on the fact that by definition $\mu_{C,O_2} = \mu_{H,O_2} = \mu_{O,CH_4} = \mu_{H,CO_2} = \mu_{C,H_2O} = 0$, one obtains the relationships

$$\begin{aligned}
Z_C &= \mu_{C,CH_4} w_{CH_4} + \mu_{C,CO_2} w_{CO_2} \\
Z_H &= \mu_{H,CH_4} w_{CH_4} + \mu_{H,H_2O} w_{H_2O} \\
Z_O &= \mu_{O,O_2} w_{O_2} + \mu_{O,CO_2} w_{CO_2} + \mu_{O,H_2O} w_{H_2O}\,.
\end{aligned}$$

For the element mass fractions in the fuel (1) and the oxidizer (2) the result is then

$$\begin{aligned}
Z_{C,1} &= \mu_{C,CH_4} = 3/4 \quad ; \quad Z_{C,2} = 0 \\
Z_{H,1} &= \mu_{H,CH_4} = 1/4 \quad ; \quad Z_{H,2} = 0 \\
Z_{O,1} &= 0 \quad ; \quad Z_{O,2} = 1\,.
\end{aligned}$$

Thus, the mixture fractions are given by the three equations

$$\begin{aligned}
\xi_C &= \frac{Z_C - Z_{C,2}}{Z_{C,1} - Z_{C,2}} = \frac{Z_C - 0}{\mu_{C,CH_4} - 0} = \frac{Z_C}{\mu_{C,CH_4}} \\
\xi_H &= \frac{Z_H - Z_{H,2}}{Z_{H,1} - Z_{H,2}} = \frac{Z_H - 0}{\mu_{H,CH_4} - 0} = \frac{Z_H}{\mu_{H,CH_4}} \\
\xi_O &= \frac{Z_O - Z_{O,2}}{Z_{O,1} - Z_{O,2}} = \frac{Z_O - 1}{0 - 1} = 1 - Z_O\,.
\end{aligned}$$

If one assumes that all species diffuse equally fast, the ratio of hydrogen and carbon does not change,

$$Z_H/Z_C = Z_{H,1}/Z_{C,1} = \mu_{H,CH_4}/\mu_{C,CH_4} \qquad \text{or} \qquad Z_H/\mu_{H,CH_4} = Z_C/\mu_{C,CH_4}\,.$$

It can be seen that $\xi_H = \xi_C$. Calculating Z_C and Z_H from ξ_C or ξ_H, one obtains $\xi_O = \xi_H = \xi_C$. Indeed, for all elements the same ξ results.

The linear relations of ξ and the mass fractions w are shown in Fig. 9.6. To draw this figure, the stoichiometric mixture fraction $\xi_{\text{stoich.}}$ has to be known. In the example above, the stoichiometric mixture consists of 1 mol CH_4 and 2 mol O_2, corresponding to an O element-mass of 64 g and a total mass of 80 g. Therefore, the element mass fraction $Z_{\text{O,stoich.}}$ is 4/5, and for the stoichiometric mixture fraction one obtains the value $\xi_{\text{stoich.}} = 1/5$. For $\xi = 0$ the mixture consists of oxygen only, for $\xi = 1$ it consists of fuel only.

At the stoichiometric mixture fraction there exist neither fuel nor oxygen, but only combustion products ($w_P = w_{CO_2} + w_{H_2O} = 1$). In the fuel-rich region (here $\xi_{\text{stoich.}} < \xi < 1$) oxygen does not exists, because excess fuel is burnt infinitely fast by the oxygen and forms the reaction products. Accordingly, there is no fuel in the fuel-lean region (here $0 < \xi < \xi_{\text{stoich.}}$).

The linear dependences of mixture fraction and mass fractions are shown in Fig. 9.6. From those dependences of w_i on ξ, one obtains for the example considered:

Fuel side ($\xi_{\text{stoich.}} < \xi < 1$)	Oxygen side ($0 < \xi < \xi_{\text{stoich.}}$)
$w_{CH_4} = (\xi - \xi_{\text{stoich.}})/(1 - \xi_{\text{stoich.}})$	$w_{CH_4} = 0$
$w_{O_2} = 0$	$w_{O_2} = (\xi_{\text{stoich.}} - \xi)/\xi_{\text{stoich.}}$
$w_P = (1 - \xi)/(1 - \xi_{\text{stoich.}})$	$w_P = \xi/\xi_{\text{stoich.}}$.

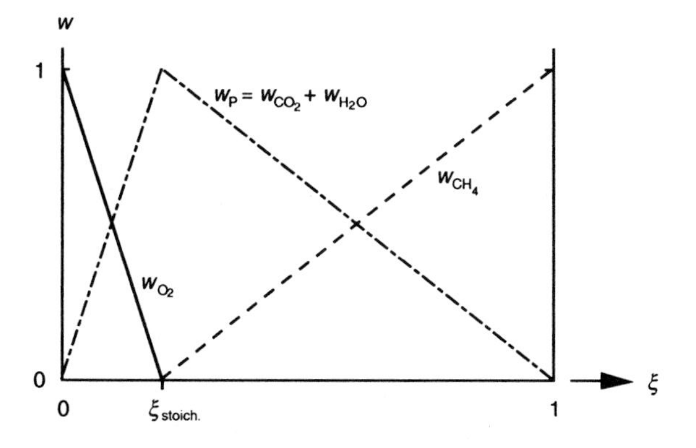

Fig. 9.6 Linear relations between mixture fraction and mass fraction in an idealized combustion system

For other systems (e. g., methane-air or partially premixed methane-air mixtures) more complex diagrams result, which, however, can be obtained by similar methods. The definition of the mixture fraction and the linear relations for $w_i = w_i(\xi)$ will be used later in the simplified description of turbulent nonpremixed flames.

If fuel and oxidizer do not react completely to the products (at equilibrium even a stoichiometric mixture contains some reactants) or if the chemical reaction has a finite rate, the linear relations no longer hold (see Fig. 9.7). Furthermore w_{Ox} and w_F

overlap in the region of the stoichiometric composition ξ_{stoich}. However, the linear relations $w_i = w_i(\xi)$ can still be used as an approximation (see Chapter 13).

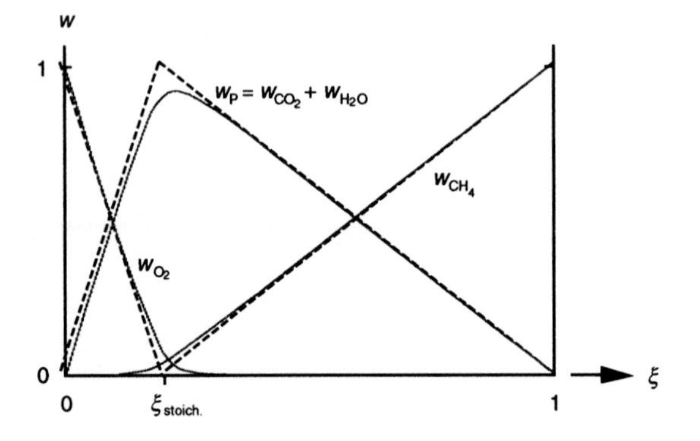

Fig. 9.7. Schematic illustration of the deviations from the linear dependence of mass fractions from the mixture fraction for finite reaction rate (coexistence of fuel and oxidizer is possible)

9.4 Exercises

Exercise 9.1: A laminar, gaseous jet of fuel flows into air, where it is ignited. The height of the flame is 8 cm. Then the jet diameter is increased by 50% and the jet velocity reduced by 50%. How does the height of the flame change? Show that the height of a nonpremixed jet flame does not depend on the pressure for constant mass flux.

Exercise 9.2: A simple acetylene-oxygen coflow nonpremixed flame shall be considered. Flow 1 consists only of oxygen (O_2), flow 2 only of acetylene (C_2H_2).

(a) Determine the mixture fractions ξ_i for the elements C, H and O prior to ignition.
(b) Determine the mixture fractions for C, H and O after the ignition. Take into account that CO_2 and H_2O are formed (assume equal diffusivities).
(c) What is the value of the mixture fraction at stoichiometric composition?

10 Ignition Processes

The discussion of premixed flames (Chapter 8) and nonpremixed flames (Chapter 9) assumed that the flames were at a steady state. The solutions are time-independent. The time-dependent process of starting with reactants and evolving in time towards a steadily burning flame is called *ignition*. Ignition processes are always time-dependent. Examples of ignition processes include induced ignition (such as occurs in gasoline engines induced by a spark), autoignition (such as occurs in Diesel engines), and photoignition caused by photolytic generation of radicals. In these cases, the ignition process is described quantitatively by addition to the time-dependent energy conservation equation (3.6) a term $\partial p/\partial t$,

$$\rho c_p \frac{\partial T}{\partial t} = \frac{\partial p}{\partial t} + \frac{\partial}{\partial z}\left(\lambda \frac{\partial T}{\partial z}\right) - \left(\rho v c_p + \sum_j j_j \, c_{p,j}\right)\frac{\partial T}{\partial z} - \sum_j h_j r_j \ . \qquad (10.1)$$

The additional term accounts for the temperature increase (or decrease) caused by compression (or expansion) of the mixture. Here it is assumed that, although the pressure p varies with time, the pressure is spatially uniform (Maas and Warnatz 1988).

The characteristic time at which pressure equilibrates is the characteristic dimension of the system divided by the sound speed. If the characteristic time of the ignition process is smaller than the pressure equilibration time, the pressure equilibration is too slow to allow the assumption of a spatially uniform pressure. In this case the conservation equations have to be extended. Such a general formulation is discussed in Chapter 11.

As the previous chapters show, a detailed simulation of a steady-state process of premixed flame propagation and of nonpremixed flames, taking into account molecular transport, chemical reactions, thermodynamics, and convection, is very demanding and nearly always requires a numerical solution. It is not surprising that the escalation to the time-dependent problem of ignition is even more demanding of computational resources. However, as with steady-state problems, a qualitative understanding can be obtained, when simplified systems are explored analytically. Here two cases are considered: homogenous ignition (Semenov 1935) and inhomogeneous ignition (Frank-Kamenetskii 1955)

If the heat exchange in the reaction system is fast in comparison to the heat exchange with the surrounding (vessel surface, etc.), the theory of Semenov is better suited. The theory of Frank-Kamenetskii is a better model if the heat exchange with the surrounding is faster than the heat exchange within the system.

10.1 Semenov's Analysis of Thermal Explosions

In his analysis of *thermal explosions*, Semenov (1935) considered a spatially homogeneous system, i. e., pressure, temperature, and composition are uniform. Furthermore, the chemistry is approximated by a one-step reaction

$$\text{Fuel (F)} \rightarrow \text{Products (P)}$$

with the first-order reaction rate

$$r = -M_F c_F A \cdot \exp(-E/RT) \tag{10.2}$$

with molar mass M_F, concentration of the fuel c_F, preexponential factor A, and activation energy E. In the early stages of ignition, fuel consumption is small. If reactant consumption is neglected ($c_F = c_{F,0}$, $\rho = \rho_0 = M_F\, c_{F,0}$, $c_{F,0}$ = initial concentration), the reaction rate

$$r = -\rho A \cdot \exp(-E/RT) \tag{10.3}$$

is obtained. For the description of the heat flux j_q to the surrounding, *Newton's law of heat exchange* is used, i. e., the heat exchange with the surrounding (vessel surface) is proportional to the temperature difference between the system and its surrounding,

$$j_q = \chi\, S \cdot \left(T - T_W\right). \tag{10.4}$$

Here T denotes the (spatially homogeneous) temperature in the system, T_w the wall temperature, S the surface area of the system, χ the *heat transfer coefficient* (W/m^2). This approach is rather simple (no differential equation in the stationary case), but on the penalty that χ is dependent on the actual conditions, especially on the geometry. The time behavior of the temperature can be calculated from an imbalance of heat production P and heat exchange (loss) L to the surrounding,

$$\rho c_p \frac{dT}{dt} = P - L = \left(h_F - h_P\right) \cdot \rho A \cdot \exp(-E/RT) - \chi\, S \cdot \left(T - T_W\right). \tag{10.5}$$

The qualitative behavior of the system can be understood, if the production term as well as the loss term are plotted (see Fig. 10.1). The loss term increases linearly with the temperature due to (10.4), whereas the heat production term increases exponentially with temperature; see (10.3). The three curves P_1, P_2, and P_3 show temperature dependences for different activation energies E and the preexponential factors A.

First curve P_3 shall be considered. Heat production and heat loss are equal when curve P_3 intersects curve L. Such intersections are called *stationary points*. Two of them occur ($T_{S,1}$ and $T_{S,2}$). If the system has a temperature $T < T_{S,1}$, heat production dominates, and the system temperature increases until heat production and heat loss compensate, i. e., until $T_{S,1}$ is reached. For temperatures $T_{S,1} < T < T_{S,2}$ heat losses dominate, and the system cools down until the steady state $T = T_{S,1}$ is reached. $T = T_{S,1}$ is thus called a *stable stationary* point, $T = T_{S,2}$ an *unstable stationary* point

If the system has a temperature $T > T_{S,2}$, heat production dominates. The temperature increases, further accelerating the chemical reaction, and an explosion occurs.

The point $T = T_M$ is a *metastable point.* Any deviation in temperature causes the system to evolve. An infinitesimal increase in temperature leads to an explosion, while an infinitesimal decrease reduces the heat production rate more that the heat loss rate, so that the systems evolves to the stable stationary point $T_{S,1}$.

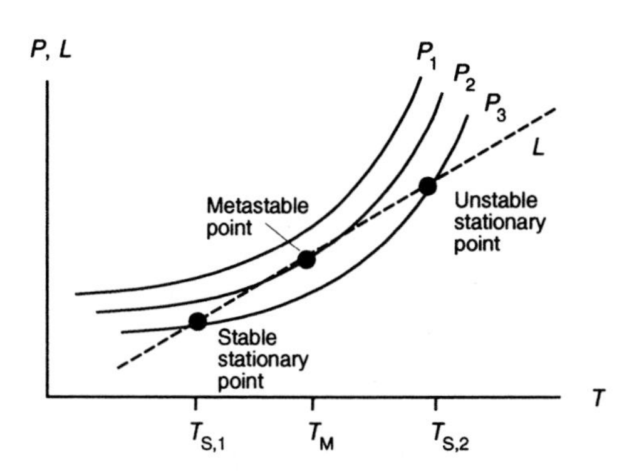

Fig. 10.1. Schematic illustration of the temperature dependences of heat production P (solid lines) and heat losses L (dotted line)

As Fig. 10.1 shows there are curves, such as curve P_1, that do not intersect with the heat transfer line L. In this case there is no stationary point, and heat production always exceeds cooling; the system explodes for all initial temperatures. For an adiabatic system, the heat transfer is zero and thus, any exothermic adiabatic system will explode. Furthermore Fig. 10.1 shows that there exists a critical heat production curve (P_2) that intersects the curve L in exactly one point.

10.2 Frank-Kamenetskii's Analysis of Thermal Explosions

Frank-Kamenetskii's (1955) analysis of thermal explosions extends Semenov's analysis by replacing the Newtonian heat transfer law (10.4) with the more realistic law of Fourier that allows for the diffusion of energy in the system to the wall. As a consequence, the temperature in the system is not uniform. For the purposes here, the treatment is restricted to one-dimensional geometries (infinite slab, infinite cylinder, or sphere). With (8.2), the energy conservation equation can then be written as

$$\frac{\lambda}{r^i}\frac{\mathrm{d}^2 r^i T}{\mathrm{d}r^2} = \rho A\cdot(h_\mathrm{P} - h_\mathrm{F})\cdot\exp(-E/RT)\ . \tag{10.6}$$

for a one-step reaction F → P. The exponent i in (10.6) allows the treatment of the three one-dimensional geometries. Here $i = 0$ for an infinite slab, $i = 1$ for cylindrical geometry (only radial dependence), and $i = 2$ for spherical geometry (only radial dependence). The different geometries are shown in Fig. 10.2.

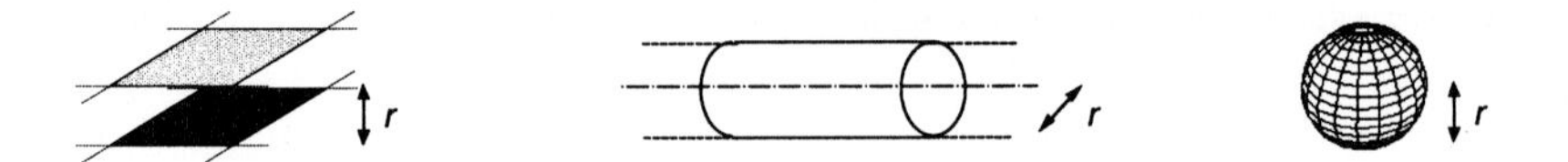

Fig. 10.2. One-dimensional geometries: infinite slab (left), infinite cylinder (center), and sphere (right)

It is instructive to recast Eq. (10.6) by introducing *dimensionless variables*. For the temperature the dimensionless variable $\Theta = (E/RT_\mathrm{W}^2)\,(T - T_\mathrm{W})$ is introduced, where T_W denotes the temperature of the wall, which is here assumed to be constant. The radius r is scaled by the overall radius r_0 of the system; thus $\tilde{r} = r/r_0$. Furthermore, let $1/\varepsilon$ denote the dimensionless activation energy ($1/\varepsilon = E/RT_\mathrm{W}$), and define δ a constant that is characteristic for the system, which is given by

$$\delta = \frac{h_\mathrm{P} - h_\mathrm{F}}{\lambda}\cdot\frac{E}{RT_\mathrm{W}^2}\cdot\rho r_0^2\, A\cdot\exp\left(-\frac{E}{RT_\mathrm{W}}\right)\ . \tag{10.7}$$

Using these definition, one obtains the rather simple differential equation

$$\frac{\mathrm{d}^2\Theta}{\mathrm{d}\tilde{r}^2} + \delta\cdot\exp\left(\frac{\Theta}{1+\varepsilon\Theta}\right) = 0 \tag{10.8}$$

with the boundary conditions $\Theta = 0$ for $\tilde{r} = 1$ (constant temperature at the surface) and $\mathrm{d}\Theta/\mathrm{d}\tilde{r} = 0$ for $\tilde{r} = 0$ (vanishing temperature gradient in the vessel center; symmetry boundary condition).

It can be shown (not here) that this differential equation has stationary solutions only when δ is smaller than a critical value δ_crit. δ_crit is dependent on the geometry with $\delta_\mathrm{crit} = 0.88$ for the infinite slab, $\delta_\mathrm{crit} = 2$ for the infinite cylinder, and $\delta_\mathrm{crit} = 3.32$ for the sphere. For $\delta > \delta_\mathrm{crit}$ explosion occurs, for $\delta < \delta_\mathrm{crit}$ one obtains a stable behavior (see Frank-Kamenetskii 1955). If the characteristic values of the reaction system are known (h_P, h_F, ρ, A, λ), the maximum wall temperatures T_W can be calculated for different vessel sizes r_0, such that the system is stable and explosion does not occur.

A weakness of Frank-Kamenetskii's analysis of thermal explosions is the assumption of no reactant consumption. Improvements of the analysis deal mainly with avoiding this assumption (e. g., Boddington et al. 1983, Kordylewski and Wach 1982). The main result, for the purposes here, is that ignition requires the chemical production of heat to be greater than the rate of heat transfer out of the system.

10.3 Autoignition: Ignition Limits

Obviously the question as to which temperatures, pressures, and compositions a mixture can be ignited, is very important (e. g., in safety considerations, ignition processes in engines, etc.) . If, e. g., a hydrogen-oxygen mixture is situated in a hot vessel, at certain values of temperature and pressure a spontaneous explosion is observed after an *ignition-delay time* sometimes called an *induction time* (which can be as long as several hours or as short as microseconds).

At other conditions only a slow reaction takes place. These phenomena are illustrated by a *p-T explosion diagram*, where the regions in which ignition takes place are separated by a curve from the regions where no ignition occurs (see Fig. 10.3). The figure shows experimental results (points) and simulations (curves) for stoichiometric mixtures of hydrogen and oxygen (Maas and Warnatz 1988).

The *explosion limits* (or *ignition limits*) were discovered in the 1920's. The detailed numerical simulation, where the complete set of time-dependent conservation equations has to be solved, has become possible during the 1980's (e. g., Maas and Warnatz 1988). The simulation shows the gas-phase reactions alone to be inadequate to explaining the measurements. Agreement is achieved by inclusion of reactions at the walls of the vessel. These surface reactions (or heterogeneous reactions) are the recombination of radicals modeled with the non-elementary reactions (see Section 6.7)

$$\mathrm{O} \rightarrow 1/2\,\mathrm{O_2} \quad , \quad \mathrm{H} \rightarrow 1/2\,\mathrm{H_2} \qquad \text{etc.}$$

Although the quantitative determination of the explosion limits is quite complicated, the processes can be understood quite easily in a qualitative way (see Fig. 10.3).

This hydrogen-oxygen system at 800 K and at low pressure ($p < 5$ mbar) does not ignite. Reactive species (radicals), which are formed in the gas phase by chemical reactions, diffuse to the vessel wall where they recombine to stable species. Due to the low pressure, diffusion is very fast, since the diffusivity is inversely proportional to the density due to (5.21). Thus, ignition does not take place; however, a slow reaction continues.

When the pressure is increased above a certain value (*first ignition limit*), spontaneous ignition is observed, because the reduced rate of diffusion of the radicals to the wall ($D \propto p^{-1}$), where they are destroyed, is below the radical production rate in the gas phase. The first explosion limit depends much on the chemical nature of the vessel surface, because the limit is a result of the concurrent processes of chain branching in the gas phase and chain termination at the surface. This surface sensitivity can be shown by noting that different wall materials (e. g., glass, iron, copper, palladium) will have different explosion limits.

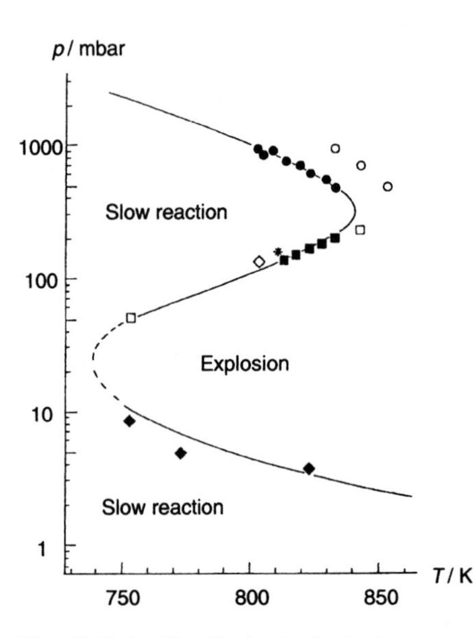

Fig. 10.3. Ignition limits in the hydrogen-oxygen system (p-T explosion diagram); points: experiments, lines: simulations (see Maas and Warnatz 1988)

Above a pressure of 100 mbar (for the same temperature 800 K), again one does not observe ignition. The *second explosion limit* is governed by the competition of chain branching and chain termination in the gas phase. At low pressures, hydrogen atoms react with molecular oxygen in a most important chain-branching step

$$\mathrm{H} + \mathrm{O_2} \rightarrow \mathrm{OH} + \mathrm{O} \ .$$

The products of this chain-branching reaction, OH and O, rapidly react with fuel to produce H which can then react in the chain-branching reaction above to yield even more radicals. The radicals increase at an exponential rate, which is the basis of an explosion. Competing with the branching reaction is the three-body reaction

$$\mathrm{H} + \mathrm{O_2} + \mathrm{M} \rightarrow \mathrm{HO_2} + \mathrm{M}$$

that produces the mildly reactive hydroperoxyl radical; this reaction (which is nearly independent of temperature) is essentially a chain termination. As with all three-body reactions, the rate increases with pressure faster than a competing two-body reaction. At some pressure, a three-body reaction rate will exceed the rate of a competing two-body reaction. This competition is the explanation for the second explosion limit.

At even higher pressures, again an explosion limit is observed. This *third explosion limit* is the *thermal explosion limit,* which is governed by the competition of *heat* production by chemical reactions $\Sigma h_j r_j$ and heat losses to the vessel wall, and has been discussed above in Sections 10.1 and 10.2. The heat production per volume increases with increasing pressure, such that at high pressures a transition to explosion is observed.

From the explanations above, it follows that the explosion limits are governed by strongly nonlinear processes; the investigation of ignition processes has contributed significantly to the overall understanding of combustion processes.

Not only in hydrogen-oxygen mixtures are ignition limits observed, but also in all hydrocarbon-air mixtures. Due to additional chemical processes (e. g., formation of peroxides) these explosion limits are much more complex (Fig. 10.4), especially in the region of the third explosion limit (see, e. g., Warnatz 1981c).

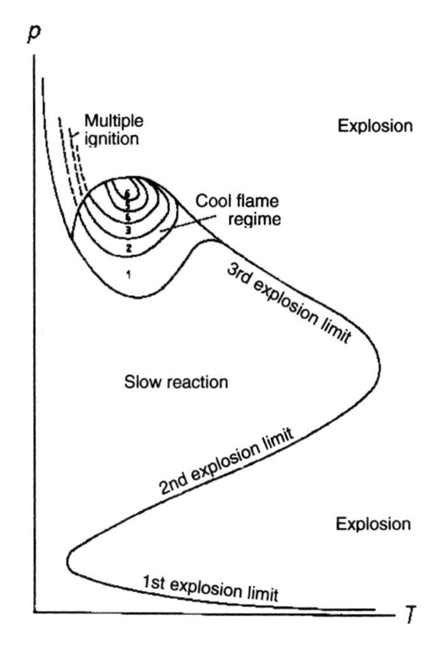

Fig. 10.4. Ignition limits (p-T explosion diagram) for hydrocarbons (schematic); see Warnatz(1981c)

Regions can be found where ignition takes place after the emission of short light pulses (*multistage ignition*) or where combustion takes place at low temperatures (*cool flames*). Here the ignition is inhibited by chemical reactions, in methane-oxygen mixtures, e. g., by the reactions

$$\bullet CH_3 + O_2 \rightleftarrows CH_3O_2\bullet \quad \text{(a)}$$

$$CH_3O_2\bullet + CH_4 \rightarrow CH_3OOH + \bullet CH_3 \quad \text{(b)}$$

$$CH_3OOH \rightarrow CH_3O\bullet + \bullet OH\ . \quad \text{(c)}$$

In principle the three reactions constitute a chain-branching mechanism which leads to ignition. However, a temperature increase shifts the equilibrium of reaction (a). At higher temperatures CH_3O_2• decomposes, and the chain-branching step (c) is no longer fed by the initial reaction (a) (This failure to branch at increasing temperature is called *degenerate branching*). Similar processes are observed for the other hydrocarbons, as it will be discussed in Chapter 16 in the context of engine knock. A comprehensive discussion of the detailed reaction mechanisms leading to explosion limits can be found in Bamford and Tipper (1977).

10.4 Autoignition: Ignition-Delay Time

Whereas in purely thermal ignition processes (see Sections 10.1 and 10.2) the temperature increases at once, it is observed in explosions of hydrogen or hydrocarbon-air mixtures that a temperature increase, and thus an explosion, takes place only after a certain *induction time* (*ignition-delay time*); see Fig. 10.5. Ignition delay is characteristic for *radical-chain explosions* (chemical reactions which are governed by a chain-branching mechanism, discussed in Section 7.5).

During the ignition-delay period, the radical-pool population is increasing at an exponential rate. Yet, the amount of fuel consumed, and hence the amount of energy liberated, is too small to be detected. Thus, important chemical reactions (chain branching, formation of radicals) take place during the induction time, whereas the temperature remains nearly constant.

Finally, the radical pool becomes large enough to consume a significant fraction of the fuel, and rapid ignition will take place. The precise definition of induction time depends on the criterion used (consumption of fuel, formation of CO, formation of OH, increase of pressure in a constant-volume vessel, increase of temperature in an adiabatic vessel, etc.)

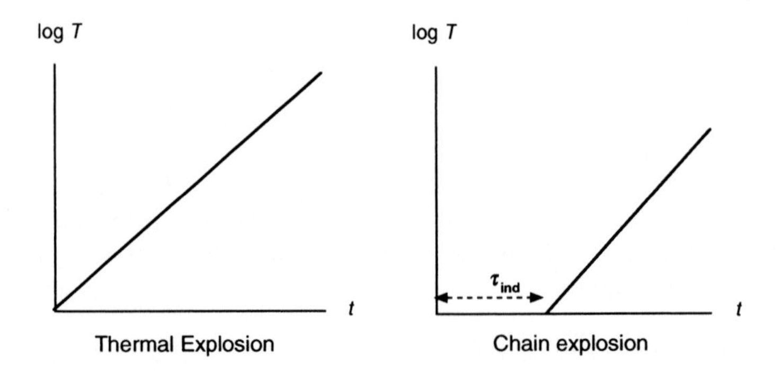

Fig. 10.5. Simplified time behavior of thermal and chain-branching explosion in an adiabatic system

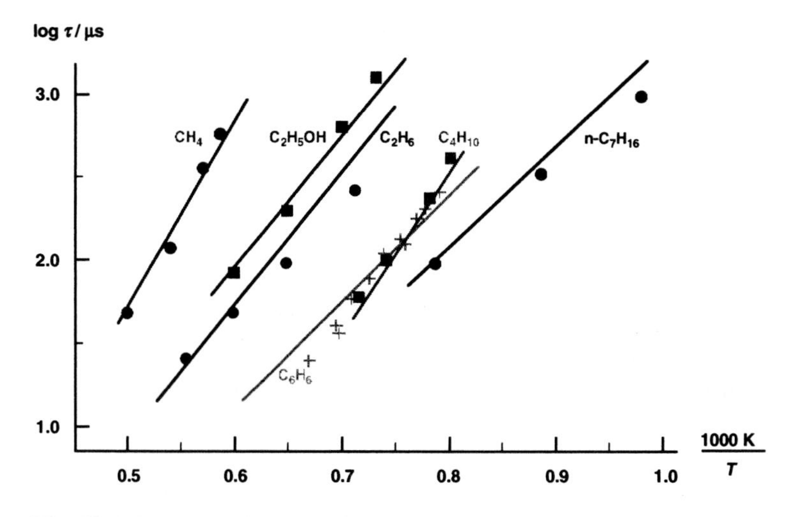

Fig. 10.6. Calculated (line) and measured (points) ignition delay times in hydrocarbon-air mixtures (see Warnatz 1993)

Due to the temperature dependence of the underlying elementary reactions, the ignition delay time depends strongly on the temperature. This is shown in Fig. 10.6 for several hydrocarbon-air mixtures (points denote experimental results, the curve denotes computational results). It can be seen that the ignition-delay time depends exponentially on the reciprocal temperature,

$$\tau = A \cdot \exp(B/T) \ ,$$

which reflects the temperature dependence (Arrhenius law) of the underlying elementary reactions occurring during the induction period.

10.5 Induced Ignition, Minimum Ignition Energies

A process where a mixture, which would not ignite by itself, is ignited locally by an ignition source is called *induced ignition*. During induced ignition, a small volume of the mixture is typically brought to a high temperature. Within this induced ignition volume, autoignition takes place with a subsequent flame propagation into the unburnt mixture. Especially for safety considerations, the understanding of the minimum ignition energy is important, i. e., the minimum energy, which is needed locally to cause ignition.

Because of the application to the ubiquitous spark-ignited engine, there is a vast literature on spark energies required to ignite fuel-air mixtures at a wide variety of pressures and temperatures (e. g., Heywood 1988). Research in spark-induced ignition is plagued by close proximity of the spark electrode which raise questions concerning surface chemistry and poorly controlled cooling rates at the nascent flame kernel. Furthermore, it is uncertain how much of the spark energy is deposited into raising the local temperature and how much energy goes into generating active radical species directly via electron impact.

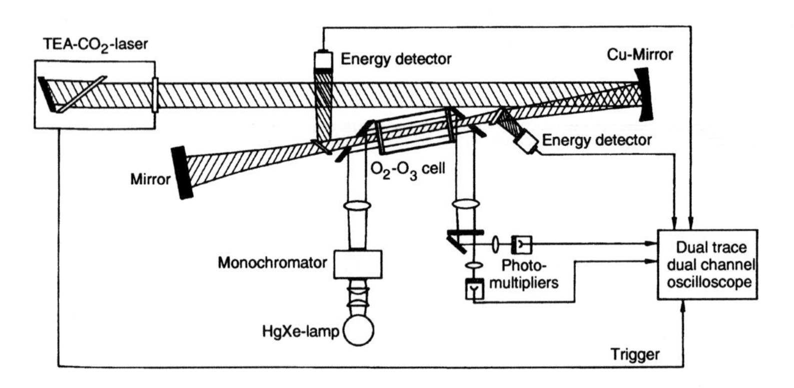

Fig. 10.7. Schematic of an experimental setup for the determination of minimum ignition energies when energy is deposited from infrared laser pulse (Raffel et al. 1985)

The laser-based research device illustrated in Fig. 10.7 is designed to study ignition without sparks or electrodes. The setup for the measurement of minimum ignition energies consists of a cylinder, where ignition is performed by a microsecond pulse from a coaxial infrared laser. The setup is (nearly) one-dimensional with a radial flame propagation. The energies of the light before and behind the cell can be measured; the difference is the ignition energy. The low energy of the infrared photons insure that the energy is delivered to thermal modes and not to direct generation of radicals. Furthermore, the flame propagation can be observed optically (Raffel et al. 1985).

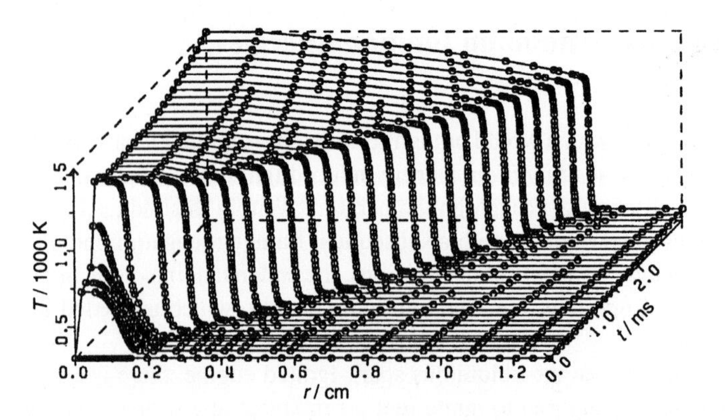

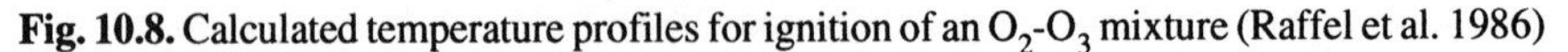
Fig. 10.8. Calculated temperature profiles for ignition of an O_2-O_3 mixture (Raffel et al. 1986)

Fig. 10.8 shows corresponding simulations for laser heating of ozone in an ozone-oxygen mixture. Upon laser heating, the ozone reacts to produce oxygen and energy. Plotted is the temperature versus the radius in the cylindrical vessel (radius = 13 mm) and the time. The laser beam, which has a diameter of about 3 mm, heats up the mixture to about 700 K in the cylinder axis (around $r = 0$). After an induction time of about 300 μs, ignition leading to a temperature of 1400 K is observed. Subsequently there is a small temperature rise in the *reactants* due to compression caused by the expanding products (details in Maas and Warnatz 1988).

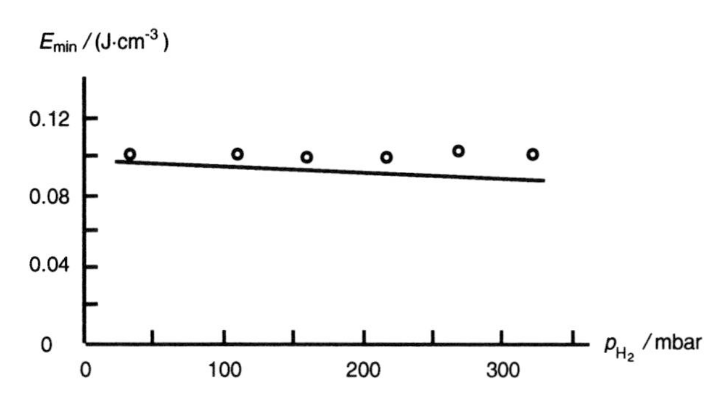

Fig. 10.9. Comparison of measured (points, Arnold et al. 1990b) and calculated (line, Maas 1990) minimum ignition energy densities in H_2-O_2-O_3 mixtures for different H_2-partial pressures; $p(O_2) = 261$ mbar, $p(O_3) = 68$ mbar

The points in Fig. 10.8 represent the spatial grid, which has been used for the numerical solution of the partial differential equation system (treated in Chapter 8). It can be seen that the mesh is adapted to the physical problem, meaning that many grid points are used where the gradients are steep (such as at the flame front) and few grid points are needed behind the flame where little action is taking place. A comparison of measurements and calculations shows (Fig. 10.9) that the minimum ignition energies differ by less than approximately 20% (Arnold et al. 1990b; Maas 1990). The level of agreement here is much better than what is typically achieved when attempting to model spark-induced ignition.

For induced ignition the concept of a *minimum ignition temperature* (corresponding to a *minimum ignition energy density*) is adequate. In order to cause a system to ignite, a small volume of the mixture has to be heated to a sufficiently high temperature. The energy needed is proportional to the pressure (change of the specific heat per volume element, see Fig. 10.10) and to the volume of the ignition source (change of the amount of mixture, which has to be heated, see Fig. 10.11), but is nearly independent of the duration of the ignition for sufficiently short ignition times. Figure 10.12 finally shows the dependence of the minimum ignition energy density on the mixture composition for a hydrogen-oxygen system. At very low, as well as at very high, hydrogen contents an ignition is not possible. Within the ignition limits the minimum ignition energy densities are nearly independent of the composition. For small ignition radii the minimum ignition energy densities increase with increasing hydrogen content, again caused by heat conduction and diffusion (fast diffusion of the light hydrogen atoms and molecules out of the ignition volume).

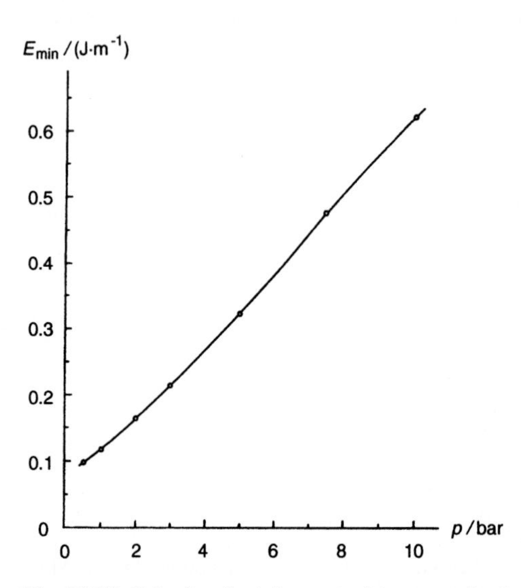

Fig. 10.10. Calculated minimum ignition energies in stoichiometric hydrogen-oxygen mixtures as function of the pressure; source time = 0.1 ms, source radius = 0.2 mm, initial temperature = 298 K, cylindrical geometry (Maas and Warnatz 1988)

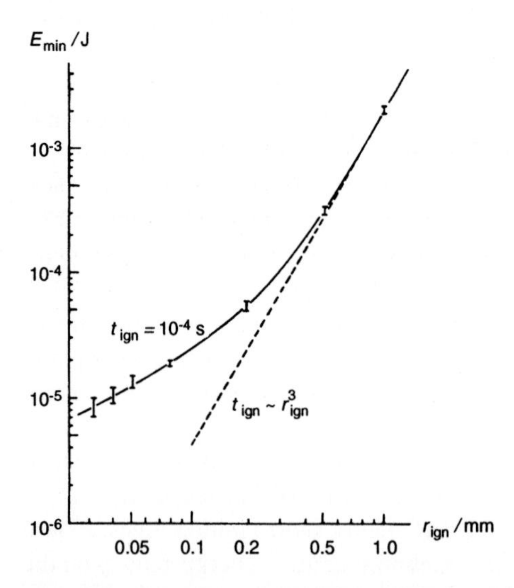

Fig. 10.11. Calculated minimum ignition energies (solid line) in stoichiometric hydrogen-oxygen mixture as function of the source radius; spherical geometry, ignition time = 0.1 ms, pressure = 1 bar, initial temperature = 298 K (Maas and Warnatz 1988)

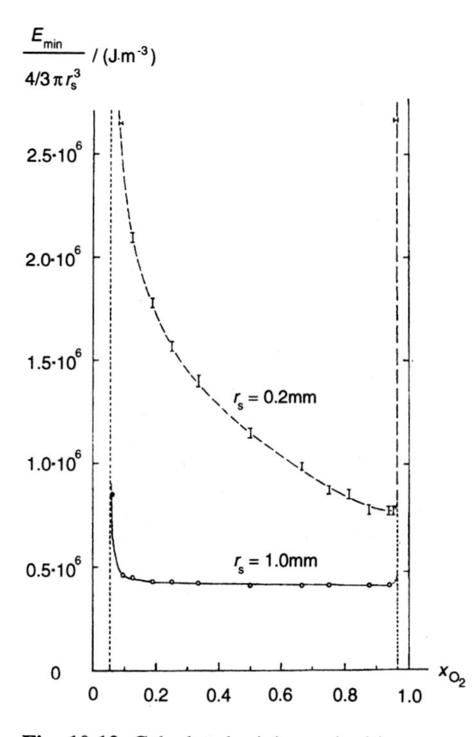

Fig. 10.12. Calculated minimum ignition energy densities in hydrogen-oxygen mixtures as function of the mixture composition (spatially homogeneous pressure, initial pressure = 1 bar, initial temperature = 298 K, source time = 0.1 ms) for two different source radii (0.2 and 1.0 mm)

Deviations from this behavior are observed for small ignition volumes, long ignition times, and low pressures, because in those cases diffusion and heat conduction become important. Consequently, the minimum ignition energies increase, due to diffusion of reactive species or heat out of the ignition volume (Maas and Warnatz 1988). The figures are based on numerical simulations of ignition processes in H_2-O_2 mixtures in spherical and cylindrical vessels. The hydrogen-oxygen chemistry described in Chapter 6 has been used in all cases.

10.6 Spark Ignition

The problems treated above (ignition limits, autoignition, and minimum ignition energy) plays an important role in spark ignition, which is an important topic, e. g., in Otto engine combustion and safety considerations (Xu et al. 1994). The sparks used are roughly cylindrically symmetric so that a 2D description should be possible.

Care is taken to provide uniform and slow discharge of the electrical power to generate well-defined conditions. Examples of studies with the help of 2D-LIF of OH radicals (see Chapter 2) are given in Fig. 10.13 (Xu et al. 1994).

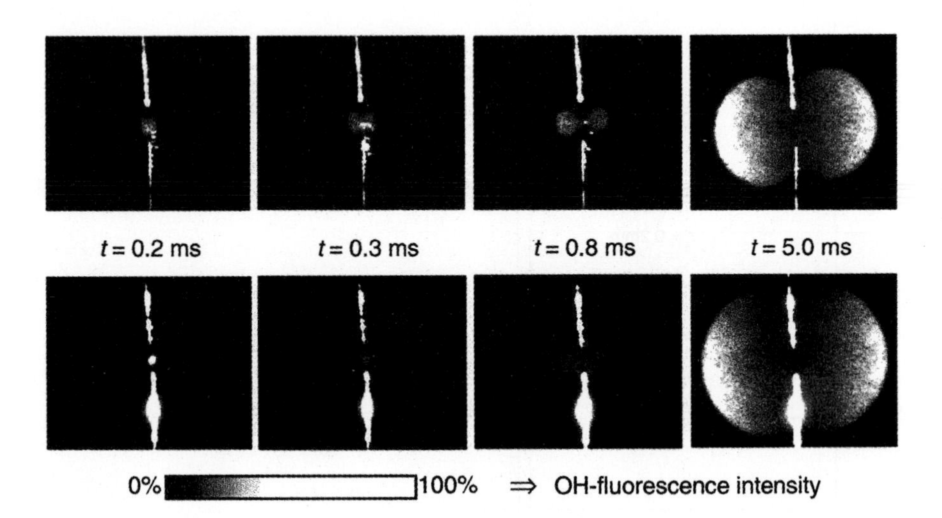

Fig. 10.13. Temporal development of the flame kernel during spark-induced ignition in a 11% CH_3OH-air mixture; E_{ign} = 1.6 mJ, t_{ign} = 35 μs, p = 600 mbar, electrode distance d = 3 mm (upper series) and d = 2 mm (lower series); the electrodes are visible due to UV fluorescence

These experiments (together with a 2D numerical evaluation; see Maas and Warnatz 1989) should allow the development of criteria for ignition/non-ignition of combustion in Otto engines for use in engine simulations. For a given mixture equivalence ratio, extinction can be caused by insufficient ignition energy or by strain of the flame fronts in the inhomogeneous flow field (discussed in Section 14.4). Furthermore, the influence of the electrode distance on the spark (due to enhanced heat removal at smaller distances) can be seen.

10.7 Detonations

Detonations shall only be outlined briefly. Detailed discussions can be found, e. g., in the book of Williams (1984). Usual flame propagation, *deflagration,* is caused by chemical reactions that sustain a gradient and molecular transport processes that propagate the gradient. In contrast, the propagation of *detonations* is caused by a pressure wave, which is sustained by the chemical reaction and the corresponding heat release. A characteristic property of detonations is propagation velocities (usually more than 1000 m/s) that are much higher than the flame speed (typically 0.5 m/s). One of the main reasons for the high propagation velocity v_D of detonations is the high value of the sound velocity in the burnt gases.

The propagation velocity v_D, the density ρ_b, and the pressure p_b in the burnt gas can be calculated according to the theory of Chapman-Jouguet (see Hirschfelder et al. 1964). They depend on pressure p_u and density ρ_u in the unburnt gas, on the specific heat of reaction q, and on γ, which is the ratio of the heat capacities at constant pressure and constant volume, $\gamma = c_p/c_V$. The approximations

$$v_D = \sqrt{2(\gamma^2 - 1)\, q} \tag{10.9}$$

$$\frac{\rho_b}{\rho_u} = \frac{\gamma + 1}{\gamma} \tag{10.10}$$

$$\frac{p_b}{p_u} = 2(\gamma - 1)\frac{q \cdot \rho_u}{p_u} \tag{10.11}$$

are the result of this treatment. A comparison of calculations and experimental results is given in Table 10.1. (p_u = 1 bar, T_u = 291 K).

Tab. 10.1. Propagation velocities, temperatures, and pressures in hydrogen-oxygen detonations (Gaydon and Wolfhard 1979)

Mixture	p_b / bar	T_b / K	v_D(calc.) / (m·s^{-1})	v_D(exp.) / (m·s^{-1})
$2\,H_2 + O_2$	18.05	3583	2806	2819
$2\,H_2 + O_2 + 5\,N_2$	14.39	2685	1850	1822

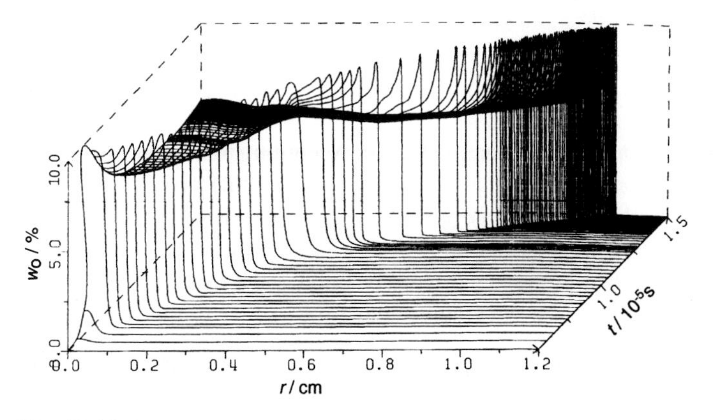

Fig. 10.14. O-atom mass fractions during the formation of a detonation in a H_2-O_2 mixture at an initial pressure of 2 bar (Goyal et al. 1990a,b). The flame propagation is induced by a small region of elevated temperature near r = 0, leading to enhanced autoignition in this area and then to flame propagation

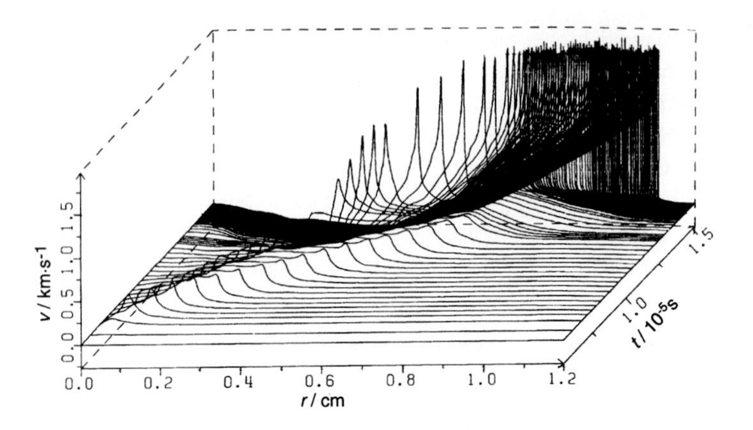

Fig. 10.15. Velocity profiles during the formation of a detonation in a H_2-O_2 mixture at an initial pressure of 2 bar (Goyal et al. 1990a,b)

The question whether a transition from a deflagration to a detonation is possible, is very important in many applications. Mathematical modeling allows the simulation of such processes for simple geometries. Figures 10.14 and 10.15 show a transition to detonation in a hydrogen-oxygen mixture. The deflagration accelerates and evolves into a detonation.

The processes that cause detonations are beyond the scope of this book. Suffice to note that detonation waves are not planar; the formation of cellular structures of the detonation fronts is observed experimentally. For an entry to the literature see Oppenheim et al. (1963), Edwards (1969), Chue et al. (1993), and He et al. (1995).

10.8 Exercises

Exercise 10.1: Consider a one-step reaction F $\rightarrow$ P. According to Semenov's theory of thermal ignition, stationary states in a reactive gas mixture in a vessel with a volume V_B exist only if the heat release rate $\dot{q}_P = M_F \cdot c_{F,0} \cdot A \cdot \exp(-E/RT) \cdot (h_F - h_P) V_B$ in the reaction equals the heat transfer rate $\dot{q}_L = \chi \cdot S \cdot (T - T_W)$ to the surrounding. Here $c_{F,0}$ denotes the concentration of the fuel prior to reaction, χ the heat transfer coefficient, T_W the wall temperature, V_B the volume, and S the surface of the vessel.

(a) Which additional condition holds for the ignition limit (i. e., for the point, where one stable state exists). Which variables are unknown?

(b) In order to determine the ignition temperature T_{ign} of a gas mixture, the latter is filled into a vessel with a variable wall temperature T_W. The wall temperature is increased and at T_W = 900 K ignition is observed. What is the value of the ignition temperature, if the activation energy is E = 167.5 kJ/mol?

11 The Navier-Stokes-Equations for Three-Dimensional Reacting Flows

In the previous chapters, the conservation equations for one-dimensional flames were developed, solution methods discussed, and results presented. In this chapter, general three-dimensional conservation equations are derived for mass, energy, and momentum; these are the Navier-Stokes equations for reactive flow.

11.1 The Conservation Equations

An arbitrarily (but reasonably) shaped element Ω of the three-dimensional space with a surface $\partial\Omega$ shall be considered (see Fig. 11.1).

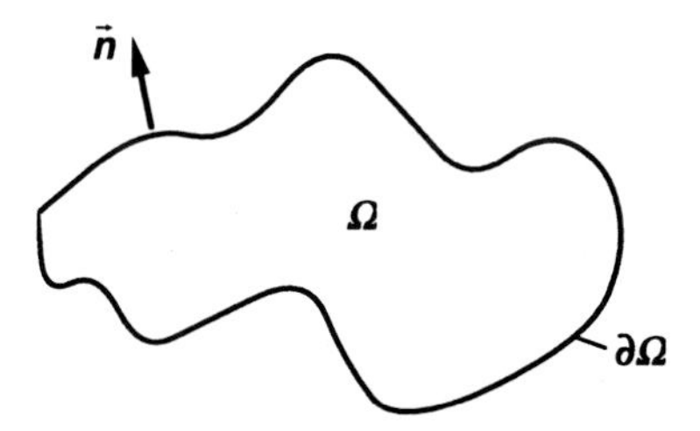

Fig. 11.1. Schematic illustration of the processes in a volume element Ω with surface $\partial\Omega$

An extensive variable $F(t)$ can be calculated from the corresponding density $f(\vec{r},t) = \mathrm{d}F/\mathrm{d}V$ by integration over the whole volume element Ω. One obtains (t = time, $\vec{r}$ = spatial location)

$$F(t) \quad = \quad \int_{\Omega} f(\vec{r},t)\mathrm{d}V \,, \qquad (11.1)$$

where $\mathrm{d}V$ denotes a differential volume element. A change of the extensive variable $F(t)$ can be caused by different processes ($\vec{n}$ = vector normal to the surface, $\mathrm{d}S$ =

differential surface element, see Fig. 11.1); a complete listing of these processes is:

1. Change caused by a *flux* $\vec{\Phi}_f \, \vec{n} \, \mathrm{d}S$ through the surface $\partial\Omega$(e. g., by diffusion, heat conduction, momentum transport, convection, etc.). The flux density $\vec{\Phi}_f$ describes the amount of F, which flows per time and unit surface.
2. Change caused by a *production* q_f (e. g., chemical reaction) within the volume element, where q_f is the amount of F formed per time and unit volume.
3. Change caused by *long-range processes* s_f (examples are radiation, gravity, coulombic attraction) from the surrounding into element Ω, where s_f is the amount of F generated per unit volume.

The overall balance of the variable F can be obtained by integration of the flux over the surface $\partial\Omega$ and integration of the production terms over the volume of the considered element Ω,

$$\int_\Omega \frac{\partial f}{\partial t} \mathrm{d}V + \int_{\partial\Omega} \vec{\Phi}_f \, \vec{n} \, \mathrm{d}S = \int_\Omega q_f \, \mathrm{d}V + \int_\Omega s_f \, \mathrm{d}V \, , \tag{11.3}$$

where use is made of the simple relation

$$\frac{\partial F}{\partial t} = \int_\Omega \frac{\partial f}{\partial t} \, \mathrm{d}V \, .$$

Using the Integral Law of Gauss (see textbooks on mathematics), the surface integral for the change of F by the flux $\vec{\Phi}_f \, \vec{n} \, \mathrm{d}S$ can be replaced by a volume integral,

$$\int_{\partial\Omega} \vec{\Phi}_f \, \vec{n} \, \mathrm{d}S = \int_\Omega \mathrm{div} \, \vec{\Phi}_f \, \mathrm{d}V \, , \tag{11.4}$$

and one obtains

$$\int_\Omega \frac{\partial f}{\partial t} \mathrm{d}V + \int_\Omega \mathrm{div} \, \vec{\Phi}_f \, \mathrm{d}V = \int_\Omega q_f \, \mathrm{d}V + \int_\Omega s_f \, \mathrm{d}V \, . \tag{11.5}$$

Now an infinitesimally small volume element is considered, and the limiting process $\Omega \rightarrow 0$ is performed with the result

$$\frac{\partial f}{\partial t} + \mathrm{div} \, \vec{\Phi}_f = q_f + s_f \, . \tag{11.6}$$

From this general equation, conservation equations for mass, energy, momentum, and species mass can be derived (Hirschfelder and Curtiss 1949, Bird et al. 1960).

11.1.1 Overall Mass Conservation

If the extensive variable F is the total mass of the system ($F = m$), then the density f is given by the *mass density* ρ. The *mass flux density* $\vec{\Phi}_f$ is given as the product of the local flow velocity $\vec{v}$ and the mass density ρ. Because mass can neither be formed nor destroyed in chemical reactions, there are no production or long-range terms, and it follows that

$$\begin{aligned} f_m &= \rho \\ \vec{\Phi}_m &= \rho\vec{v} \\ q_m &= 0 \\ s_m &= 0 \ . \end{aligned}$$

Insertion into (11.6) then leads to

$$\frac{\partial \rho}{\partial t} + \operatorname{div}(\rho\vec{v}) = 0 \ . \tag{11.7}$$

This equation is usually called the *mass conservation equation* or the *continuity equation* (Hirschfelder and Curtiss 1949, Bird et al. 1960).

11.1.2 Species Mass Conservation

If the masses of different species m_i are considered, the density f is given by the *partial mass density* ρ_i of species i. The local flow velocity $\vec{v}_i$ of species i is composed of the *mean flow velocity* $\vec{v}$ of the center of mass and the *diffusion velocity* $\vec{V}_i$ of species i (relative to the center of mass). In analogy to the total mass, one has no long-range process. However, because species are formed and consumed in chemical reactions, one obtains a *production term* $q_{m,i}$, which is given by the product of the molar mass M_i and the molar rate of formation ω_i (in mol/m^3·s). The result is

$$\begin{aligned} f_{m,i} &= \rho_i &&= w_i\,\rho \\ \vec{\Phi}_{m,i} &= \rho_i\,\vec{v}_i &&= \rho_i\,(\vec{v}+\vec{V}_i) \\ q_{m,i} &= M_i\,\omega_i \\ s_{m,i} &= 0 \ . \end{aligned}$$

If $\rho_i\vec{V}_i = \vec{j}_i$ denotes the *diffusion flux*, insertion into (11.6) yields the conservation equations (Hirschfelder and Curtiss 1949, Bird et al. 1960)

$$\frac{\partial \rho_i}{\partial t} + \operatorname{div}(\rho_i\,\vec{v}) + \operatorname{div}\vec{j}_i = M_i\,\omega_i \ . \tag{11.8}$$

11.1.3 Momentum Conservation

Considering the conservation of momentum $m\vec{v}$, which is a vector, the *momentum density* $f_{m\vec{v}}$ is given by the momentum density $\rho\vec{v}$. The *momentum flux density* $\vec{\Phi}_{m\vec{v}}$ consists of a convective part $\rho\vec{v}\otimes\vec{v}$ and a part $\overline{\overline{p}}$, which describes the momentum change due to viscous dissipation and pressure (treated in Section 11.2). There is no production term, but a long-range term, namely the *gravity*. The result is

$$
\begin{aligned}
f_{m\vec{v}} &= \rho\vec{v} \\
\vec{\Phi}_{m\vec{v}} &= \rho\vec{v}\otimes\vec{v} + \overline{\overline{p}} \\
q_{m\vec{v}} &= 0 \\
s_{m\vec{v}} &= \rho\vec{g} \ .
\end{aligned}
$$

Here $\overline{\overline{p}}$ denotes the *pressure tensor* (described in Section 11.2), $\otimes$ denotes the *dyadic product* of two vectors (a short summary of definitions and laws from vector- and tensor analysis is given in Section 11.3), $\vec{g}$ is the gravitational acceleration.

Insertion into (11.6) yields the *momentum conservation equation* (see, e. g., Hirschfelder and Curtiss 1949, Bird et al. 1960), which together with the mass conservation equation (11.7) forms the well-known and often used Navier-Stokes equations for nonreactive flow,

$$
\frac{\partial(\rho\vec{v})}{\partial t} + \operatorname{div}(\rho\vec{v}\otimes\vec{v}) + \operatorname{div}\overline{\overline{p}} = \rho\vec{g} \ . \tag{11.9}
$$

11.1.4 Energy Conservation

The conservation equation for the internal energy or the enthalpy results from a separate consideration of the potential, the kinetic, and the total energy. For the total energy one obtains the relations

$$
\begin{aligned}
f_e &= \rho e \\
\vec{\Phi}_e &= \rho e\vec{v} + \overline{\overline{p}}\vec{v} + \vec{j}_q \\
q_e &= 0 \\
s_e &= q_{\mathrm{r}} \ ,
\end{aligned}
$$

where e denotes the *total specific energy*. The *energy flux density* $\vec{\Phi}_e$ is composed of a convective term $\rho e\vec{v}$, a term $\overline{\overline{p}}\vec{v}$, which describes energy changes due to pressure and viscous dissipation, and a term which accounts for heat conduction ($\vec{j}_q$ = heat flux density). There are no production terms, but *radiation* is a long-range effect (q_{r} = heat generation due to radiation, in J/(m^3·s), e. g.).

Noting that the total energy density e is given as the sum of internal, kinetic, and potential energy,

$$
\rho e = \rho u + \frac{1}{2}\rho|\vec{v}|^2 + \rho G \ , \tag{11.10}
$$

with G = gravitational potential, $\vec{g} = \operatorname{grad} G$, u = specific internal energy, one obtains the *energy conservation equation* (Hirschfelder and Curtiss 1949, Bird et al. 1960) in the form

$$\frac{\partial(\rho u)}{\partial t} + \operatorname{div}(\rho u \vec{v} + \vec{j}_q) + \overline{\overline{p}} : \operatorname{grad} \vec{v} = q_r , \qquad (11.11)$$

where : denotes the contraction of two tensors, which leads to a scalar (Section 11.3). Together with (p = pressure) $\rho h = \rho u + p$, (11.11) can be rewritten as a conservation equation for the specific enthalpy (Hirschfelder and Curtiss 1949, Bird et al. 1960),

$$\frac{\partial(\rho h)}{\partial t} - \frac{\partial p}{\partial t} + \operatorname{div}(\rho \vec{v} h + \vec{j}_q) + \overline{\overline{p}} : \operatorname{grad} \vec{v} - \operatorname{div}(p \vec{v}) = q_r . \qquad (11.12)$$

11.2 The Empirical Laws

When the number of equations is equal to the number of variables, the system of equations is *closed.* The conservation equations described in Section 11.1 are closed by specifying laws, which describe the flux densities $\vec{j}_q$ and $\vec{j}_i$, as well as the pressure tensor $\overline{\overline{p}}$ as functions of known physical properties of the system. *Empirical laws* of Newton, Fourier, and Fick are used. The transport coefficients in these laws are modified by today's improved understanding derived from the kinetic theory of dilute gases and irreversible thermodynamics as discussed in Chapter 5 (see Hirschfelder et al. 1964).

11.2.1 Newton's Law

Many experimental investigations yield for the pressure tensor (see Section 11.3)

$$\overline{\overline{p}} = p\overline{\overline{E}} + \overline{\overline{\Pi}} . \qquad (11.13)$$

Here $\overline{\overline{E}}$ denotes the unit tensor (explained in Section 11.3) and p the hydrostatic pressure. The first term in (11.13) describes the *hydrostatic* part of $\overline{\overline{p}}$, the second term the *viscous* part.

The rigorous kinetic theory of dilute gases yield the relation (derived for example by Hirschfelder et al. 1964)

$$\overline{\overline{\Pi}} = -\mu\left[(\operatorname{grad} \vec{v}) + (\operatorname{grad} \vec{v})^{\mathrm{T}}\right] + \left(\frac{2}{3}\mu - \kappa\right)(\operatorname{div} \vec{v})\overline{\overline{E}} , \qquad (11.14)$$

where μ denotes the *mean dynamic viscosity* of the mixture. The *volume viscosity* κ describes viscous dissipation due to normal shear stress, which occurs during the expansion of a fluid (caused by relaxation between internal degrees of freedom and translation).

For monoatomic gases there are no internal degrees of freedom, and one obtains $\kappa = 0$. If volume viscosity is negligible (this is usually assumed) the result is

$$\overline{\overline{\Pi}} = -\mu\left[(\text{grad}\,\vec{\upsilon})+(\text{grad}\,\vec{\upsilon})^{\text{T}}-\frac{2}{3}(\text{div}\,\vec{\upsilon})\overline{\overline{E}}\right]. \tag{11.15}$$

11.2.2 Fourier's Law

The heat flux (Section 11.1.4) is given by three different parts (Hirschfelder et al. 1964),

$$\vec{j}_q = \vec{j}_q^{\,\text{c}} + \vec{j}_q^{\,\text{D}} + \vec{j}_q^{\,\text{d}}, \tag{11.16}$$

where $\vec{j}_q^{\,\text{c}}$ denotes flux caused by *heat conduction,* $\vec{j}_q^{\,\text{D}}$ denotes flux caused by the *Dufour effect*, and $\vec{j}_q^{\,\text{d}}$ denotes flux caused by diffusion (treated in Chapter 5),

$$\vec{j}_q^{\,\text{c}} = -\lambda\,\text{grad}\,T \tag{11.17}$$

$$\vec{j}_q^{\,\text{D}} = \overline{M}RT\sum_i\sum_{j\neq i}\frac{D_i^{\text{T}}}{\rho D_{ij}M_iM_j}\left(\frac{w_j}{w_i}\vec{j}_i-\vec{j}_j\right) \tag{11.18}$$

$$\vec{j}_q^{\,\text{d}} = \sum_i h_i\,\vec{j}_i \tag{11.19}$$

with λ = mixture thermal conductivity (see Section 5.2), T = temperature, M_i = molar mass, R = gas constant, D_i^{T} = coefficient of thermal diffusion, D_{ij} = binary diffusion coefficients, w_i = mass fractions, and h_i = specific enthalpy of species i. Usually the Dufour effect is negligible in combustion processes, such that one can write

$$\vec{j}_q = -\lambda\,\text{grad}\,T + \sum_i h_i\,\vec{j}_i\,. \tag{11.20}$$

11.2.3 Fick's Law and Thermal Diffusion

Diffusion is caused by three effects, which are a part $\vec{j}_i^{\,\text{d}}$, describing ordinary diffusion, a part $\vec{j}_i^{\,\text{T}}$ caused by thermal diffusion, and a contribution $\vec{j}_i^{\,\text{p}}$ caused by *pressure diffusion*,

$$\vec{j}_i = \vec{j}_i^{\,\text{d}} + \vec{j}_i^{\,\text{T}} + \vec{j}_i^{\,\text{p}} \tag{11.21}$$

$$\vec{j}_i^{\,\text{d}} = \rho_i\vec{V}_i = \frac{\rho M_i}{\overline{M}^2}\sum_{j\neq i}D_{ij}^{\text{mult}}M_j\,\text{grad}\,x_j \tag{11.22}$$

$$\vec{j}_i^{\,\text{T}} = -D_i^{\text{T}}\,\text{grad}(\ln T) \tag{11.23}$$

$$\vec{j}_i^{\,\text{p}} = \frac{\rho M_i}{\overline{M}^2}\sum_{j\neq i}D_{ij}^{\text{mult}}M_j\left(x_j-w_j\right)\text{grad}(\ln p) \tag{11.24}$$

with $\vec{V}_i$ = diffusion velocity of species i, x_i = mole fraction, p = pressure. The D_{ij}^{mult} are *multicomponent diffusion coefficients*, which depend on the concentrations, and can be calculated from the binary diffusion coefficients (Waldmann 1947, Curtiss and Hirschfelder 1949, Ern and Giovangigli 1996).

Usually pressure diffusion can be neglected in combustion processes. As shown in Chapter 5, the diffusion flux density can be approximated in most applications by

$$\vec{j}_i = -D_i^{\text{M}} \rho \frac{w_i}{x_i} \operatorname{grad} x_i - D_i^{\text{T}} \operatorname{grad}(\ln T) . \tag{11.25}$$

Here D_i^{M} (see Section 5.4) denotes a mean diffusion coefficient for the diffusion of species i into the mixture of the remaining species, which can be calculated from the binary diffusion coefficients D_{ij} by (Stefan 1874)

$$D_i^{\text{M}} = \frac{1 - w_i}{\sum_{j \neq i} x_j / D_{ij}} . \tag{11.26}$$

11.2.4 Calculation of the Transport Coefficients from Molecular Parameters

The transport coefficients λ, μ, D_i^T and D_{ij}, which are needed for the calculation of the fluxes, can be computed from molecular parameters (Chapter 5). Thus, the conservation equations for mass, momentum, energy, and species masses are now complete.

11.3 Appendix: Some Definitions and Laws from Vector- and Tensor-Analysis

Some definitions and laws from vector- and tensor-analysis, used in the previous sections, shall be given in the following. Details can, e. g., be found in the books of Bird et al. (1960) or Aris (1962). Here S denotes a scalar, $\vec{v}$ a vector, and $\overline{\overline{T}}$ a tensor.

The *dyadic product* $\vec{v} \otimes \vec{v}'$ of two vectors $\vec{v}$ and $\vec{v}'$ leads to a tensor $\overline{\overline{T}}$,

$$\vec{v} \otimes \vec{v}' = \begin{pmatrix} v_x v_x' & v_x v_y' & v_x v_z' \\ v_y v_x' & v_y v_y' & v_y v_z' \\ v_z v_x' & v_z v_y' & v_z v_z' \end{pmatrix} \quad \text{with} \quad \overline{\overline{T}} = \begin{pmatrix} T_{xx} & T_{xy} & T_{xz} \\ T_{yx} & T_{yy} & T_{yz} \\ T_{zx} & T_{zy} & T_{zz} \end{pmatrix} .$$

The *transposed tensor* $\overline{\overline{T}}^{\text{T}}$ is obtained by interchanging rows and columns of $\overline{\overline{T}}$,

$$\overline{\overline{T}}^{\text{T}} = \begin{pmatrix} T_{xx} & T_{yx} & T_{zx} \\ T_{xy} & T_{yy} & T_{zy} \\ T_{xz} & T_{yz} & T_{zz} \end{pmatrix} .$$

Furthermore, the *unit tensor* $\overline{\overline{E}}$ is defined as

$$\overline{\overline{E}} = \begin{pmatrix} 1 & 0 & 0 \\ 0 & 1 & 0 \\ 0 & 0 & 1 \end{pmatrix}.$$

The *contraction* $\overline{\overline{T}}:\overline{\overline{T'}}$ of two tensors $\overline{\overline{T}}$ and $\overline{\overline{T'}}$ yields a scalar by means of

$$\overline{\overline{T}}:\overline{\overline{T'}} = \sum_i \sum_j T_{ij} T'_{ji} = S .$$

The *gradient* of a scalar yields a vector of the form

$$\operatorname{grad} S = \begin{pmatrix} \dfrac{\partial S}{\partial x} \\ \dfrac{\partial S}{\partial y} \\ \dfrac{\partial S}{\partial z} \end{pmatrix}.$$

The *gradient* of a vector yields a tensor of the form

$$\operatorname{rad} \vec{v} = \begin{pmatrix} \dfrac{\partial v_x}{\partial x} & \dfrac{\partial v_y}{\partial x} & \dfrac{\partial v_z}{\partial x} \\ \dfrac{\partial v_x}{\partial y} & \dfrac{\partial v_y}{\partial y} & \dfrac{\partial v_z}{\partial y} \\ \dfrac{\partial v_x}{\partial z} & \dfrac{\partial v_y}{\partial z} & \dfrac{\partial v_z}{\partial z} \end{pmatrix}.$$

The *divergence* of a vector yields a scalar

$$\operatorname{div} \vec{v} = \frac{\partial v_x}{\partial x} + \frac{\partial v_y}{\partial y} + \frac{\partial v_z}{\partial z} .$$

The *divergence* of a tensor yields a vector

$$\operatorname{div} \overline{\overline{T}} = \begin{pmatrix} \dfrac{\partial T_{xx}}{\partial x} + \dfrac{\partial T_{yx}}{\partial y} + \dfrac{\partial T_{zx}}{\partial z} \\ \dfrac{\partial T_{xy}}{\partial x} + \dfrac{\partial T_{yy}}{\partial y} + \dfrac{\partial T_{zy}}{\partial z} \\ \dfrac{\partial T_{xz}}{\partial x} + \dfrac{\partial T_{yz}}{\partial y} + \dfrac{\partial T_{zz}}{\partial z} \end{pmatrix}.$$

11.4 Exercises

Exercise 11.1. Write the pressure tensor (see the definitions in Section 11.3)

$$\overline{\overline{p}} = p\overline{\overline{E}} - \mu\left[(\operatorname{grad}\vec{\upsilon}) + (\operatorname{grad}\vec{\upsilon})^{\mathrm{T}} - \frac{2}{3}(\operatorname{div}\vec{\upsilon})\overline{\overline{E}}\right]$$

in matrix notation in Cartesian coordinates. What is the momentum conservation equation for a one-dimensional viscous flow?

Exercise 11.2. A thin pipe that is 150 cm long is attached to two 1-liter tanks. Both tanks contain a gaseous mixture of equal amounts of He and Xe, $x_{\mathrm{Xe}} = x_{\mathrm{He}} = 0.5$, at a pressure of 1 bar. The temperatures are initially the same at 300K. At time zero, one tank is rapidly heated to 400 K, with a constant temperature gradient in the pipe.

(a) What is the molar flux density $\vec{j}^{\,*}_{\mathrm{He}}$ of the helium into the pipe from the colder tank (assume that $V_{\mathrm{pipe}} << V_{\mathrm{chambers}}$).
(b) What is the mole fraction of helium in each tank at steady state (i. e., after a sufficiently long time)?
(c) How much helium has then diffused through the pipe?

Remark: The molar flux in a mixture consisting of two species is defined analogous to the mass flux as

$$\vec{j}_i^{\,*} = -D_{12}\, c \operatorname{grad} x_i - D_{12}^{\mathrm{T}}\, c \operatorname{grad}(\ln T) \ .$$

The coefficient of thermal diffusion based on the molar amount of the component is given by

$$D_{12}^{\mathrm{T}} = D_{12}\, \alpha \cdot x_1\, x_2 \ ,$$

where α is positive for the heavy component, and negative for the light component. D_{12} and α are given as

$$D_{\mathrm{He,Xe}} = 0.71 \frac{\mathrm{cm}^2}{\mathrm{s}} \quad , \quad \alpha_{\mathrm{He}} = -0.43 \, .$$

Exercise 11.3. Assume the x-component of the velocity field in an inviscid, incompressible, stationary flow is given by $\upsilon_x(x,y) = -x$ (let $\rho = 1$ for the density.).

(a) Which condition holds for the y-component $\upsilon_y(x,y)$, such that the continuity equation is fulfilled ($\upsilon_y(x,y) = 0$ at $x = 0$, $y = 0$)?
(b) Determine the streamlines in the velocity field under consideration. What kind of flow is it?
(c) What is the pressure distribution, if the pressure is p_0 at $x = 0$, $y = 0$? The pressure tensor in this case is given by the expression

$$\overline{\overline{p}} = \begin{pmatrix} p & 0 \\ 0 & p \end{pmatrix} \, .$$

Exercise 11.4. Derive the momentum equation for an inviscid two-dimensional flow using a small surface element. Assume that only pressure forces are present.

12 Turbulent Reacting Flows

In previous chapters, premixed and nonpremixed reacting flows have been studied on the assumption that the underlying fluid flow is laminar. In most combustion equipment, e. g., engines, boilers, and furnaces, the fluid flow is usually *turbulent*. In turbulent flows, mixing is greatly enhanced. As a consequence, the combustion chamber is, for example, much smaller than possible with laminar flows. In spite of the widespread use of turbulent combustion, many questions are still open here.

Much remains to be investigated about turbulent fluid flow by itself, and the addition of chemical kinetics with energy release only further complicates an already difficult problem. Because of this complexity, the mathematical models for turbulent combustion are much less developed than corresponding models for laminar flames. Rather than review the vast subject of turbulence, the goal of this chapter is to present the salient features of various approaches taken to add combustion chemistry to models of turbulent flow. An in-depth review of the subject of turbulent combustion can be found in the book by Libby and Williams (1994) and Peters (2000). Reviews on turbulence without combustion are given in the texts by Libby (1996) and Pope (2000).

12.1 Some Fundamental Phenomena

In laminar flows, velocity and scalars have well-defined values. In contrast, turbulent flows are characterized by continuous *fluctuations* of velocity, which can lead to fluctuations in scalars such as density, temperature, and mixture composition. These fluctuations in velocity (and then scalars) are a consequence of vortices generated by shear within the flow. Figure 12.1 shows the vortex generation and growth when two fluid streams of different velocities (assume fuel on bottom and air on top) are brought together following a splitter plate. Two key features of Fig. 12.1 are noteworthy:

- First, note that fluid from the top is *convected* (not diffused) transversely to the flow across the layer while fluid from the bottom is convected toward the top. This convection process, brought about by the vortex motion, greatly accelerates the mixing process.

- Second, note that the interfacial area between the two fluids is greatly increased and thus the overall rate of molecular mixing is greatly increased as well. The rate of molecular mixing is further accelerated by steepened gradients generated when the interface is stretched (see Fig. 3.4).

The growth of these vortices is the result of competition between the (nonlinear) generation process and the destruction process caused by viscous dissipation. The generation term exceeds the viscous damping term when a critical value of *Reynolds number Re* is exceeded; transition from a *laminar* to a *turbulent* flow is taking place. The Reynolds number $Re = \rho \upsilon l/\mu = \upsilon l/\nu$ is a ratio of a destabilizing momentum and a stabilizing (or damping) viscous effect. Here ρ denotes the density, υ the velocity (or better: a velocity difference), μ the viscosity of the fluid, and l a characteristic length of the system ($\nu = \mu/\rho$; see Eq. 5.20). The length l is geometry-dependent (for pipe flow, e. g., the diameter is used for l). With this definition it can be found by experiments that the critical Reynolds number for pipe flow is about 2000.

There are several examples of turbulent flows, which are important for the theoretical understanding as well as for practical applications. Here, only some simple examples shall be presented (see, e. g., Hinze 1972, Sherman 1990):

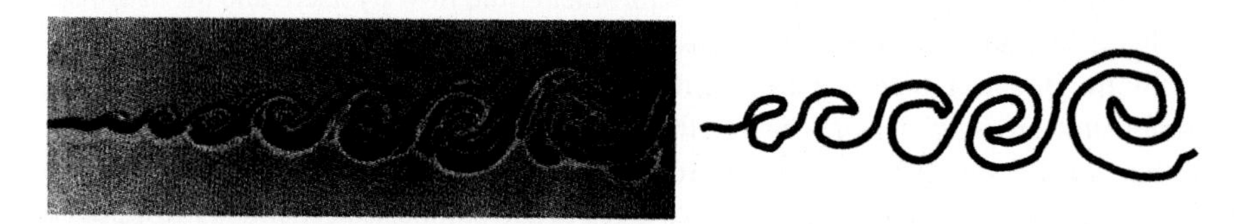

Fig. 12.1. Formation of a turbulent shear layer (Roshko 1975)

Shear Flow: Two parallel flowing fluid layers mix in a growing layer behind a splitter plate. Right behind the plate the flow is still laminar. *Vortices* (loosely: *eddies*) are generated in the steep velocity gradient, the *shear,* between the two flows. The vortices grow (see Fig. 12.1) and, in time, become three-dimensional, indicating the full transition to turbulence. The characteristic length l used for this flow is distance downstream from the splitter plate or distance across the flow. With either definition, the Reynolds number increases with downstream distance and the flow will develop into a fully turbulent flow (Oran and Boris, 1993). The shear flow is the simplest flow that has the basic ingredients of turbulent flow and, thus, has received the most research interest (see, e. g., Dimotakis and Miller 1990).

Pipe Flow: Here, turbulence is caused by shear as a consequence of the zero velocity at the wall and nonzero velocity on the pipe centerline. Above a Reynolds number of 2000, the viscous forces are unable to dampen the instabilities in the momentum with the result that the flow transits from laminar to turbulent. With the transition, a profound increase of axial and radial mixing is observed. The increase of momentum transfer is manifested by a larger pressure drop for turbulent flow than for laminar flow at the same volumetric rate.

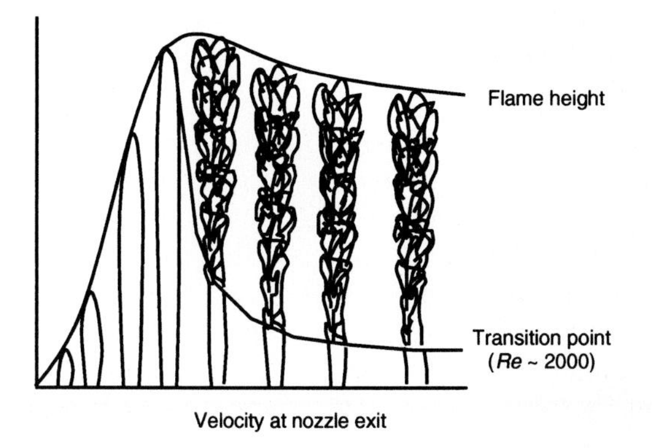

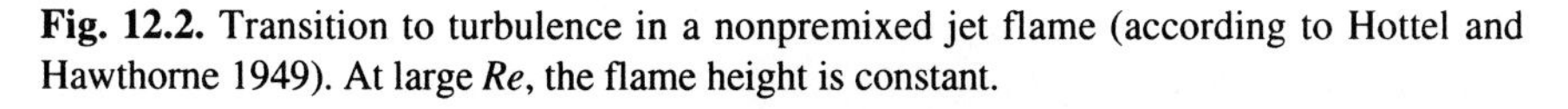
Fig. 12.2. Transition to turbulence in a nonpremixed jet flame (according to Hottel and Hawthorne 1949). At large Re, the flame height is constant.

Turbulent Premixed Flame: Turbulent premixed flames are the main mode of combustion in spark-ignited engines and in the afterburners of jet engines. These flames can be stabilized in, or at the exit of, a premixed flow in a pipe. At low flow velocities, these Bunsen flames are laminar with a well-defined flame front that is stationary in time. Above a certain velocity, the pipe flow is turbulent and combustion is accompanied by a roaring sound. The flame appears to have a broad, diffuse flame front. However, pictures with high temporal resolution show wrinkled and corrugated flame fronts, as shown in Fig. 14.1, 14.2, and 14.3.

Turbulent Nonpremixed Jets: When fuel exits from a pipe into the surrounding air, the resulting nonpremixed flame is laminar at low velocities and then becomes turbulent at high velocities. In the first few jet diameters, this flow is well described as an axisymmetric shear layer. Beyond a few diameters, the central-core fluid that formed one side of the shear layer is now constantly being diluted by mixing with the surrounding air. As Fig 12.2 illustrates, the flame length increases with velocity until turbulent flow causes the air entrainment to increase at the same rate as the fuel input, with the result that the turbulent flame length is independent of jet velocity.

12.2 Direct Numerical Simulation

Just as the Navier-Stokes equations were solved for laminar flame problems, there is in principle no reason that one cannot solve the Navier-Stokes equations for any of the preceding turbulent examples and for turbulent flows in general (*direct numerical simulations* or *DNS*, Reynolds 1989). In practice, solutions of the Navier-Stokes equations for turbulent flows demand a prohibitive amount of computational time.

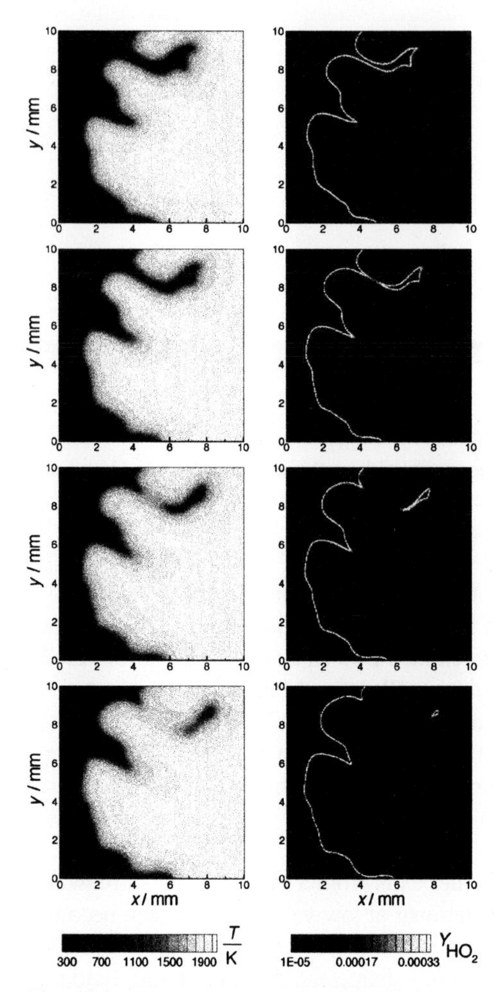

Fig. 12.3. Direct numerical simulation of temperature T and HO_2 mass fraction Y_{HO_2} in a premixed hydrogen-air flame (Lange et al. 1998). The times of interaction with the turbulent flow field (beginning at the top) are 0.90 ms, 0.95 ms, 1.00 ms, and 1.05 ms. The initial turbulence intensity is described by the Reynolds number $Re_l = 175$ at $t = 0$.

This is mainly due to the fact that resolution of the small scales in turbulent flows demands far more grid points than does the analogous laminar flow. An appreciation of the large computation problem can be seen from a simple calculation: The ratio between the largest and the smallest turbulent length scales is (see Section 12.10)

$$\frac{l_0}{l_K} \approx R_l^{3/4} , \tag{12.1}$$

where R_l is a turbulence Reynolds number, which will be defined in Section 12.10 and has the property $R_l < Re$. The *integral length scale* l_0 denotes the largest length scale and is governed by the dimension of the system. l_K is the *Kolmogorov length scale*, describing the size of the smallest turbulent structures (see Section 12.10). For a typical turbulent flow with $R_l = 500$ one has $l_0/l_K \approx 100$. Thus, about 1000 grid points in one dimension are needed, and for three-dimensions 10^9 grid points are necessary in order to resolve the smallest turbulent eddies.

Furthermore, unlike laminar flow solutions, the Navier-Stokes solution to turbulent flows is itself time-dependent in any case; there is not a steady solution. If one estimates that at least 1000 time steps are needed to mimic a turbulent combustion process, the number of computational operations needed in the calculation easily exceeds 10^{14} (assuming 100 operations per grid point). (Yet, one further problem is caused by the fact that the maximum time steps are inversely proportional to the square of the grid point distance.) As a consequence, the overall time for the computation increases with the fourth power of the Reynolds number.

Despite these problems, DNS is possible for low R_l (i. e., $R_l < 1000$ presently) in very small 3-dimensional domains with one or two chemical reactions or in 2-dimensional domains with detailed chemical reactions (Poinsot 1996, Thévenin et al.1996, Chen and Hong 1998; an example is given in Fig. 12.3). These low-R_l solutions are far from practical interest, but are of great interest to researchers probing the details of turbulent flows.

For practical purposes, a solution of the Navier-Stokes equations for turbulent reacting flows is not yet possible. As a consequence there is a long history of approximate solutions obtained in a wide variety of approaches. Before these are discussed, several concepts will be developed in Sections 12.3 to 12.5.

12.3 Concepts for Turbulence Modeling: Probability Density Functions (PDFs)

Even if one had a DNS solution of a flow of practical interest, one would be overwhelmed with a vast amount of details, in time and in space, that would be of little practical interest. One would likely average the time-dependent output to obtain what is typically desired: the average fuel consumption, the average power, the average pollutant formation, etc. It is natural to seek time-independent equations that describe the mean of these quantities. Progress has been made by assuming the flow to be a random, chaotic process that can be adequately described by its statistics.

The probability that the fluid at the spatial location $\vec{r}$ has a density between ρ and $\rho+\mathrm{d}\rho$, a velocity in x-direction between v_x and $v_x+\mathrm{d}v_x$, a velocity in y-direction between v_y and $v_y+\mathrm{d}v_y$, a velocity in z-direction between v_z and $v_z+\mathrm{d}v_z$, a temperature between T and $T+\mathrm{d}T$, and a local composition, which corresponds to mass fractions between w_i and $w_i+\mathrm{d}w_i$, is given by (see, e. g., Libby and Williams 1980, 1994)

$$P(\rho, v_x, v_y, v_z, w_1, ..., w_{S-1}, T; \vec{r})\,\mathrm{d}\rho\ \mathrm{d}v_x\ \mathrm{d}v_y\ \mathrm{d}v_z\ \mathrm{d}w_1, ..., \mathrm{d}w_{S-1}\ \mathrm{d}T\ ,$$

where P is called the *probability density function, PDF* ($w_S = 1 - \sum_{i=1}^{S-1} w_i$ is known).

A *normalization condition* for the PDF is obtained from the fact that the overall probability for the system to be somewhere in the whole configuration space, which is spanned by the coordinates ρ, v_x, v_y, v_z, w_1, . . . , w_{S-1}, T has to be 1,

$$\int_0^\infty \int_{-\infty}^\infty \int_{-\infty}^\infty \int_{-\infty}^\infty \int_0^\infty \ldots \int_0^\infty \int_0^\infty P(\rho, v_x, v_y, v_z, w_1, \ldots, w_{S-1}, T; \vec{r}) \cdot d\rho \, dv_x \, dv_y \, dv_z \, dw_1 \ldots dw_{S-1} \, dT = 1 . \quad (12.2)$$

If the PDF $P(\vec{r})$ is known at some point $\vec{r}$, the *means* of the local properties can be calculated very easily. For the mean density or the mean momentum density in i-direction (in order to simplify the notation, $\int$ denotes all the integrations), one obtains

$$\bar{\rho}(\vec{r}) = \int \rho \, P(\rho, \ldots, T; \vec{r}) \, d\rho \ldots dT$$

$$\overline{\rho v_i}(\vec{r}) = \int \rho \, v_i \, P(\rho, \ldots, T; \vec{r}) \, d\rho \ldots dT .$$

This is an *ensemble-averaging*. A sufficiently large number of different realizations is considered and averaged. The statistical weight of any realizable state is contained in the probability density function. In experiments, mean values are obtained in an analogous way, by averaging over a large number of time- and space-resolved measurements, obtained at constant experimental boundary conditions.

12.4 Concepts for Turbulence Modeling: Time- and Favre-Averaging

Using *time-averaging*, a mean can be obtained which is equal to the ensemble-average. Time-averaging is explained using the example of a *statistically stationary* process that is illustrated in Fig. 12.4. If the time behavior of a variable is observed, e. g., the density ρ, it can be seen that the value fluctuates about an average. The time-average is obtained by integration over a long (ideally infinitely long) time interval,

$$\bar{\rho}(\vec{r}) = \lim_{\Delta t \to \infty} \frac{1}{\Delta t} \int_0^{\Delta t} \rho(\vec{r}, t) \, dt . \quad (12.3)$$

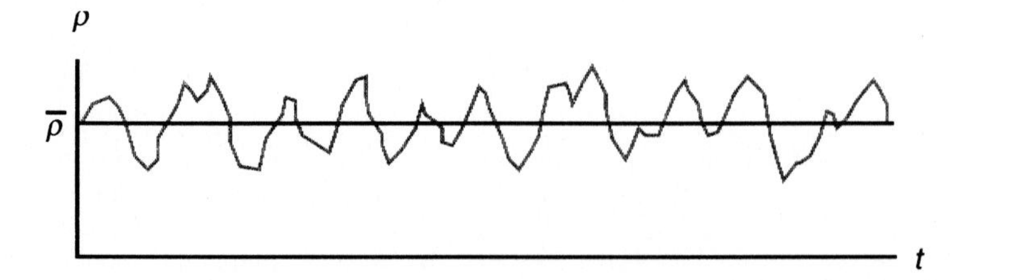

Fig. 12.4. Temporal fluctuations and time-average in a statistically stationary process

If the average itself is changing in time, local time-averages can be calculated in non-stationary systems, if the temporal fluctuations are fast compared to the time-behavior of the mean value (see Fig. 12.5). In this case the result for the time-average at t' ($t_1 < t' < t_2$) is

$$\bar{\rho}(\vec{r},t') = \frac{1}{t_2 - t_1}\int_{t_1}^{t_2}\rho(\vec{r},t)\,\mathrm{d}t \quad ; \quad t_1 \le t' \le t_2 \; . \tag{12.4}$$

However, one can see from Fig. 12.4 that the choice of the time interval $[t_1, t_2]$ has a strong influence on the results.

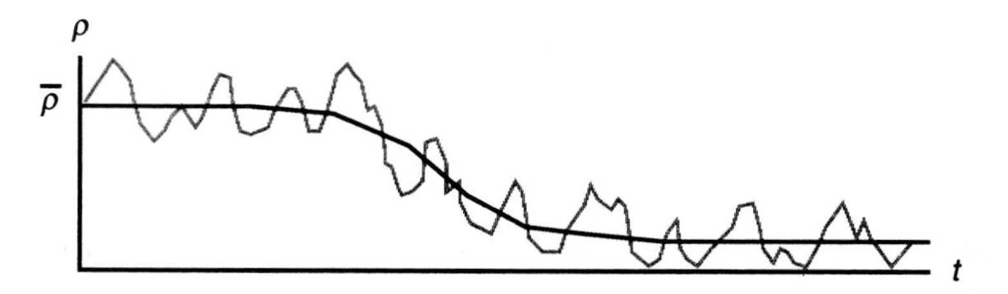

Fig. 12.5. Temporal fluctuations and time-average in a statistically non-stationary process

Now, the value of a function q shall be split into its mean and the fluctuation (indicated by the apostrophe),

$$q(\vec{r},t) = \bar{q}(\vec{r},t) + q'(\vec{r},t) \; . \tag{12.5}$$

If q is averaged, the important result is that the mean of the fluctuations is zero,

$$\overline{q'} = 0 \; . \tag{12.6}$$

Large density variations are typical for combustion processes. Thus it is useful (see below) to introduce another average, namely the *Favre average* (also called *density-weighted average*), which is, for an arbitrary property q, given by

$$\tilde{q} = \frac{\overline{\rho q}}{\bar{\rho}} \quad \text{or} \quad \bar{\rho}\,\tilde{q} = \overline{\rho q} \; . \tag{12.7}$$

As in (12.5), any property q again can be split into its mean value and the fluctuation,

$$q(\vec{r},t) = \tilde{q}(\vec{r},t) + q''(\vec{r},t) \; , \tag{12.8}$$

and the result for the average of the *Favre fluctuation* (which is characterized by two apostrophes) is

$$\overline{\rho q''} = 0 \; . \tag{12.9}$$

Introduction of (12.5) into the definition (12.7) leads to a relation which formally allows the calculation of the Favre average from the average of a variable q,

$$\tilde{q} = \frac{\overline{\rho q}}{\bar{\rho}} = \frac{\overline{(\bar{\rho} + \rho')(\bar{q} + q')}}{\bar{\rho}} = \frac{\overline{\bar{\rho}\,\bar{q}} + \overline{\bar{\rho}\,q'} + \overline{\rho'\bar{q}} + \overline{\rho' q'}}{\bar{\rho}}$$

or

$$\tilde{q} = \bar{q} + \frac{\overline{\rho' q'}}{\bar{\rho}} . \tag{12.10}$$

However, the *correlation* $\overline{\rho' q'}$ of the fluctuation of the density with the fluctuation of the variable q must be known. Ideally, the correlation is computed from a conservation equation or from an empirically derived equation.

In the following, additional relations for the averages are derived that will be used in the next section. The mean of the square of a variable q is easily calculated from (12.5),

$$\overline{q^2} = \overline{(\bar{q} + q')(\bar{q} + q')} = \overline{\bar{q}\,\bar{q}} + \overline{\bar{q}\,q'} + \overline{q'\,\bar{q}} + \overline{q'\,q'} = \bar{q}\,\bar{q} + 2\,\bar{q}\,\overline{q'} + \overline{q'\,q'}$$

or

$$\overline{q^2} = \bar{q}^2 + \overline{q'^2} . \tag{12.11}$$

The density weighted mean of the correlation between two variables u and υ can be calculated according to

$$\begin{aligned} \overline{\rho u \upsilon} &= \overline{(\bar{\rho} + \rho')(\bar{u} + u')(\bar{\upsilon} + \upsilon')} \\ &= \overline{\bar{\rho}\,\bar{u}\,\bar{\upsilon} + \bar{\rho}\,\bar{u}\,\upsilon' + \bar{\rho}\,u'\,\bar{\upsilon} + \bar{\rho}\,u'\,\upsilon' + \rho'\,\bar{u}\,\bar{\upsilon} + \rho'\,\bar{u}\,\upsilon' + \rho'\,u'\,\bar{\upsilon} + \rho'\,u'\,\upsilon'} \\ &= \bar{\rho}\,\bar{u}\,\bar{\upsilon} + \bar{\rho}\overline{u'\,\upsilon'} + \bar{u}\overline{\rho'\,\upsilon'} + \bar{\upsilon}\overline{\rho'\,u'} + \overline{\rho'\,u'\,\upsilon'} \quad . \end{aligned} \tag{12.12}$$

Splitting into Favre average and Favre fluctuation leads to

$$\overline{\rho u \upsilon} = \overline{\rho(\tilde{u} + u'')(\tilde{\upsilon} + \upsilon'')} = \overline{\rho \tilde{u} \tilde{\upsilon} + \rho \tilde{u} \upsilon'' + \rho u'' \tilde{\upsilon} + \rho u'' \upsilon''}$$

or

$$\overline{\rho u \upsilon} \quad = \bar{\rho}\,\tilde{u}\,\tilde{\upsilon} + \overline{\rho\,u''\,\upsilon''} . \tag{12.13}$$

A comparison of (12.12) and (12.13) shows that a much more compact formulation, with fewer (unknown) correlations, is possible, if Favre averaging is used. This is the main reason for the use of the Favre average.

12.5 Averaged Conservation Equations

The Navier-Stokes equations, which were derived in Chapter 11, allow the simulation of reacting flows. If one is only interested in the mean values of the turbulent flow, but not in the fluctuations, one can derive *Reynolds-averaged* conservation equations from the Navier-Stokes equations (RANS), using the methods discussed in Section 12.4 (see, e. g., Libby and Williams 1980, 1994).

For the conservation of mass (11.7), after averaging with the aid of (12.7), one obtains the rather simple equation

$$\frac{\partial \bar{\rho}}{\partial t} + \operatorname{div}(\bar{\rho}\,\tilde{\vec{v}}) = 0\,. \tag{12.14}$$

In the same way, one obtains for the conservation of the species masses (11.8), using the approximation $\vec{j}_i = -D_i \rho \operatorname{grad} w_i$ and (12.7) and (12.13),

$$\frac{\partial(\bar{\rho}\,\tilde{w}_i)}{\partial t} + \operatorname{div}\left(\bar{\rho}\,\tilde{\vec{v}}\,\tilde{w}_i\right) + \operatorname{div}\left(-\overline{\rho\, D_i \operatorname{grad} w_i} + \overline{\rho \vec{v}''\, w_i''}\right) = \overline{M_i\, \omega_i}\ . \tag{12.15}$$

For the momentum conservation (11.9), averaging leads to

$$\frac{\partial\left(\bar{\rho}\,\tilde{\vec{v}}\right)}{\partial t} + \operatorname{div}\left(\bar{\rho}\,\tilde{\vec{v}} \otimes \tilde{\vec{v}}\right) + \operatorname{div}\left(\bar{\bar{\bar{p}}} + \overline{\rho \vec{v}'' \otimes \vec{v}''}\right) = \bar{\rho}\,\bar{\vec{g}}\,, \tag{12.16}$$

and for the energy conservation equation (11.12) one obtains together with the approximation $\vec{j}_q = -\lambda \operatorname{grad} T$

$$\frac{\partial\left(\bar{\rho}\,\tilde{h}\right)}{\partial t} - \frac{\partial \bar{p}}{\partial t} + \operatorname{div}\left(\bar{\rho}\,\tilde{\vec{v}}\,\tilde{h}\right) + \operatorname{div}\left(-\overline{\lambda \operatorname{grad} T} + \overline{\rho v''\, h''}\right) = \overline{q_{\mathrm{r}}}\ . \tag{12.17}$$

Here the terms $\bar{\bar{p}} : \operatorname{grad} \vec{v}$ und $\operatorname{div}(p\vec{v})$ are not considered, because they are only important if extreme pressure gradients, i. e., shock waves or detonations, occur in the system. Like in the original (not averaged) equations, an equation of state (the ideal gas law) is needed. Averaging of $p = \rho RT\, \Sigma(w_i/M_i)$ leads to

$$\bar{p} = R \sum_{i=1}^{S} \left(\bar{\rho}\,\tilde{T}\,\tilde{w}_i + \overline{\rho T''\, w_i''}\right) \frac{1}{M_i}\ . \tag{12.18}$$

If the molar masses do not differ too much, one can use the approximation that fluctuations in the mean molar mass are negligible. Then averaging leads to the approximation

$$\bar{p} = \bar{\rho} R \tilde{T} / \overline{M}\,, \tag{12.19}$$

where $\overline{M}$ denotes the averaged mean molar mass of the mixture (see Section 1.2).

The treatment of the source terms (i. e., chemical reaction rates) in the species conservation equations is usually very difficult. Thus it is advantageous to derive *element conservation equations*. Elements are neither created nor destroyed in chemical reactions, and thus the source terms vanish in the element conservation equations. One introduces the *element mass fraction* (Williams 1984)

$$Z_i = \sum_{j=1}^{S} \mu_{ij}\, w_j \quad \text{for} \quad i = 1, \ldots, M\,, \tag{12.20}$$

where S denotes the number of species and M the number of different elements in the mixture considered. The μ_{ij} denote the mass ratio of element i in species j (see Section 9.3).

If it is assumed (as an approximation), that all diffusion coefficients D_i in (12.15) are equal, the species conservation equations (11.8) can be multiplied by μ_{ij} and summed; the result is the simple relation

$$\frac{\partial(\rho Z_i)}{\partial t} + \mathrm{div}(\rho \vec{v} Z_i) - \mathrm{div}(\rho D \,\mathrm{grad}\, Z_i) = 0 . \qquad (12.21)$$

Because of the element conservation $\Sigma \mu_{ij} M_i \omega_i = 0$, this equation does not contain a reaction term. This is very often quite useful (see Chapter 13). Averaging of (12.21) leads to an equation without source terms,

$$\frac{\partial(\bar{\rho} \tilde{Z}_i)}{\partial t} + \mathrm{div}\left(\bar{\rho} \tilde{\vec{v}} \tilde{Z}_i\right) + \mathrm{div}\left(\overline{\rho \vec{v}'' Z_i''} - \overline{\rho D \,\mathrm{grad}\, Z_i}\right) = 0 . \qquad (12.22)$$

12.6 Turbulence Models

The Navier-Stokes equations are closed when the empirical laws for the flux densities are used. The averaged conservation equations are not closed until analogous terms of the form $\overline{\rho \vec{v}'' q''}$ are specified. These new terms generated in the averaging process are not explicitly known as functions of the dependent (averaged) variables. Consequently, there are more unknowns than equations (*closure problem* in turbulence theory).

In order to solve the closure problem, models are proposed describing the *Reynold stresses* $\overline{\rho \vec{v}'' q''}$ in terms of the dependent variables. Current turbulence models (see, e. g., Launder and Spalding 1972, Jones and Whitelaw 1985) interpret $\overline{\rho \vec{v}'' q''}$ ($q = w_i$, $\vec{v}$, h, Z_i) in (12.14)-(12.17), and (12.22) as *turbulent transport* and model it in analogy to the laminar case (see Chapter 11), using a *gradient-transport* assumption which states that the term is proportional to the gradient of the mean value of the property,

$$\overline{\rho \vec{v}'' q_i''} = -\bar{\rho}\, \nu_{\mathrm{T}} \,\mathrm{grad}\, \tilde{q}_i , \qquad (12.23)$$

where ν_{T} is called *turbulent exchange coefficient*. This gradient-transport assumption is the source of many controversial discussions. Indeed, experiments show that turbulent transport can even proceed against the gradient of the mean values (Moss 1979).

Usually it is argued that turbulent transport is much faster than the laminar transport process. Thus, the averaged laminar transport terms in (12.14)-(12.17) can be neglected very often, and then the conservation equations simplify to

$$\frac{\partial \bar{\rho}}{\partial t} + \operatorname{div}(\bar{\rho}\,\tilde{\vec{v}}) = 0 \tag{12.24}$$

$$\frac{\partial(\bar{\rho}\,\tilde{w}_i)}{\partial t} + \operatorname{div}\left(\bar{\rho}\,\tilde{\vec{v}}\,\tilde{w}_i\right) - \operatorname{div}\left(\bar{\rho}\,\nu_{\mathrm{T}}\,\mathrm{grad}\,\tilde{w}_i\right) = \overline{M_i\,\omega_i} \tag{12.25}$$

$$\frac{\partial\left(\bar{\rho}\,\tilde{\vec{v}}\right)}{\partial t} + \operatorname{div}\left(\bar{\rho}\,\tilde{\vec{v}}\otimes\tilde{\vec{v}}\right) - \operatorname{div}\left(\bar{\rho}\,\nu_{\mathrm{T}}\,\mathrm{grad}\,\tilde{\vec{v}}\right) = \overline{\rho\,\vec{g}} \tag{12.26}$$

$$\frac{\partial\left(\bar{\rho}\,\tilde{h}\right)}{\partial t} - \frac{\partial \bar{p}}{\partial t} + \operatorname{div}\left(\bar{\rho}\,\tilde{\vec{v}}\,\tilde{h}\right) - \operatorname{div}\left(\bar{\rho}\,\nu_{\mathrm{T}}\,\mathrm{grad}\,\tilde{h}\right) = \bar{q}_{\mathrm{r}} \tag{12.27}$$

$$\frac{\partial(\bar{\rho}\,\tilde{Z}_i)}{\partial t} + \operatorname{div}\left(\bar{\rho}\,\tilde{\vec{v}}\,\tilde{Z}_i\right) - \operatorname{div}\left(\bar{\rho}\,\nu_{\mathrm{T}}\,\mathrm{grad}\,\tilde{Z}_i\right) = 0\,. \tag{12.28}$$

If the turbulent exchange coefficient ν_{T} (which most likely has different values for the different equations) is known, the equations can be solved numerically. There exist many models determining the turbulent exchange coefficient; examples include:

Zero-Equation Models: These (now obsolete) models yield explicit algebraic expressions for the turbulent exchange coefficient. Examples are models determining ν_{T} based on the mixing length formula of Prandtl (1925). In this case the turbulent transport term is given by

$$\overline{\rho \vec{v}''q''} = -\bar{\rho}\,l^2 \left|\frac{\partial \tilde{v}}{\partial z}\right| \frac{\partial \tilde{q}}{\partial z}\,, \tag{12.29}$$

where l is a characteristic length, which is problem-dependent. For the turbulent exchange coefficient one then obtains

$$\nu_{\mathrm{T}} = l^2 \left|\frac{\partial \tilde{v}}{\partial z}\right|. \tag{12.30}$$

If one considers a turbulent shear flow (see Fig. 12.6), l is a function of the thickness δ of the shear layer, calculated from empirical (geometry-dependent) formulas like

$$\delta = \begin{cases} 0.115\,x & \text{for a 2D-jet from a slit} \\ 0.085\,x & \text{for a cylinder-symmetric jet} \end{cases}\,.$$

Furthermore, one has to take into account whether the considered location is in the inner or outer part of the shear layer. For the mixing length one obtains

$$l = \begin{cases} \kappa z & \text{for } z \le z_{\mathrm{c}} \quad \text{(inner boundary layer)} \\ \alpha\delta & \text{for } z_{\mathrm{c}} \le z \le \delta \quad \text{(outer boundary layer)} \end{cases}\,.$$

Prandtl's formula for the mixing length has been extended by the so-called mixing-length formula (von Karman 1930)

$$l \propto \left| \frac{\partial \tilde{\upsilon}}{\partial z} \Big/ \frac{\partial^2 \tilde{\upsilon}}{\partial z^2} \right| , \tag{12.31}$$

where l is now computed, not specified. Another problems with this formula are the singularities in the points of inflection of the profile $\tilde{\upsilon}$, where the formula becomes meaningless.

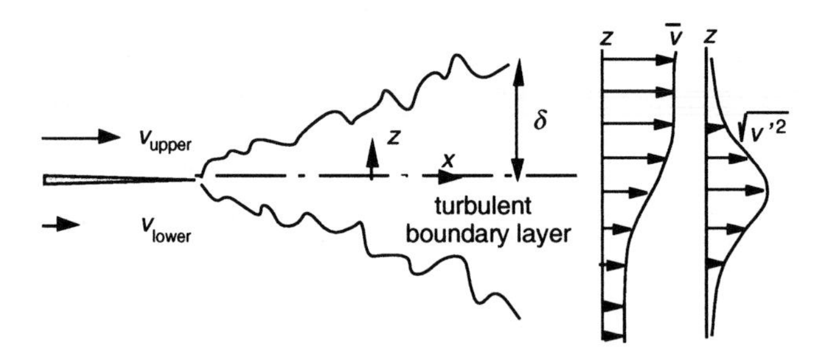

Fig. 12.6. Schematic illustration of a turbulent shear layer

The coefficients α and κ, as well as the thickness z_c of the inner boundary layer, are determined from a variety of experiments for typical conditions. They are $\kappa = 0.4$, $\alpha = 0.075$, and $z_c = 0.1875\ \delta$. Using these coefficients, one obtains a dependence of the mixing length on the location in the shear layer, which is shown in Fig. 12.7.

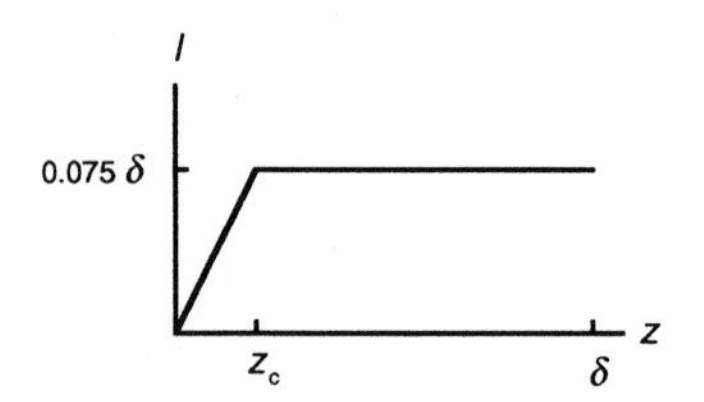

Fig. 12.7. Plot of the dependence of mixing length l vs. location z in the boundary layer.

One-Equation Models: In one-equation models (also obsolete), the turbulent exchange coefficient ν_T is calculated from one additional partial differential equation (this is the reason for the name), e. g., that for the turbulent kinetic energy (TKE)

$$\tilde{k} = \frac{1}{2} \frac{\overline{\rho \sum \upsilon_i''^2}}{\bar{\rho}} . \tag{12.32}$$

The turbulent exchange coefficient is calculated from the kinetic energy according to (Prandtl 1945)

$$\nu_T = l \sqrt{\tilde{k}} . \tag{12.33}$$

The mixing length l is still calculated using algebraic equations, as used in zero-equation models.

Two-Equation Models: The two-equations models, which are usually applied today, use two partial differential equations for the determination of the turbulent exchange coefficient ν_T. One of the equations is usually for the turbulent kinetic energy $\tilde{k}$. The second variable z is of the form $z = \tilde{k}^m \cdot l^n$ (m, n constant). Then the viscosity hypothesis reads

$$\nu_T \propto z^{\frac{1}{n}} \tilde{k}^{\frac{1}{2}-\frac{m}{n}} . \qquad (12.34)$$

The most widely used turbulence model is the *k-ε turbulence model* (Launder and Spalding 1972, Jones and Whitelaw 1985), which uses an equation for the turbulent kinetic energy which can be derived like a conservation equation. The constants n and m have the values -1 and 3/2, respectively, and one obtains for z, which is in this case called *dissipation rate* $\tilde{\varepsilon}$ of the kinetic energy,

$$\tilde{\varepsilon} = \frac{\tilde{k}^{3/2}}{l} \left(= \frac{\tilde{k}}{l/\tilde{k}^{1/2}} = \frac{\text{energy}}{\text{time}} \right). \qquad (12.35)$$

For $\tilde{\varepsilon}$, which is given by

$$\tilde{\varepsilon} = \nu \, \overline{\operatorname{grad} \vec{v}''^{\mathrm{T}} : \operatorname{grad} \vec{v}''} \qquad (12.36)$$

with $\nu = \mu/\rho$ = laminar kinematic viscosity, a differential equation is formulated empirically. The two differential equations read (e. g., Kent and Bilger 1976)

$$\frac{\partial(\bar{\rho}\tilde{k})}{\partial t} + \operatorname{div}(\bar{\rho}\tilde{\vec{v}}\tilde{k}) - \operatorname{div}(\bar{\rho}\nu_T \operatorname{grad}\tilde{k}) = G_k - \bar{\rho}\tilde{\varepsilon} \qquad (12.37)$$

$$\frac{\partial(\bar{\rho}\tilde{\varepsilon})}{\partial t} + \operatorname{div}(\bar{\rho}\tilde{\vec{v}}\tilde{\varepsilon}) - \operatorname{div}(\bar{\rho}\nu_T \operatorname{grad}\tilde{\varepsilon}) = (C_1 G_k - C_2\bar{\rho}\tilde{\varepsilon})\frac{\tilde{\varepsilon}}{\tilde{k}} . \qquad (12.38)$$

The turbulent exchange coefficient ν_T can be calculated from (12.34); the result is the simple relation

$$\nu_T = C_\nu \frac{\tilde{k}^2}{\tilde{\varepsilon}} . \qquad (12.39)$$

Here $C_\nu = 0.09$ is an empirically determined constant; C_1 and C_2 are two further empirical constants of the model. The term G_k is a complicated function of the stress tensor, which results from (12.38),

$$G_k = -\overline{\rho \vec{v}'' \otimes \vec{v}''} : \operatorname{grad} \tilde{\vec{v}} . \qquad (12.40)$$

The constants of the k-ε model depend on the geometry and on the nature of the problem considered. Another shortcoming of the model is the gradient transport assumption (12.23). However, this model is in wide use, e. g., in the program packages like PHOENICS™, FLUENT™, FIRE™, NUMECA™, STAR-CD™, and KIVA™ developed for the simulation of chemically reacting turbulent flows (see Rosten and Spalding 1987) in engines, turbines, furnaces, and chemical reactors, due to the lack of better, but nevertheless simple, models.

12.7 Mean Reaction Rates

In order to solve the averaged conservation equations (12.24-12.28), one still has to specify the averaged chemical reaction rates $\overline{\omega}_i$ which will then lead to computation of the density and pressure. In order to demonstrate the impending problems, two simple examples will be considered (Libby and Williams 1994).

The first example is a reaction A + B → Products at constant temperature, but with variable concentrations for A and B:

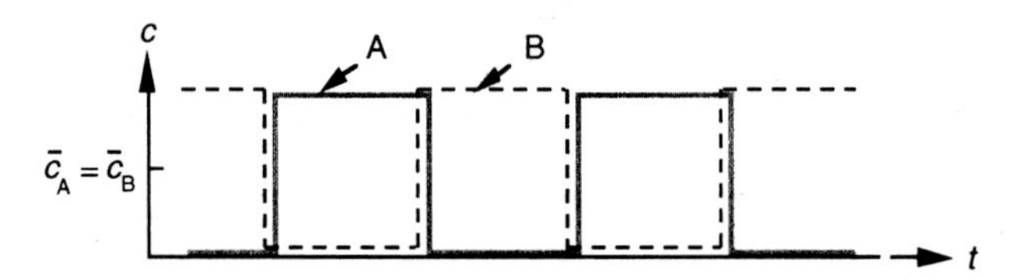

Fig. 12.8. Hypothetical time behavior of the concentrations in a reaction A + B → Products

Here a rather hypothetical (but one resembling turbulent nonpremixed combustion) time behavior, illustrated in Fig. 12.8, will be assumed, where the concentrations c_A and c_B are never nonzero at the same time. Then one obtains (in order to avoid confusion with the notation for the turbulent kinetic energy k, the rate coefficient k has here a subscript R)

$$\omega_A = -k_R\, c_A\, c_B = 0 \quad \text{and} \quad \overline{\omega}_A = 0 \,.$$

This demonstrates that the mean reaction rate cannot be calculated from the means of the concentrations, i. e.,

$$\overline{\omega}_A = -\overline{k_R \cdot c_A \cdot c_B} = -k_R \cdot \overline{c_A} \cdot \overline{c_B} - k_R \cdot \overline{c'_A \cdot c'_B} \quad \neq \quad -\overline{k}_R \cdot \overline{c_A} \cdot \overline{c_B} \,. \qquad (12.41)$$

Thus, there is no circumstance where it is valid to calculate the averaged reaction rates by simply replacing the average of the concentrations product by the product of the average concentrations. (The right-hand-side of (12.41) is correct if the term $k_R\, \overline{c'_A\, c'_B}$ is retained. For this example, this term is equal to the preceding term, as it must be.)

The second example is a reaction with varying temperature (for constant concentrations; this is resembling turbulent premixed combustion), where a sine-curve temperature variation is assumed (see Fig. 12.9).

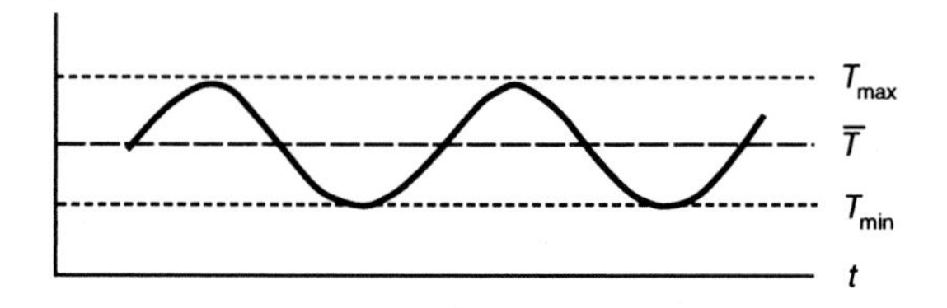

Fig. 12.9. Hypothetical time behavior of the temperature in a reaction A + B → Products

As a result of the strong non-linearity of the rate coefficients $k_R = A \cdot \exp(-T_a/T)$, $\overline{k_R}$ is completely different from $k_R(\overline{T})$. This shall be explained by a simple example. For $T_{min} = 500$ K and $T_{max} = 2\,000$ K one obtains $\overline{T} = 1\,250$ K. If the reaction rate coefficient for an activation temperature of $T_a = 50\,000$ K ($T_a = E_a/R$) is calculated, the result is

$$k_R(T_{max}) = 1.4 \cdot 10^{-11}\,A$$

$$k_R(T_{min}) = 3.7 \cdot 10^{-44}\,A$$

$$k_R(\overline{T}) = 4.3 \cdot 10^{-18}\,A$$

and after time-averaging (e. g., by numerical integration)

$$\overline{k_R} = 7.0 \cdot 10^{-12}\,A.$$

This behavior is especially interesting in the case of the nitrogen oxide formation, which is strongly temperature-dependent due to the high activation temperature (T_a = 38 000 K, see Chapter 17). Thus, NO is mainly formed at peak values of the temperature. Calculation of the NO production using an average temperature is thus completely useless; temperature fluctuations have to be accounted for!

One (on the first glance attractive) proposal to account for temperature fluctuations is to replace the temperature T by $\tilde{T} + T''$ and derive a power series for the exponential function (Libby and Williams 1980, 1994),

$$k_R = A\exp\left(-T_a/\tilde{T}\right)\left\{1+\left(\frac{T_a}{\tilde{T}^2}\right)T'' + \left[\left(\frac{T_a^{\,2}}{2\tilde{T}^4}\right)-\left(\frac{T_a}{\tilde{T}^3}\right)\right]T''^2 + \ldots\right\}. \quad (12.42)$$

Favre averaging and neglecting the term $T_a/\tilde{T}^3$ leads to an expression for the mean reaction rate coefficient,

$$\tilde{k}_R = \frac{\overline{\rho k_R}}{\overline{\rho}} = A \cdot \exp\left(-\frac{T_a}{\tilde{T}}\right)\left[1+\frac{T_a^{\,2}}{2\tilde{T}^4}\frac{\overline{\rho T''^2}}{\overline{\rho}}+\ldots\right]. \quad (12.43)$$

The series may be truncated after the second term if the condition

$$\frac{T_a \cdot T''}{\tilde{T}^2} \ll 1$$

is fulfilled. Usually $T_a > 10\,\tilde{T}$; thus, for $T_a T''/\tilde{T}^2 = 0.1$ the temperature fluctuations must not exceed 1%. In turbulent premixed flames fluctuations between burnt and unburned gas occur, and thus, fluctuations of 70% are possible for $T_u = 300$ K, $T_b =$ 2000 K. For this reason, this *moment method* is not practical.

An attractive method that avoids the mean reaction rate problem is the statistical approach using the probability density function (PDF). If the PDF is known, the mean reaction rate can be determined by integration. For the example A + B → Products one obtains (Libby and Williams 1994)

$$\overline{\omega} = -\int_0^1 \ldots \int_0^1 \int_0^\infty \int_0^\infty k_R\, c_A\, c_B\, P(\rho, T, w_1, \ldots, w_{S-1}; \vec{r}) \mathrm{d}\rho \mathrm{d}T \mathrm{d}w_1 \ldots \mathrm{d}w_{S-1}$$

$$= -\frac{1}{M_A M_B} \int_0^1 \ldots \int_0^1 \int_0^\infty \int_0^\infty k_R(T) \rho^2 w_A w_B\, P(\rho, T, w_1, \ldots, w_{S-1}; \vec{r}) \mathrm{d}\rho \mathrm{d}T \mathrm{d}w_1 \ldots \mathrm{d}w_{S-1}\,. \quad (12.44)$$

The main problem in, this case, is that the PDF P has to be known. There are different methods to determine the PDF, which are used in different applications:

PDF-Transport Equations (see, e. g., Dopazo and O'Brien 1974, Pope 1986, Chen et al. 1989): The most elegant way is the solution of the PDF-transport equation. A transport equation for the time behavior of the PDF can be derived from the conservation equations for the species masses. The main advantage of this method is the fact that the chemistry can be treated exactly (while molecular transport has still to be modelled).

For the numerical solution of the transport equation the PDF is approximated by a large number of stochastic particles, which represent different realizations of the flow. Then Monte-Carlo methods can be used to solve the transport equations (see Section 13.4). The numerical solution is very time-consuming and restricted to systems with only a small number of chemical species. Thus, reduced reaction mechanisms are strongly favored (see Section 7.4).

Empirical Construction of PDFs: In this method, PDFs are constructed using empirical knowledge about the shape. Here the observation that major features of turbulent flame calculations are not sensitive to the exact shape of the PDF is consequently used.

A simple way to construct multidimensional PDFs is to assume statistical independence of the different variables. In this case, the PDF can be written as a product of one-dimensional PDFs (Gutheil and Bockhorn 1987),

$$P(\rho, T, w_1, \ldots, w_{S-1}) = P(\rho) \cdot P(T) \cdot P(w_1) \cdot \ldots \cdot P(w_{S-1})\,. \quad (12.45)$$

Of course, this separation is not correct, because T and ρ, e. g., are in general not independent because of the ideal gas law. Thus, additional correlations between the variables have to be accounted for.

One-dimensional PDFs can be obtained from experiments. In the following, some results for simple geometries are presented (Libby and Williams 1994).

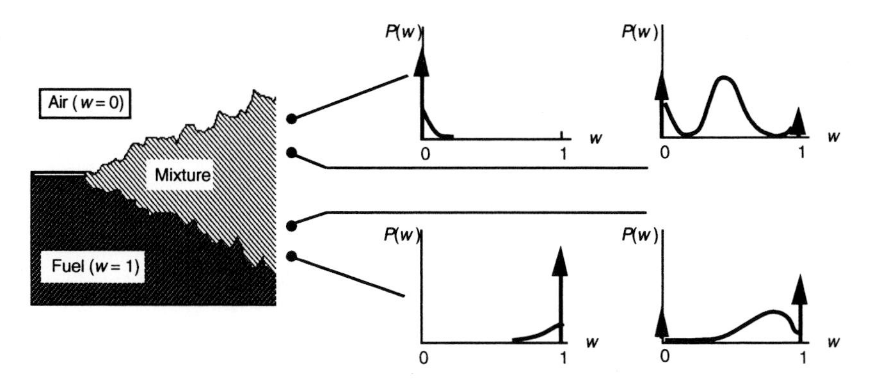

Fig. 12.10. Schematic illustration of probability density functions for the mass fraction of the fuel in a turbulent mixing layer

Examples of PDFs for the mass fraction of the fuel are shown in Fig. 12.10 for different points in a turbulent mixing layer. In the outer flow, the probability that pure fuel or oxidizer are present is very high (arrows in the figure), whereas a mixture of fuel and air has only a small probability. In the mixing layer, the probability to find mixed compounds is high. Thus, the PDF has a maximum for a certain mixture fraction between 1 and 0.

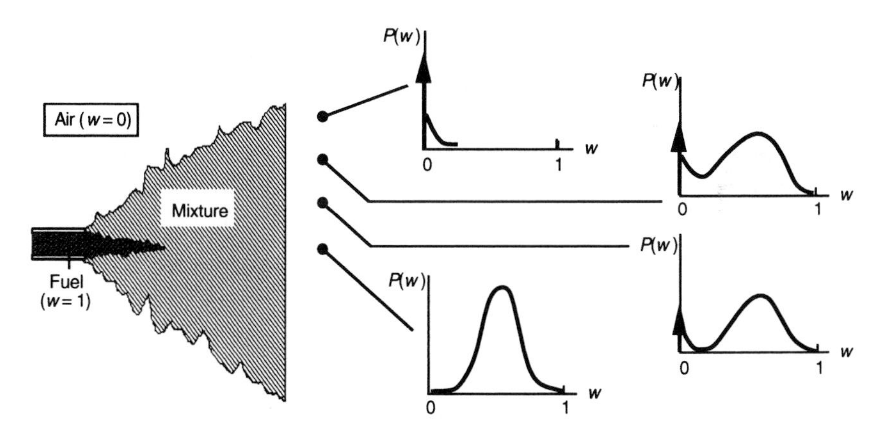

Fig. 12.11. Schematic plot of probability density functions, of the mass fraction of fuel, at various locations in a turbulent jet

Similar results are obtained for a turbulent jet, which can be regarded as a mixing between a pure fluid and one that is constantly being diluted with the pure fluid (see Fig. 12.11).

In a turbulent reactor (Fig. 12.12), the PDF corresponds approximately to a Gaussian distribution. The probability of a complete mixing increases with increasing distance from the inflow boundary. The width of the Gaussian distribution decreases asymptotically, approaching a *Dirac delta-function* (the probability of complete mixing goes to one).

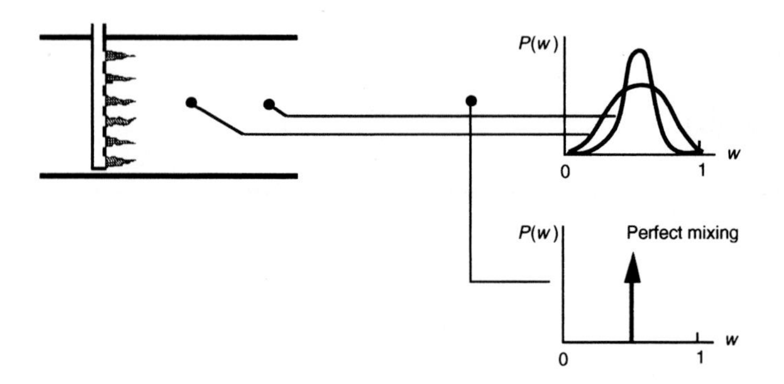

Fig. 12.12. Schematic plot of probability density functions, of the mass fraction of fuel, at three locations in a multiple jet turbulent reactor

Usually, because of their simplicity, *truncated Gaussian functions* (or *clipped Gaussian functions*) or *beta-functions* are used for analytical approximation to the one-dimensional PDFs.

The *truncated Gaussian function* (see Fig. 12.13) consists of a Gaussian distribution and two Dirac δ-functions for the description of *intermittency peaks* (Gutheil and Bockhorn 1987).

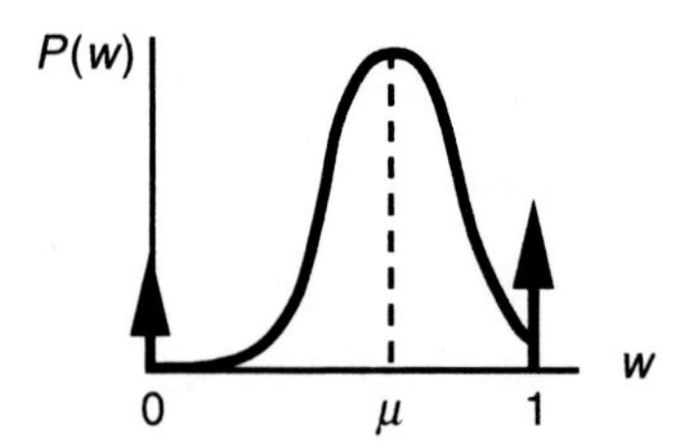

Fig. 12.13. Truncated Gaussian function with intermittency peaks

An analytical representation of this frequently used function is given by (Williams 1984)

$$P(Z) = \alpha \cdot \delta(Z) + \beta \cdot \delta(1-Z) + \gamma \cdot \exp\left[-(Z-\zeta)^2 / \left(2\sigma^2\right)\right]. \qquad (12.46)$$

Here ζ and σ characterize the position and the width of the Gaussian distribution, respectively ($Z = w_i, T, ...$). The normalization constant γ is obtained (for given α and β) from the relation

$$\gamma = \frac{(1-\alpha-\beta)\sqrt{\frac{2\sigma}{\pi}}}{\operatorname{erf}\left(\frac{1-\zeta}{\sqrt{2\sigma}}\right)+\operatorname{erf}\left(\frac{\zeta}{\sqrt{2\sigma}}\right)}, \tag{12.47}$$

where "erf" denotes the *error function* tables of which are readily available (e. g., in integral tables).

The *beta function* (*β-function*, shown in Fig. 12.14) has the advantage of having only two parameters (α, β) and yet can assume a wide variety of shapes; see Fig. 12.14 (Rhodes 1979),

$$P(Z) = \gamma \cdot Z^{\alpha-1} \cdot (1-Z)^{\beta-1} \quad \text{with} \quad \gamma = \frac{\Gamma(\alpha+\beta)}{\Gamma(\alpha)\cdot\Gamma(\beta)}. \tag{12.48}$$

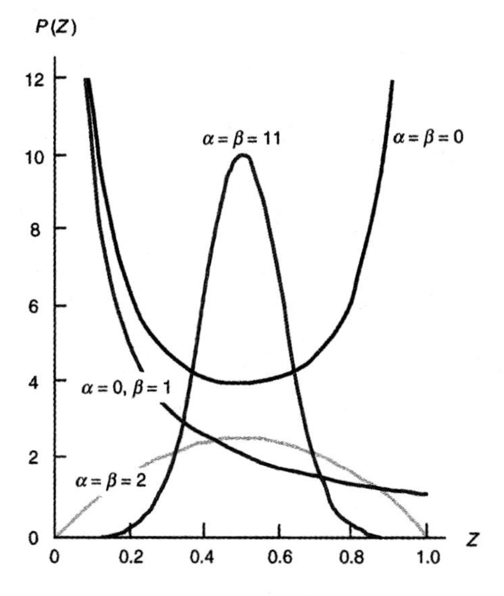

Fig. 12.14. Behavior of the β-function for different sets of parameters α and β; for simplicity, the normalization constant γ is taken to be 1 (Libby and Williams 1994)

The third parameter is obtained from the normalization condition $\int P(Z)\,\mathrm{d}Z = 1$. (Note that in mathematics the integral $B(\alpha,\beta) = \int_0^1 t^{\alpha-1}(1-t)^{\beta-1}\mathrm{d}t$ is usually called β-function). The attractivity of this function is that the constants α and β can be obtained directly from mean and variance of Z,

$$\overline{Z} = \frac{\alpha}{\alpha+\beta} \quad \text{and} \quad \overline{Z'^2} = \frac{\overline{Z}(1-\overline{Z})}{1+\alpha+\beta}. \tag{12.49}$$

It is easily seen that the β-function is a flexible and easy-to-use two-parameter function.

12.8 Eddy-Break-Up Models

Eddy-Break-Up models (see Peters 2000, e. g.) are empirical models for the mean reaction rate in the case of fast chemistry. In this case the reaction rate is governed by the rate of turbulent dissipation ("*mixed is burnt*"). The reaction zone is described as a mixture of unburned and burnt regions.

In analogy to the decay of turbulent energy, a formulation by Spalding (1970) describes the rate which governs the breakup of domains of unburned gas into smaller fragments. These fragments are in sufficient contact with already burnt gases, thus are at sufficiently high temperature, and thus react. For the reaction rate (F = fuel, C_F is an empirical constant of the order of 1) one obtains (Spalding 1970)

$$\overline{\omega}_F = -\frac{\overline{\rho}\, C_F}{\overline{M}} \sqrt{\overline{w_F''^{\,2}}}\, \frac{\tilde{\varepsilon}}{\tilde{k}} \quad , \quad \overline{w_F''^{\,2}} \le \overline{w}_F \cdot (1 - \overline{w}_F); \text{ see Section 17.5.} \qquad (12.50)$$

12.9 Large-Eddy Simulation (LES)

Large Eddy Simulation (*LES*, Reynolds 1989, Pope 2000) means the simulation of a turbulent flow field using direct numerical simulation except that grid points are not extended to the smallest scale. Instead, there are fewer grid points so that the *large scales* are determined with DNS, but the unresolved scales, the *subgrid scales,* are modelled as isotropic turbulence using any of the turbulence models, such as the k-ε model and, more recently, the linear eddy model of Kerstein 1992 (LES-LE). Applications of LES include automotive engine calculations (e. g., with the program KIVA™, Amsden et al. 1989) or weather forecast calculations.

12.10 Turbulent Scales

As mentioned above, turbulent processes occur at different length scales. The largest length scales correspond to the geometrical dimensions of the system (*integral length scale* l_0). Perturbations with long wave lengths (low frequency) are associated with large eddies. These eddies interact and fission into smaller and smaller eddies (smaller wave length means higher frequency). Thus, an *energy cascade,* from large eddies to many small eddies, is observed. The major part of the kinetic energy is in the motion of the large eddies. The energy cascade terminates, as the kinetic energy of the many small eddies (at or below the Kolmogorov-length scale l_K) is dissipated by viscosity into thermal energy (i. e., molecular motion).

The distribution of the turbulent kinetic energy (*TKE*) among the spectrum of eddies with diameters l is described by the *turbulent energy spectrum* (Fig. 12.15). The energy density $e(k)$ describes the dependency of the turbulent kinetic energy (here called q) on the wave number $k = 1/l$, i. e., the reciprocal value of the turbulent length scale,

$$TKE = q(\vec{r},t) = \int_0^\infty e(k;\vec{r},t)\mathrm{d}k \ . \tag{12.51}$$

The energy spectrum begins at the *integral length scale* l_0 (governed by the characteristic dimensions of the experiment) and ends at the *Kolmogorov length-scale* l_K

At the Kolmogorov length l_K, the time for an eddy to rotate 1/2 revolution is equal to the diffusion time across the diameter l_K. Below l_K, diffusion (and in general molecular transport) is faster than turbulent transport; hence turbulence does not extend below l_K.

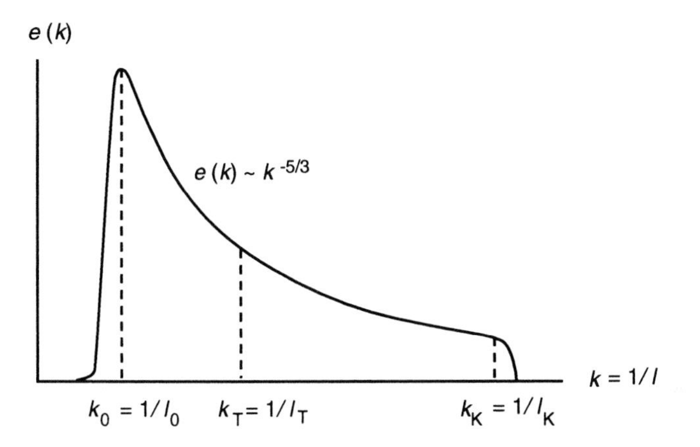

Fig. 12.15. Turbulent energy spectrum showing the energy cascade, left to right

For the special case of fully developed isotropic turbulence, Kolmogorov (1942) derived that the relation

$$e(k) \propto k^{-5/3} \tag{12.52}$$

is valid for completely developed turbulence. This result was verified later by measurements. In an attempt to universally describe the degree of turbulence, the *turbulent Reynolds number* has been proposed (see Williams 1984, Libby and Williams 1994),

$$R_l = \frac{\bar{\rho}\sqrt{2q}\, l_0}{\bar{\mu}} \ , \tag{12.53}$$

which is defined by the integral length scale l_0 and the turbulent kinetic energy q, instead of the mean velocity. R_l is an improvement over the traditional Reynolds number *Re* that is explicitly geometry-dependent (see Section 12.1). The Kolmogorov length l_K can be calculated from the turbulent Reynolds number according to the relation

$$R_l = \left(\frac{l_0}{l_K}\right)^{4/3} . \tag{12.54}$$

Thus, the turbulent Reynolds number is a measure for the ratio between integral length scale and Kolmogorov length scale, and it is evident that the turbulent Reynolds number characterizes turbulent flows better than the Reynolds number *Re*.

One further length scale that is used very often in the description of the dissipation is the *Taylor length-scale* $l_T = l_0/R_l^{1/2}$. A turbulent length scale based on the Taylor length can be defined (see (12.53)) by

$$R_T = \frac{\bar{\rho}\sqrt{2q}\, l_T}{\bar{\mu}} , \tag{12.55}$$

where as a simple relation between R_T and R_l,

$$R_T = \sqrt{R_l} , \tag{12.56}$$

is obtained. In the steady state, the *dissipation rate* of the turbulent kinetic energy (on the right hand side of the spectrum) has to equal the rate of formation of turbulent energy on the left hand side (e. g., by shear processes in the boundary layers, which are the cause of the turbulence).

An analysis of dimensions shows that the dissipation rate depends on the energy q of the spectrum and the integral length scale l_0,

$$\varepsilon = (2q)^{3/2}/l_0 . \tag{12.57}$$

The energy cascade model has contributed greatly to the development of the popular k-ε model.

12.11 Exercises

Exercise 12.1. A plate, heated to the temperature $T_P = 500$ °C is in a gas flow with temperature $T_0 = 0$ °C. The flow is turbulent. For point 2, the temperature is measured 30 times (see table).

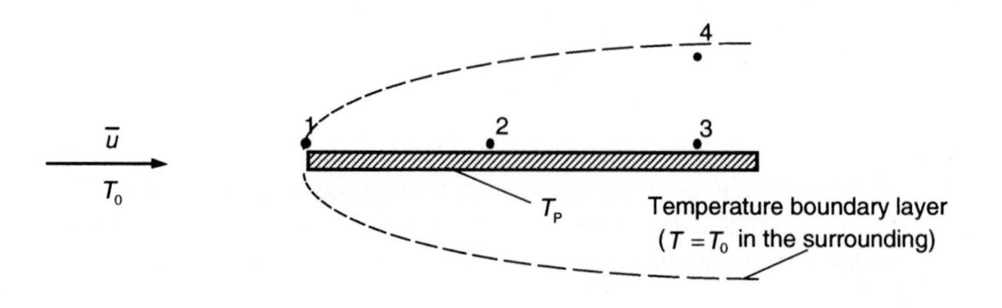

no.	1	2	3	4	5	6	7	8	9	10	11	12	13	14	15
T [°C]	400	392	452	410	363	480	433	472	402	350	210	490	351	421	279

no.	16	17	18	19	20	21	22	23	24	25	26	27	28	29	30
T [°C]	403	404	221	445	292	430	444	370	482	102	412	409	302	480	308

(a) Determine the mean temperature and the probability that the temperature is in the interval 0 - 100 °C, 100 - 200 °C, 200 - 300 °C, 300 - 400 °C, 400 - 500 °C, respectively. Plot the resulting probability density function (replace differentials by differences).

(b) The probability density function at point 2 shall be given by the function

$$P(T) = \frac{1}{100[1-\exp(-S)]} \cdot \exp\left(\frac{T-500^0}{100^0}\right) \qquad (T \text{ in } °\mathrm{C})$$

Show that this function fulfills the normalization condition, and determine the mean temperature. Add the plot of this function to the previous plot of part a).

(c) Plot the PDFs for the points 1, 2, and 4 schematically.

Exercise 12.2. Consider a turbulent pipe flow. What are the values for the Kolmogorov length scale l_K and the specific turbulent energy q for a pipe diameter of 200 mm and a Reynolds number of $Re = 15\,000$? Measurements (Hinze 1972) show that $\sqrt{\upsilon'^2}/\overline{\upsilon}$ is 5% on a pipe centerline. Show the dependency of the spectral density $e(k)$ on the specific turbulent energy (use $\overline{\nu} = \overline{\mu}/\overline{\rho} = 20\ \mathrm{mm^2/s}$).

Exercise 12.3. Turbulent mixing consists of two simultaneously occurring processes: (1) stirring, which increases interfacial surface area, and (2) diffusion, which smooths out interfaces. An appreciation for turbulent mixing is obtained by considering the process depicted in the figure below, where at each "eddy turnover time" a cube (edge length l) is subdivided into 8 cubes. This is analogous to the fission process where large eddies cascade into smaller ones (see Section 12.10).

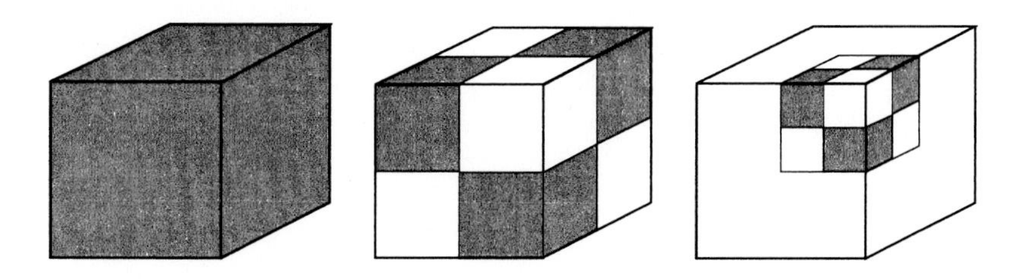

(a) Convince yourself that after N turnovers, the surface area has increased from $6l^2$ to $2^N \cdot 6\,l^2$. Also, the characteristic cube dimension has decreased from l to $l/2^N$. Thus, assuming that the turnover time τ is 1 ms, the area is increased by a factor of a thousand in 10 ms and a 1 cm cube is reduced to many 10 μm cubes.

(b) As noted in the text, the Kolmogorov length scale l_K is the cube dimension at which diffusion obliterates the cube in the same time as the "eddy turnover" process. The length l_K can be calculated from a room temperature gas-phase diffusivity of $D = 0.1\ \text{cm}^2/\text{s}$ and an eddy-turn-over time of 1 ms. Compute the Kolmogorov length, $l_K \approx \sqrt{Dt}$ (see Eq. 3.14).

(c) Compute the time required to reduce a 1 cm cube to Kolmogorov length scale; at this time, the property considered is completely homogeneous. (This example suggests that complete mixing and chemical reaction in less than 10 ms is not unreasonable.)

(d) Assuming only molecular diffusion, no stirring, compute a characteristic mixing time (at atmospheric pressure).

13 Turbulent Nonpremixed Flames

Turbulent nonpremixed flames are of interest in practical applications. They appear in jet engines, Diesel engines, steam boilers, furnaces, and hydrogen-oxygen rocket motors. Except for the turbulent premixed combustion in many spark-ignited engines (Otto cycles), most combustion is turbulent nonpremixed.

Nonpremixed flames are safer to handle than premixed flames, because fuel and oxidizer are mixed in the combustor itself. The widespread use of nonpremixed combustion is the major motivation for the numerous model approaches to their numerical simulation.

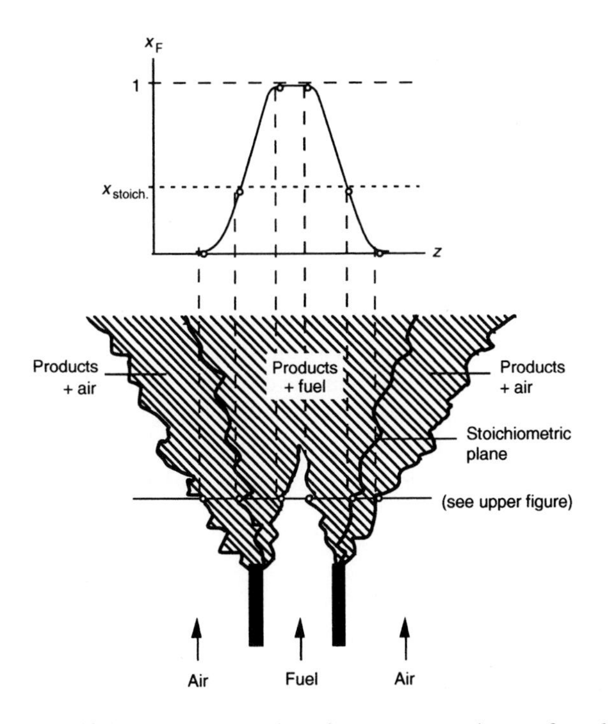

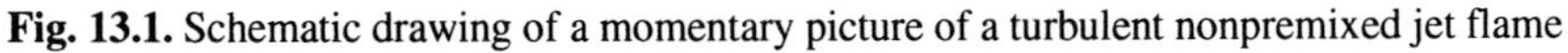
Fig. 13.1. Schematic drawing of a momentary picture of a turbulent nonpremixed jet flame

As shown below, the understanding of laminar nonpremixed flames, discussed in Chapter 9, forms the basis for the understanding of turbulent nonpremixed flames. As noted already in Chapter 9, these flames were historically called *diffusion flames* in the recognition that typically the diffusion of fuel and air toward the flame zone was slow in comparison to the reaction rate between fuel and oxidizer at the flame zone. Thus diffusion dominates in diffusion flames and, by default, one might incorrectly reason that chemical kinetics, not the diffusion process, dominates in premixed flames.

As shown in Chapter 8, the premixed flame speed is a consequence of diffusion along a gradient that is sustained by chemical reaction (e. g,. Liñán and Williams 1993). Since diffusion is essential to both flames, the distinction of premixed and nonpremixed is an improved description.

This chapter will start with an idealized model that illuminates many of the macroscopic features of nonpremixed flames. The shortcomings of the idealized model will then be examined. Next will be a discussion of model improvements. As always, a complete description of the combustion, heat transfer, and pollutant formation is intended. It will be seen that much progress has been made and that much still remains to be done. Reviews that chronicle this progress can be found in the literature (e. g,. Bilger 1976 and 1980, Peters 1987, Libby and Williams 1994, Dahm et al. 1995, Pope 1986, and Takeno 1995).

13.1 Nonpremixed Flames with Equilibrium Chemistry

Much understanding of nonpremixed flames is obtained by assuming that the chemicals react to equilibrium as fast as they mix. With this assumption, all that remains is to compute how fuel mixes with oxidizer. An example of turbulent mixing is shown in Fig. 13.1. The prediction of mixing in isothermal non-reacting turbulent jets is a formidable problem; the additional considerations of variable density and volumetric expansion due to heat release from combustion further complicates the problem.

The mixing problem is greatly simplified when it is assumed that the diffusivities of all scalars are equal. Then all species mix alike, and one can focus on the mixing of a single variable. Because some molecules are consumed, it is better to track the mixing of elements since they are unchanged by chemical reaction. To track elements, one generates a scalar called the *mixture fraction* ξ, as was done for the laminar flame in Section 9.3, by

$$\xi = \frac{Z_i - Z_{i2}}{Z_{i1} - Z_{i2}}. \tag{13.1}$$

By computing the mixing of ξ, the mixing of everything can be computed. For example, a jet flame into air can be viewed as a two-stream problem with the element mass fractions Z_{i1} and Z_{i2} in the two streams. (ξ does not depend on the choice of the element i ($i = 1, \ldots, M$) and ξ depends linearly on the mass fractions w_j because of

(13.1) and $Z_i = \Sigma\mu_{ij}w_j$; see Section 12.5.) In stream 1 the boundary condition $\xi = 1$ and in stream 2 the boundary condition $\xi = 0$ shall be assumed. As mixing proceeds, ξ takes on values between 0 and 1. At any point in the flow, ξ can be regarded as the mass fraction of the fluid material that originated from stream 1 and $1-\xi$ can be regarded as the mass fraction of the fluid material that originated from stream 2.

Using (12.21) and (13.1), a conservation equation for the mixture fraction ξ can be derived,

$$\frac{\partial(\rho\xi)}{\partial t} + \mathrm{div}\left(\rho\vec{v}\xi\right) - \mathrm{div}\left(\rho D \cdot \mathrm{grad}\ \xi\right) = 0\ . \tag{13.2}$$

It should be noted that ξ does not have a chemical source or sink term, ξ is conserved during chemical reaction and hence it is often called a *conserved scalar*. If one further assumes that energy diffuses at the same rate for all species, i. e., Lewis number $Le = \lambda/(D\rho c_p) = 1$, and no heat transfer occurs, then the enthalpy field, as well as the temperature field, can also be uniquely described by ξ (the kinetic energy of the flow is here considered negligible, thus pressure is constant) via

$$\xi = \frac{h - h_2}{h_1 - h_2}\ . \tag{13.3}$$

Thus, assuming (a) equilibrium ("fast") chemistry, (b) equal diffusivity and $Le = 1$ and, (c) no heat loss, all scalar variables (temperature, mass fractions, and density) are known functions of the mixture fraction only. The known function is the equilibrium composition. The turbulent nonpremixed flame problem is now reduced to tracking the turbulent mixing of ξ. This tracking can be done from wide variety of levels including DNS (Reynolds 1989), LES (McMurtry et al. 1992), the Lagrangian integral method (LIM; Dahm et al. 1995) and, quite often, from a PDF description (Pope 1991).

The time average of ξ is obtained (see (12.28)) after time averaging and using the (rather dubious) gradient transport assumption (12.23) with turbulent transport ν_T,

$$\mathrm{div}\left(\bar{\rho}\tilde{\vec{v}}\tilde{\xi}\right) - \mathrm{div}\left(\bar{\rho}\nu_\mathrm{T}\ \mathrm{grad}\tilde{\xi}\right) = 0\ . \tag{13.4}$$

If the PDF of the mixture fraction is known, the mean values of the scalars can be calculated. In this way the coupled system of the averaged conservation equations can be solved, because the mean density is used in (12.24)-(12.28). Ideally, the PDF should be calculated from its own set of conservation equations and associated boundary conditions (Chen et al. 1989). Much simplification is achieved by constraining the PDF to a certain generic shape that is described by two parameters such as the mean and variance of ξ (for example Gaussian function or a β-function, discussed in Section 12.7) Then instead of a conservation equation for the PDF, one needs only a conservation equation for the mean and variance of ξ. An equation for the Favre variance $\widetilde{\xi''^2} = \overline{\rho\xi''^2}/\bar{\rho}$ can be derived from (13.4) by multiplication of (13.4) with $\tilde{\xi}$ and subsequent averaging. One obtains (Bilger 1980)

$$\mathrm{div}(\bar{\rho}\,\tilde{\vec{v}}\,\widetilde{\xi''^2}) - \mathrm{div}(\bar{\rho}\,\nu_\mathrm{T}\,\widetilde{\xi''^2}) = 2\,\bar{\rho}\,\nu_\mathrm{T}\,\mathrm{grad}^2\tilde{\xi} - 2\,\overline{\rho D\,\mathrm{grad}^2\xi''}\ , \tag{13.5}$$

where $\text{grad}^2\xi$ denotes the square of the absolute value of the gradient, $(\text{grad}\,\xi)^T \cdot \text{grad}\,\xi$. The last term in this equation is called *scalar dissipation rate* χ. The scalar dissipation rate dissipates fluctuations in scalars just as viscous dissipation dissipates fluctuations in velocity. This term χ has to be modelled in terms of "known" variables, e. g., by the gradient transport assumption

$$\tilde{\chi} = \overline{2\rho D \text{grad}^2 \xi''} / \bar{\rho} \approx 2D \,\text{grad}^2 \tilde{\xi} \,. \tag{13.6}$$

A PDF $P(\xi;\vec{r})$ can be synthesized from $\tilde{\xi}$ and $\widetilde{\xi''^2}$ (e. g., a β-function). Then all the averages can be calculated, because ρ, w_i and T are known functions (due to the equilibrium assumption) of ξ,

$$
\begin{aligned}
\tilde{w}_i(\vec{r}) &= \int_0^1 w_i(\xi)\, \tilde{P}(\xi;\vec{r})\, \mathrm{d}\xi \\
\tilde{T}(\vec{r}) &= \int_0^1 T(\xi)\, \tilde{P}(\xi;\vec{r})\, \mathrm{d}\xi \\
\widetilde{w_i''^2}(\vec{r}) &= \int_0^1 \left[w_i(\xi) - \tilde{w}_i(\vec{r})\right]^2 \tilde{P}(\xi;\vec{r})\mathrm{d}\xi \\
T''^2(\vec{r}) &= \int_0^1 \left[T(\xi) - \tilde{T}(\vec{r})\right]^2 \tilde{P}(\xi;\vec{r})\mathrm{d}\xi \,.
\end{aligned}
\tag{13.7}
$$

$\tilde{P}$ is a Favre-averaged probability density function, which is obtained from the conventional one by integration over the density,

$$\widetilde{P}(\xi;\vec{r}) = \frac{1}{\bar{\rho}} \int_0^\infty \rho\, P(\rho, \xi;\vec{r})\, \mathrm{d}\rho \,. \tag{13.8}$$

Thus, in review, the equation system includes the conservation equations for density and velocity field (e. g., using the additional equations of the k-ε model), as well as the conservation equations for Favre average $\tilde{\xi}$ and Favre variance $\widetilde{\xi''^2}$ of the mixture fraction ξ. From $\tilde{\xi}$ and $\widetilde{\xi''^2}$ one can synthesize P(ξ). Because of the unique relationship between ξ and all scalars (i. e., the equilibrium chemistry relationship), one can compute the statistics of any scalar. This system of equations allows for investigations of flame length and spatial variation of major species such as temperature and concentration of fuel, oxygen, and water.

However, this flame will never extinguish because of the fast chemistry assumption. Furthermore, soot (Chapter 18) is not predicted, and nitric oxide, an important pollutant discussed in Chapter 17, is erroneously predicted. These are important aspects of flames that one wishes to command.

Accordingly, model enhancements will be explored in the following that include the consequences of finite rate, as opposed to infinite rate chemical kinetics (i. e., equilibrium chemistry).

13.2 Finite-Rate Chemistry in Nonpremixed Flames

If one wishes to relax the fast chemistry assumption, then, in addition to a conservation equation for total mass, energy, and momentum, each species will have a conservation equation with a chemical source term $M_i\omega_i$,

$$\frac{\partial(\rho w_i)}{\partial t}+\mathrm{div}(\rho\vec{v}\,w_i)+\mathrm{div}(\rho D\cdot\mathrm{grad}\;w_i) = M_i\omega_i \quad , \quad i=1,\ldots,S\;, \tag{13.9}$$

where the source term is the sum of all chemical kinetic reactions that involve species i. These kinetic rates depend on other species and, more importantly here, have a nonlinear dependence on both species and temperature as described in Section 12.7. Thus, it is unclear how to form the time average of (13.9).

In principle, if the PDFs of the species mass fractions w_i are known, these equations can be averaged and solved as was done in the preceding section (Gutheil and Bockhorn 1987). This is an enormous undertaking that rapidly exceeds computational limits, as predictions are attempted with greater detail in chemistry, i. e., more species.

As the mixing rate increases, one chemical process will emerge at first to depart from chemical equilibrium. Increasing the mixing rate further will result in another process departing from equilibrium. One by one, processes will depart from equilibrium until the main energy releasing reactions are competing with the mixing rate. As the mixing rate increases further, the temperature begins to depart from equilibrium solution.

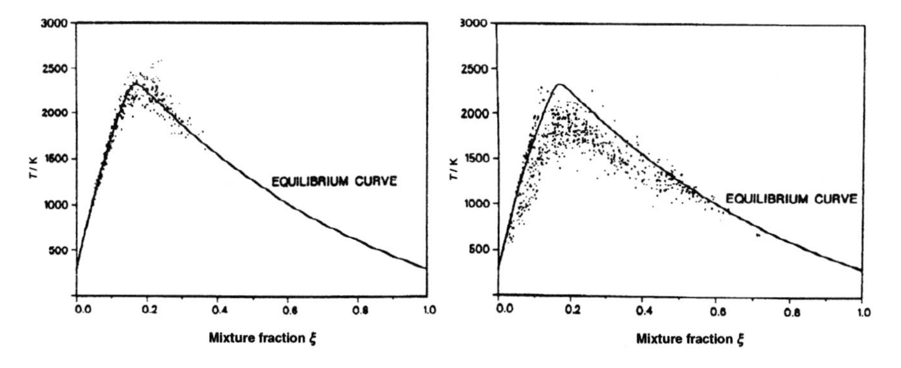

Fig. 13.2. Laser Raman-scatter plots of simultaneous measurement of mixture fraction and temperature in a hydrogen turbulent nonpremixed jet flame where the jet velocity is increased by a factor of three going from the left drawing to the right one (Magre and Dibble 1988)

An example of this is shown in Fig. 13.2 . Moderate departure of temperature from chemical equilibrium is demonstrated here. Left and right scatter plots are from the same flame except for a factor of three increase in hydrogen jet velocity on the right.

The laser Raman-scattering device measures both mixture fraction and temperature. Each microsecond laser-pulse leads to a dot on the figure. As can be seen, the measurements group around the equilibrium line in the left figure. On the right, the fall in temperature shows that the mixing rate, which is movement from right to left on the abscissa, is competing with the heat-releasing chemical rates, which is vertical movement on these figures. The measurements are clearly below the equilibrium line. A further increase in jet velocity leads to sudden global flame extinction.

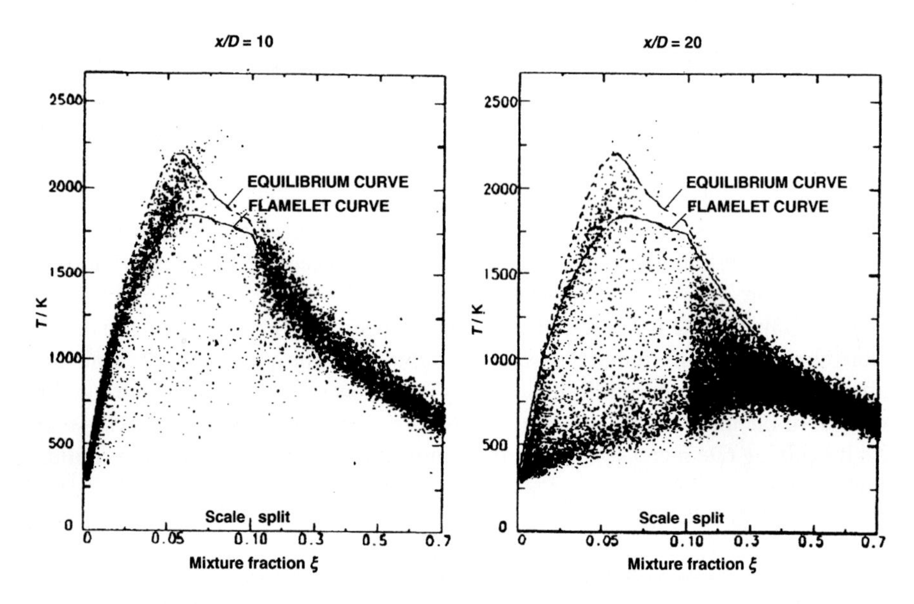

Fig. 13.3. Laser Raman-scatter plots of simultaneous measurement of mixture fraction and temperature in a methane turbulent nonpremixed jet flame at different heights over the burner (Dibble et al. 1987); note the scale change on the ξ-axis

A different behavior is shown in Fig. 13.3. These scatter plots of mixture fraction and temperature show evidence of local flame extinction, internal to the flame. On the left, a methane flame at low mixing rates is shown. On the right is the same flame, but at a different location where air is rapidly mixing with the hydrocarbon fuel. Local flame extinction is manifested by numerous data points being far from the equilibrium line. A further increase in jet velocity results in flame extinction.

Accordingly, a first-level improvement to the equilibrium model is to compute the rate of the first non-equilibrium process of interest only and to assume that the remaining (faster) processes are in equilibrium. This process will depart further from equilibrium as the mixing rate is increased. A parameter is needed that characterizes this departure.

The laminar opposed flow flames of Chapter 9 have solutions that increasingly depart from equilibrium as the mixing rate is increased (as demonstrated in Fig. 3.4).

The mixing rate is characterized by the scalar dissipation rate $\chi = 2\,D\,(\text{grad}\;\xi)^2$, which is related to the strain rate a by (Dahm and Bish 1993, Bish and Dahm 1995)

$$a \;=\; 2\pi D\left[\frac{\text{grad}\xi\cdot\text{grad}\xi}{\left(\xi^+-\xi^-\right)^2}\right]\cdot\exp 2\left\{\text{erf}^{-1}\left[\frac{\xi-\frac{1}{2}\left(\xi^++\xi^-\right)}{\frac{1}{2}\left(\xi^+-\xi^-\right)}\right]\right\}^2 \tag{13.10}$$

in a locally two-dimensional flow. (For the Tsuji geometry in Fig. 9.1, e. g., the strain rate is usually approximated by the potential flow solution $a = 2V/R$.) This equation reflects correctly that, at any strain rate a, the scalar dissipation can be large or small if the difference between maximum ξ^+ and minimum ξ^- is large or small.

It is thus natural to use the scalar dissipation rate as the parameter describing the departure from chemical equilibrium. Accordingly, the improvement to the equilibrium turbulent flame model is to use this unique relationship instead of the equilibrium relationship. This is tantamount to assuming that the turbulent nonpremixed flame is an ensemble of laminar nonpremixed flamelets, each with the same scalar dissipation rate χ. This model shows great improvement. Non-equilibrium amounts of CO, NO, and others are predicted.

The model is further improved, if one admits that the ensemble of laminar flamelets should have a distribution of scalar dissipation rates as a consequence of spatial and temporal distributions in velocities, caused by the underlying rotation of eddies. A model for this distribution will follow.

For a given scalar dissipation rate χ, the concentration of species at any point in the flamelet is a unique function of ξ,

$$w_i = w_i^{(\mathrm{F})}(\xi)\;,\quad \frac{\partial w_i}{\partial t} = \frac{\partial w_i^{(\mathrm{F})}}{\partial \xi}\frac{\partial \xi}{\partial t} \quad\text{and}\quad \text{grad}\; w_i = \frac{\partial w_i^{(\mathrm{F})}}{\partial \xi}\,\text{grad}\;\xi\;.$$

Insertion into the conservation equation for w_i yields (only if the assumptions above are made) the equation (Bilger 1980, Peters 1987)

$$\frac{\partial w_i^{(\mathrm{F})}}{\partial \xi}\left[\frac{\partial(\rho\xi)}{\partial t}+\text{div}(\rho\vec{v}\xi)-\text{div}\left(\rho D\,\text{grad}\xi\right)\right]-\rho D\left(\text{grad}\xi\right)^2\frac{\partial^2 w_i^{(\mathrm{F})}}{\partial \xi^2} \;=\; M_i\omega_i\;.$$

According to (13.4), the term in the square bracket vanishes, and one obtains the *flamelet* equation

$$-\rho\,D\,\text{grad}^2\xi\;\frac{\partial^2 w_i^{(\mathrm{F})}}{\partial \xi^2} \;=\; M_i\omega_i\;. \tag{13.11}$$

Favre averaging leads to the mean reaction rate

$$\overline{M_i\omega_i} \;=\; -\frac{1}{2}\,\bar{\rho}\int_0^1\int_0^\infty \chi\,\frac{\partial^2 w_i^{(\mathrm{F})}}{\partial \xi^2}\;\tilde{P}(\chi,\xi)\,\mathrm{d}\chi\,\mathrm{d}\xi\;, \tag{13.12}$$

where $\chi = 2\,D\,(\text{grad}\;\xi)^2$ is the scalar dissipation rate as a measure of the mixing rate.

The dependence of the mass fractions w_i on the mixture fraction, which has to be known in order to calculate $\partial^2 w_i^{(F)}/\partial\xi^2$, is usually obtained from simulations of laminar nonpremixed flames or from experiment (see Chapter 9). Using (13.12), the equation system (12.24)-(12.28) can be solved, if the PDF $\tilde{P}(\chi,\xi)$ is known.

It is expedient to assume that χ and ξ are *statistically independent*, such that a factorization can be used, $\tilde{P}(\chi,\xi)=\tilde{P}_1(\chi)\cdot\tilde{P}_2(\xi)$ (Peters 1987). Following the idea of Kolmogorov, a log-normal distribution is used for $\tilde{P}_1(\chi)$ (see, e. g., Liew et al. 1984, Buch and Dahm 1996, 1998), whereas a β-function is chosen for $\tilde{P}_2(\xi)$ (see Section 12.7).

The method described so far requires the solution of all the averaged species conservation equations (13.9), where the right hand side of (13.9) is computed using (13.12), and is very time-consuming. Furthermore, the mass fractions, the temperature, and the density are (according to the assumptions above) unique functions of the mixture fraction and the scalar dissipation rate χ. Thus, a more elegant way is to compute the density, the mass fractions, and the temperatures using the PDF for χ and ξ such that – in analogy to (13.7) – one obtains the relations

$$
\begin{aligned}
\bar{\rho}(\vec{r}) &= \int_0^1\int_0^\infty \rho^{(F)}(\chi,\xi)P(\chi,\xi;\vec{r})\,\mathrm{d}\chi\,\mathrm{d}\xi \\
\tilde{w}_i(\vec{r}) &= \int_0^1\int_0^\infty w_i^{(F)}(\chi,\xi)\tilde{P}(\chi,\xi;\vec{r})\,\mathrm{d}\chi\,\mathrm{d}\xi \\
\tilde{T}(\vec{r}) &= \int_0^1\int_0^\infty T^{(F)}(\chi,\xi)\tilde{P}(\chi,\xi;\vec{r})\,\mathrm{d}\chi\,\mathrm{d}\xi \qquad (13.13)\\
\widetilde{w_i''^2}(\vec{r}) &= \int_0^1\int_0^\infty \left[w_i^{(F)}(\chi,\xi)-\tilde{w}_i^{(F)}(\vec{r})\right]^2\tilde{P}(\chi,\xi;\vec{r})\,\mathrm{d}\chi\,\mathrm{d}\xi \\
\widetilde{T''^2}(\vec{r}) &= \int_0^1\int_0^\infty \left[T^{(F)}(\chi,\xi)-\tilde{T}^{(F)}(\vec{r})\right]^2\tilde{P}(\chi,\xi;\vec{r})\,\mathrm{d}\chi\,\mathrm{d}\xi\,.
\end{aligned}
$$

Here $\tilde{P}(\chi,\xi)$ is again a Favre-averaged probability density function, i. e.,

$$
\tilde{P}(\chi,\xi;\vec{r}) = \frac{1}{\bar{\rho}}\int_0^\infty \rho^{(F)}P\left(\rho^{(F)},\chi,\xi;\vec{r}\right)\mathrm{d}\rho^{(F)}\,. \qquad (13.14)
$$

If it is assumed as an approximation that the density $\rho^{(F)}$ depends only on the mixture fraction ξ, the relation between conventional and Favre-PDF

$$
\tilde{P}(\chi,\xi;\vec{r}) = \frac{\rho^{(F)}(\xi)}{\bar{\rho}(\vec{r})}P(\chi,\xi;\vec{r}) \qquad (13.15)
$$

holds. A prerequisite for the use of (13.13) is that the dependences $\rho^{(F)}=\rho^{(F)}(\chi,\xi)$, $w_i^{(F)}=w_i^{(F)}(\chi,\xi)$, $T^{(F)}=T^{(F)}(\chi,\xi)$ are known from calculations of laminar nonpremixed flame fronts. Thus, *libraries* of flame structures $\rho^{(F)}=\rho^{(F)}(\xi)$, $w_i^{(F)}=w_i^{(F)}(\xi)$, $T^{(F)}=T^{(F)}(\xi)$ are needed for different dissipation rates χ. The computation of such libraries with hundreds of different laminar flame structures (e. g., for different pressures and unburned gas temperatures) is very time-consuming, but has to be done only once. Then the calculation of the averages according to (13.13) is quite easy and straightforward (see Rogg et al. 1987, Gill et al. 1994).

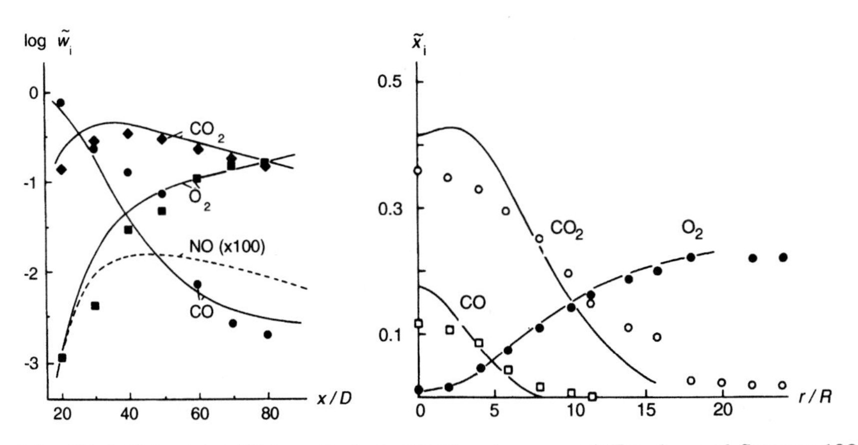

Fig. 13.4. Calculated (Behrendt et al. 1987) and measured (Razdan and Stevens 1985) concentration profiles in a turbulent CO-air jet diffusion flame; shown are radial mean mass fraction (left) and axial mean mole fraction (right) profiles; $D = 2R$ = nozzle diameter

The model described above yields quite good results at a small expense, if the flamelet assumption is fulfilled. In order to demonstrate this, Fig. 13.4 shows measured (Razdan and Stevens 1985) and calculated (using the flamelet model, Behrendt et al. 1987) concentration profiles in a turbulent jet nonpremixed flame of CO in air. The CO-air system has the advantage that the assumption of equal diffusivities is fulfilled quite well and that the temperature is not decreased by radiation from soot.

13.3 Flame Extinction

Laminar counterflow nonpremixed flames had been discussed in Chapter 9. It was explained that characteristic parameters such as flame temperatures depend strongly on the scalar dissipation rate χ, describing the mixing rate which is related to the strain rate a by (13.10).

If χ is high enough, laminar nonpremixed flames extinguish. This behavior is shown in Fig. 13.5 (Tsuji and Yamaoka 1967). The flame is "blown out" above a critical dissipation rate χ_q (corresponding to a critical flow velocity V of the air). f_w is a dimensionless outflow parameter, which can be calculated from the velocity V of the air, the outflow velocity of the fuel v_w , the Reynolds number Re and the cylinder radius R. The strain rate is approximately given by $a = 2V/R$ (discussed in Chapter 9).

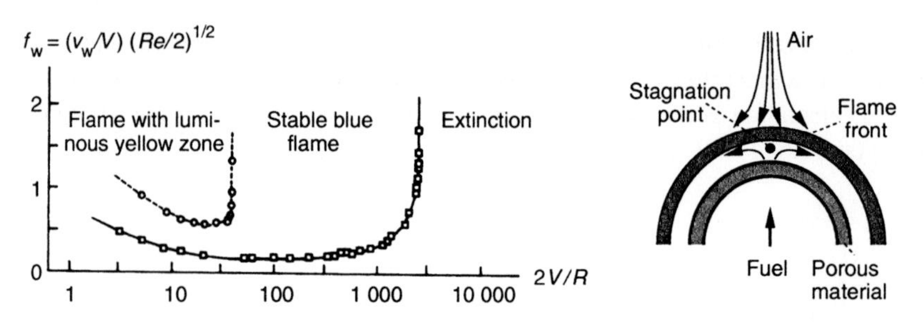

Fig. 13.5. Stability diagram of a laminar counterflow diffusion flame measured by Tsuji and Yamaoka 1967 (left) and burner configuration used (right)

Figure 13.6 shows calculated temperature profiles for some scalar dissipation rates χ in a counterflow nonpremixed flame. The maximum flame temperature decreases with increasing scalar dissipation rate. For scalar dissipation rates larger than a critical χ_q (here $\chi_q = 20.6\ s^{-1}$; q stands for "quenching"), extinction is observed (Rogg et al. 1987).

The temperature is dropping because the convective-diffusive heat removal rate is increasing while, at the same time, the rate of heat generation is decreasing due to the reduced reaction rate and to the reduced residence time in the flame zone. The abrupt extinction is entirely analogous to the ignition-extinction analysis of Chapter 10.1 (see especially Fig. 10.1). As for nonpremixed flames, anemic flames near extinction are sensitive to Lewis numbers $Le = D\rho c_p/\lambda$, i. e., the ratio of thermal diffusivity to mass diffusivity (Tsuji and Yamaoka 1967; Peters and Warnatz 1982).

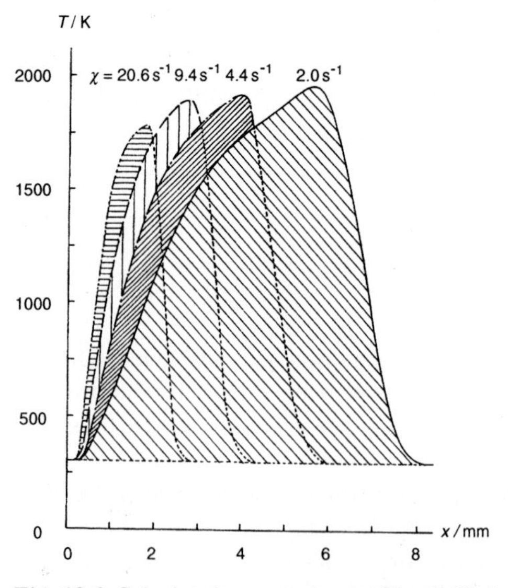

Fig. 13.6. Calculated temperature profiles (left) in a methane-air counterflow nonpremixed flame (right) for different scalar dissipation rates χ (Rogg et al. 1987); extinction occurs at $\chi > 20.6\ s^{-1}$; unburnt gas temperatures are $T = 298$ K on fuel and oxidizer side; $p = 1$ bar

The lift-off of turbulent flames which is shown in Fig. 13.7 can be explained by extinction due to scalar dissipation. The scalar dissipation is highest near the nozzle, where the scalar ξ take on its maximum value ξ^+ and minimum value ξ^- and the strain rate is largest. Thus, extinction is observed at this location. The mean luminescent flame contour shows a lift-off, which increases with increasing jet velocity. The practical importance of this fact lies in the possibility to optimize extinction (e. g., of burning oil wells) by cooling of the lower part of the flame, where the tendency to extinguish is highest due to the high strain.

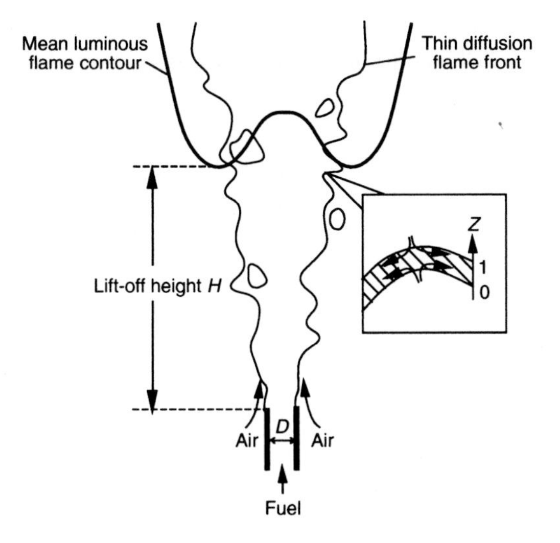

Fig. 13.7. Schematic illustration of the lift off behavior of a turbulent jet nonpremixed flame; the inset box depicts how the laminar opposed-jet flame front is mapped into the turbulent flow field

Modelling turbulent nonpremixed flames, extinction processes can be accounted for, if the integration over the scalar dissipation rates for the determination of the means of density, temperature, and mass fractions is only performed over the interval, where no extinction occurs, e. g.,

$$\tilde{T}(\vec{r}) = \int_0^1 \int_0^{\chi_q} T^{(F)}(\chi,\xi)\tilde{P}(\chi,\xi;\vec{r})\mathrm{d}\chi\mathrm{d}\xi + \int_0^1 \int_{\chi_q}^{\infty} T_u(\chi,\xi)\tilde{P}(\chi,\xi;\vec{r})\mathrm{d}\chi\mathrm{d}\xi \ . \qquad (13.16)$$

Analogous expressions can easily be written for the other properties evaluated in Eqs. (13.13).

Extinction in nonpremixed flames is followed by local premixing of reactants. This leads to the very complex case of a partially premixed turbulent flame, where a further variable in the PDF is needed to describe the degree of premixedness (Rogg et al. 1987). The processes in turbulent premixed flames are the subject of Chapter 14.

13.4 PDF-Simulations of Turbulent Non-Premixed Flames

It was pointed out in Chapter 12 that the *closure problem* for the chemical source term would be solved, if the joint PDF of the scalar variables were known. Thus, in one approach, the shape of the PDF is a prescribed analytic function. Examples include the "clipped" Gaussian and the β-functions. These functions are determined by the mean and the variance of a variable. From the Navier-Stokes equations, a conservation equation for each of these two moments is generated and then solved. Much progress has been made with analytic PDFs (e. g., Libby and Williams 1994); however, the actual PDF often has features that are poorly represented by two-parameter analytic functions. In principle, any PDF can be represented by a sum of weighted moments of the PDF. In practice, generating, and then solving, additional conservation equations for the higher moments has been impractical.

The shape of the joint scalar PDF is a consequence of the fluid mixing and reaction and hence the PDF, in principle, is generated by solving the Navier-Stokes equations. Starting from these equation, a conservation equation for the *joint PDF of velocities and scalars* can be derived (Pope 1986). Let the one-point joint PDF

$$f\left(\upsilon_x, \upsilon_y, \upsilon_z, \Psi_1, \ldots, \Psi_n; x, y, z, t\right) \mathrm{d}\upsilon_x \mathrm{d}\upsilon_y \mathrm{d}\upsilon_z \mathrm{d}\Psi_1 \ldots \mathrm{d}\Psi_n$$

denote the probability at time t and at the one spatial point x, y, z, that the fluid has velocity components in the range between υ_i and $\upsilon_i + \mathrm{d}\upsilon_i$ and values of the scalars (mass fractions, density, enthalpy) in the range between Ψ_α and $\Psi_\alpha + \mathrm{d}\Psi_\alpha$. Then the conservation equation, which describes the evolution of the PDF, reads (Pope 1986, Pope 1991)

$$\rho(\vec{\Psi})\frac{\partial f}{\partial t} + \rho(\vec{\Psi})\sum_{j=1}^{3}\left(\upsilon_j \frac{\partial f}{\partial x_j}\right) + \sum_{j=1}^{3}\left(\left[\rho(\vec{\Psi})g_j - \frac{\partial \bar{p}}{\partial x_j}\right]\frac{\partial f}{\partial \upsilon_j}\right) + \sum_{\alpha=1}^{n}\left(\frac{\partial}{\partial \psi_\alpha}\left[\rho(\vec{\Psi})S_\alpha(\vec{\Psi})f\right]\right)$$

(13.17)

$$= \sum_{j=1}^{3}\left(\frac{\partial}{\partial \upsilon_j}\left[\left\langle \frac{\partial p'}{\partial x_j} - \sum_{i=1}^{3}\frac{\partial \tau_{ij}}{\partial x_i}\middle| \vec{\upsilon}, \vec{\Psi}\right\rangle f\right]\right) + \sum_{\alpha=1}^{n}\left(\frac{\partial}{\partial \psi_\alpha}\left[\sum_{i=1}^{3}\left\langle \frac{\partial J_i^\alpha}{\partial x_i}\middle| \vec{\upsilon}, \vec{\Psi}\right\rangle f\right]\right),$$

where the x_i denote the x, y, and z coordinates, respectively, the g_i denote the gravitational forces in x, y, and z direction, $\vec{\Psi}$ denotes the n-dimensional vector of the scalars, υ_j the components of the velocity vector $\vec{\upsilon}$, S_α the source terms for the scalars (e. g., the chemical source terms), τ_{ij} the components of the stress tensor, and J_i^α the components of the molecular flux (e. g., diffusion or heat conduction) of scalar α in i-direction. The terms $\langle q|\vec{\upsilon}.\vec{\Psi}\rangle$ denote *conditional expectations* of the variable q. Thus, $\langle q|\vec{\upsilon}.\vec{\Psi}\rangle$ is the average of q provided that the velocity and the composition have the values $\vec{\upsilon}$ and $\vec{\Psi}$ respectively. In physical terms, these conditional expectations determine the averages of the molecular fluxes for given values of the velocity and the scalars.

The first term on the left hand side describes the change of the PDF with time, the second describes convection which is transport in physical space, the third describes the transport in velocity space due to gravitation and pressure gradients, and the forth term describes the transport in the composition space due to source terms (e. g., chemical reaction). It is important to note that all terms on the left hand of the equation appear in closed form. In particular, the main advantage of the PDF approach is that the chemical reaction is treated exactly by the method.

However, the conditional expectations $\langle q | \vec{v}. \vec{\Psi} \rangle$ of the molecular flux terms on the right hand side do not appear in closed form; they have to be modeled. One is forced to postulate a functional relationship between the molecular flux and the "known" (i. e., calculated) quantities. The need for this model is an artifact of assuming that the one-point PDF is a sufficient description of the flow. Such one-point descriptions are potentially limited in that they carry no information about spatial correlations.

Even with the simplification afforded by neglect of spatial correlations, the resulting single-point PDF evolution equation (13.17) is not readily solved with present computers. The problem in solving (13.17) stems from its high dimensionality. Whereas in the Navier-Stokes equations only the spatial coordinates and the time are independent variables, all scalars and the velocity components enter the PDF evolution equation (13.17) as independent variables. Thus, the difficulties in solving the Navier-Stokes equations will be greatly increased when solving the more numerous PDF transport equations.

The *Monte-Carlo method* has emerged as a solution to this problem. In this approach, the PDF is represented by a large number (e. g., 100 000 for two-dimensional systems) of *stochastic particles*. These particles evolve in time according to convection, chemical reaction, molecular transport, and body forces, and thus mimic the evolution of the PDF (see Pope 1986 for details).

In practical applications, the joint PDF for scalars and velocity $f(\vec{v}, T, w_i, \rho)$ is reduced to a PDF for scalars to treat chemical reaction exactly, and the velocity field is computed by a standard turbulence model (e. g., the k-ε model) based on the averaged Navier-Stokes equations (the *turbulent flow model*).

The two models are coupled via the density ρ. The PDF model supplies a density field, which enters the flow model. Then a new flow field is calculated and the information is passed back to the PDF-code. This procedure is repeated until the solution has converged (Nau et al. 1996).

Such hybrid PDF/turbulent flow simulations offer a realistic description of turbulent flames. As an example, Fig. 13.8 shows a comparison of experimental results, obtained from a recirculating methane-air nonpremixed flame, with a simulation based on a combined PDF/turbulent flow model, where an ILDM model for detailed chemical reaction (described in Section 7.4) is used. The agreement is quite good. The model is also a clear improvement over an *eddy-dissipation model* (sophisticated EBU model, see Section 12.8), which assumes reaction to be much faster than mixing, and hence only computing the mixing rate (the *eddy-break-up rate*). A study of Fig. 13.8 shows that the fast chemistry assumption grossly overpredicts the amount of product formation and, consequently, overpredicts the temperature increase. Thus, the predicted nitric oxide NO will be far to large.

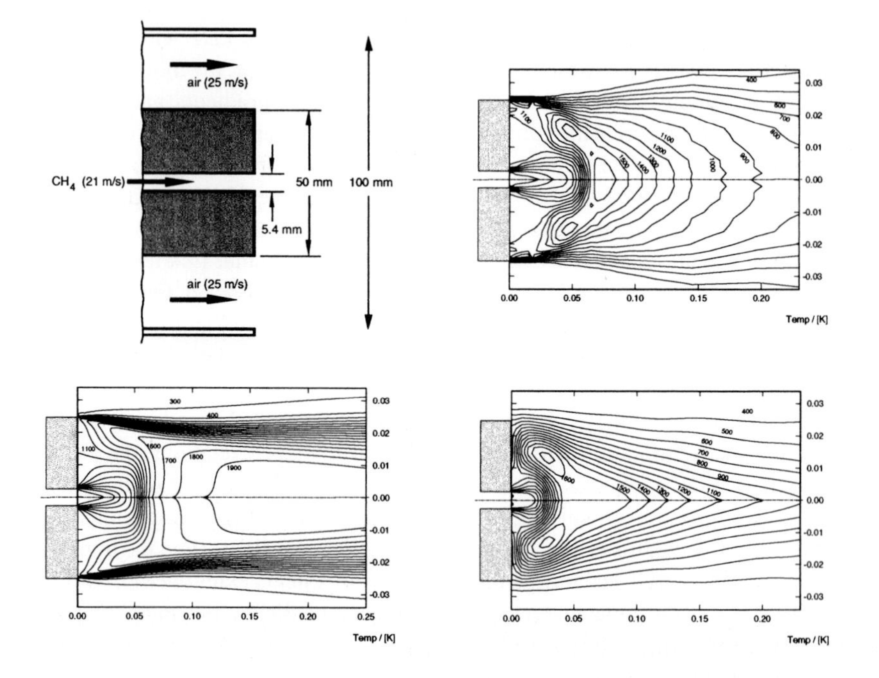

Fig. 13.8. Nonpremixed CH_4-air jet flame (Nau et al. 1996); (top left) configuration; (top right) experimental temperature, $T_{max} \approx 1600$ K (Perrin et al. 1990); (bottom left) eddy-dissipation model, $T_{max} \approx 1900$ K; (bottom right) combined PDF/turbulent flow model, $T_{max} \approx 1600$ K

13.5 Exercises

Exercise 13.1. Consider a cylinder, open at both ends, where oxidizer O is flowing along one face, and fuel F along the other., reacting according to F + 2 O → 3 P.

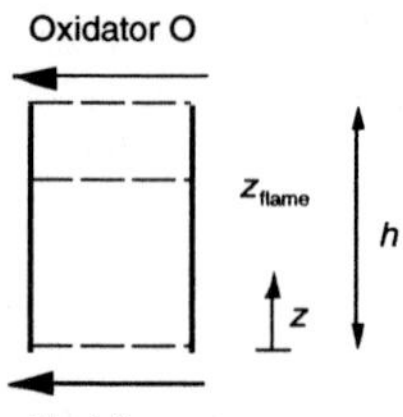

(a) Which condition determines the position of the flame front in the laminar case, where is the flame front? Draw the profiles of fuel, oxidizer, and product over the mixture fraction and over the height z.
(b) Now the combustion shall be assumed to be turbulent. Which diagram from a) is still valid, and which can no longer be used?

14 Turbulent Premixed Flames

This chapter will discuss *turbulent premixed flames*. The distinction between premixed flames and nonpremixed flames is made clear by reviewing the ideal case of each. The ideal nonpremixed flame has fast (equilibrium) chemistry that rapidly adjusts to the local mixture fraction; the mixture fraction is constantly changing. The unburnt gas in an ideal premixed flame is completely mixed before chemistry begins. Then the ideal premixed flame has a delta function PDF for mixture fraction with chemistry that suddenly evolves from unburnt to burnt at the interface between reactants and products; the interface propagates with a speed v_L.

The distinction between premixed and nonpremixed is not always clear when mixing times and chemical kinetic times become competitive. For example, local flame extinction in a nonpremixed flame may allow fuel and air to mix, before being ignited by an adjacent nonpremixed flame zone (leading to *partially premixed combustion*). Other distinctions can be made.

Nonpremixed flames establish themselves at the interface between fuel and oxidizer; the flame is sustained by diffusion on each side. The flame does not propagate and moves only as fuel and air are convected by, sometimes turbulent, fluid motion. In contrast, premixed flames have reactants on one side of the flame only (burnt products on the other side). The product water appears on one side of the premixed flame whereas water appears on both sides of a nonpremixed flame.

The motion of the premixed flame is a superposition of flame propagation and, sometimes turbulent, fluid convection. Simply stated, the consequence of this superposition is that turbulent premixed flame modeling is significantly more challenging than modeling of turbulent nonpremixed flames.

14.1 Classification of Turbulent Premixed Flames

An illustration of a *premixed flame* in *turbulent flow* is shown in Fig. 14.1. Premixed fuel and oxidizer is flowing upward. A premixed flame is stabilized by the recirculation of hot gas behind a bluff-body. The flame propagates from this bluff-body into the oncoming unburnt fuel-air mixture. If the oncoming flow were laminar, the premixed flame would form a planar "V". The framework of the previous chapters allows one to calculate a laminar flame speed that can be used with (1.8) to predict the angle of the "V". However, if the oncoming stream is turbulent, then the flame angle changes depending on the local approach velocity of the reactants. As a consequence, the premixed flame takes on a shape as depicted in Fig. 14.1.

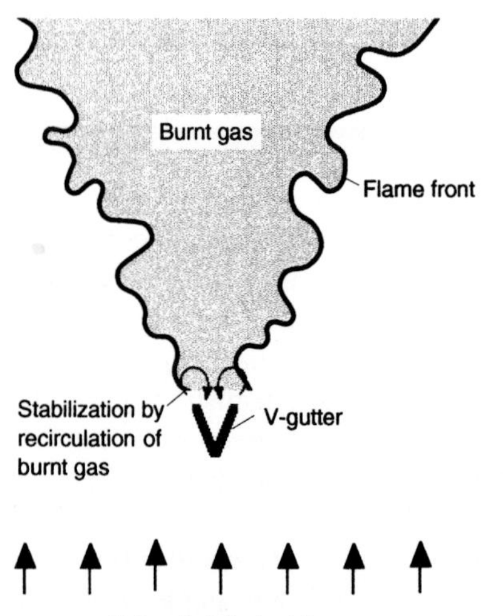

Fig. 14.1. Schematic illustration of an instantaneous picture of a "V-shaped" turbulent premixed flame stabilized by a bluff-body

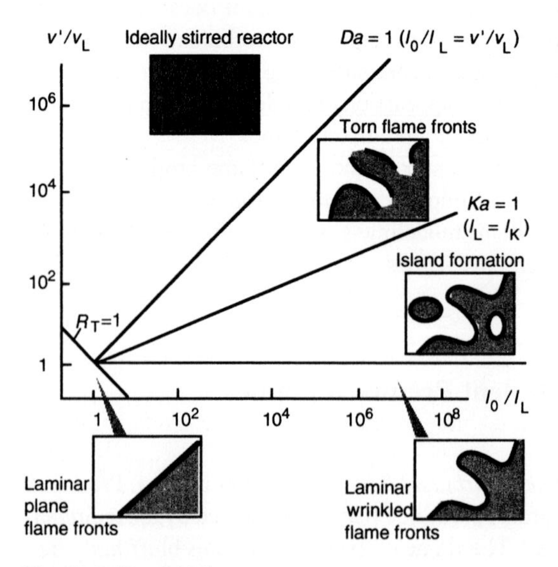

Fig. 14.2. Borghi Diagram

The departure of the turbulent premixed flame front from a plane to an increasingly three-dimensional structure is the subject of the *Borghi diagram* (Borghi 1984; Candel et al. 1994; Poinsot et al. 1991) presented in Fig. 14.2 in a double-logarithmic form. Plotted is υ'/υ_L which is the oncoming turbulence intensity υ' normalized by the laminar burning velocity υ_L, versus l_0/l_L, which is the largest turbulent eddy length scale l_0, normalized by the laminar flame thickness l_L. (Recall from Chapter 12 that the velocity fluctuations υ' are a consequence of vortical motion in the flow and that $\upsilon' = \sqrt{2k_0/\bar{\rho}_0}$, where k_0 is the turbulent kinetic energy and ρ_0 is the density of the oncoming stream.)

Several lines divide the diagram into different domains with different flame behavior. When the turbulent Reynolds number $R_l = \upsilon' \, l_0/\nu$ defined in Eq. (12.53) is smaller than one, $R_l = R_T^2 < 1$, laminar combustion is observed. The domain of turbulent combustion ($R_l = R_T^2 > 1$) is further divided into three zones. For this, two dimensionless ratios are valuable, namely the *turbulent Karlovitz number Ka* and the *turbulent Damköhler number Da*.

The turbulent Karlovitz number *Ka* denotes the ratio between the time scale t_L of the laminar flame ($t_L = l_L/\upsilon_L$) and the Kolmogorov time scale t_K,

$$Ka = \frac{t_L}{t_K} \qquad \text{with} \qquad t_K = \sqrt{\frac{\nu}{\tilde{\varepsilon}}} \,, \tag{14.1}$$

where ν is a characteristic kinematic viscosity ($\nu = \mu/\rho$) and $\tilde{\varepsilon}$ is the dissipation rate of turbulent energy (12.35). At the Kolmogorov scale, the time for an eddy of size l_K to rotate is equal to the time to diffuse across the eddy. For dimensions below l_K, the flow is laminar (Peters 1987). When the flame thickness is less that the Kolmogorov scale, the system is described as a locally laminar premixed flame embedded in a turbulent flow. On the Borghi diagram, this *flamelet regime* is below the $Ka = 1$ line.

The turbulent Damköhler number *Da* describes the ratio between the macroscopic time scales and the time scale of the chemical reaction,

$$Da = \frac{t_0}{t_L} = \frac{l_0 \, \upsilon_L}{\upsilon' \, l_L} \,. \tag{14.2}$$

When $Da < 1$, the time needed for chemical change is greater that the time needed for fluid motion induced change. In this regime, nearly all of the turbulent eddies are embedded in the reaction zone, which is so broad that the term "flame front" is not useful. On the Borghi diagram, this regime is above the line defined by $Da = 1$. This zone, called the *well-mixed reactor* or *perfectly stirred reactor* regime, will be discussed later.

Between the stirred reactor zone and the flamelet zone is the *distributed reaction zone*, where a fraction of the eddies are embedded in flame front, i. e., eddies that have length scales less than l_K. In any given turbulent flow, there is a wide range of dissipation $\tilde{\varepsilon}$; it appears that $\tilde{\varepsilon}$ is log-normally distributed (Buch and Dahm 1996). Thus, a turbulent premixed flame is not represented by a point on the Borghi diagram, but by a zone that may cross boundaries.

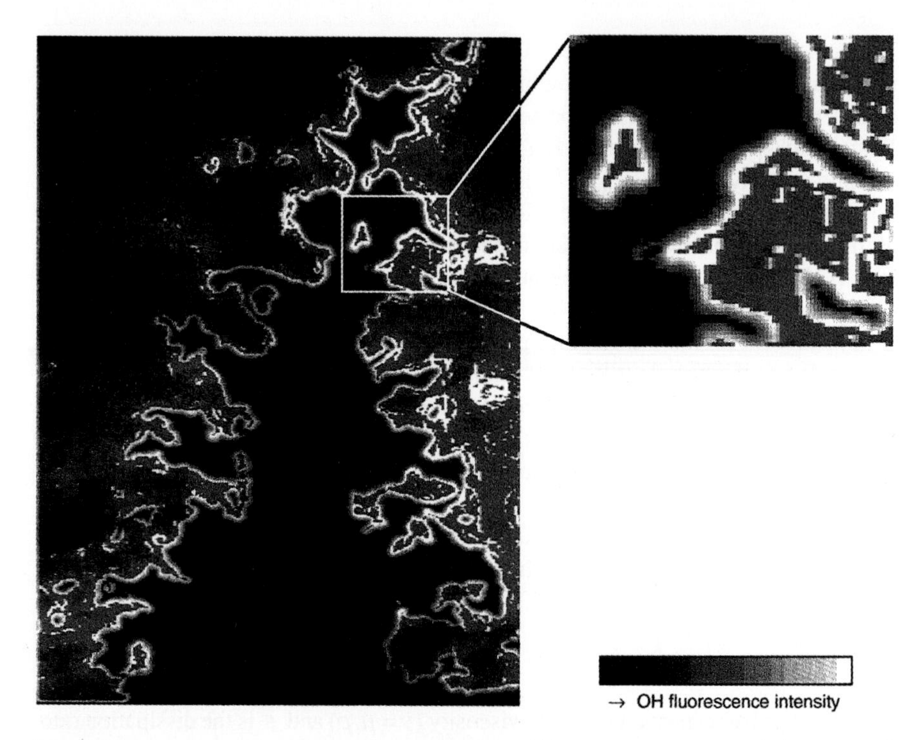

Fig. 14.3. Laser light-sheet LIF measurement of the OH-concentration in a turbulent natural gas-air premixed jet flame stabilized on a nozzle with 3 cm diameter (Dinkelacker et al. 1993); the black interior shows the region of incoming unburnt turbulent mixture ($\Phi = 0.8$, $R_l = 857$, $Ka = 0.07$)

14.2 Flamelet Models

The tools of the first eight chapters allows prediction, from first principles, of properties of laminar premixed flame. The predicted properties include profiles of temperature and species, including pollutants, as well as flame speed. The laminar flame problem is one-dimensional in space and stationary in time. As the previous section demonstrates, the turbulent premixed flame is unsteady in time and is three-dimensional. The analogous first-principle calculation would be a direct numerical simulation (DNS). As indicated in Section 12.2, such calculations far exceed present day computational ability. The practical alternative has been, and continues to be, the development of models that allow prediction of essential features. Much computation is eliminated by including physical insight in the form of a submodel, an example of which is the *flamelet model.*

The flamelet model of premixed flames is analogous to the flamelet model of non-premixed flames. It describes a turbulent premixed flame as a laminar premixed flame imbedded in a turbulent flow field. As the turbulent Reynolds number R_l goes to zero, this model correctly evolves to a laminar premixed flame. There is general agreement that the flamelet concept is applicable in the zone of large Damköhler numbers with turbulent scales larger than the flame thickness; this is the lower right part of the Borghi diagram (Section 14.1).

In the flamelet regime, the central problem is quantifying flame speed. As noted in Chapter 8, the flame speed is a consequence of diffusion sustained by chemical reaction. Thus one needs to know the chemical reaction rate. Usually a *reaction progress variable* c is used, describing the progress of combustion in a premixed flame front and (like the mixture fraction) has values between 0 and 1 (Bray and Libby 1976, Bray and Moss 1977, Bray 1980). Temperature is often used as an indicator of reaction progress, however, other scalars can be used, e. g., the formation of a final product, such as CO_2,

$$w_{CO_2} = c \cdot w_{CO_2,b} , \tag{14.3}$$

where the index b denotes the burnt gas. The scalar profile used must not have a maximum between c_{max} and c_{min} because the definition of c would not be unique in this case. The scalars, such as OH, O_2, CO, CO_2, etc., at any point in the flow can be characterized by the reaction progress variable c and, if necessary, by the local dissipation of c.

Laminar premixed flames with prescribed dissipation rates can be studied in an opposed-flow configuration experimentally (Law 1989) and numerically (Stahl and Warnatz 1991) using 1D conservation equation. Analogous to turbulent nonpremixed flames, the hope is that these turbulent premixed flames can be described as an ensemble of these laminar premixed flames.

Encouragement of use of the flamelet assumption in turbulent premixed combustion can be seen from laser light sheet experiments. Figure 2.7 shows measurements in an Otto engine of OH-concentrations (about 0.3 mol-%) in the flame front, which fluctuates due to vortical motion in the velocity field. The locally wrinkled flame fronts can clearly be seen. Note especially that the flame front is nearly everywhere the same.

A second example is presented by Fig. 14.3. In this turbulent Bunsen flame, the flamelet assumptions, again, appears justified. The figure shows a 2D-LIF measurement of the OH-concentration in a turbulent natural gas/air premixed flame stabilized on a semi-industrial scale burner (Dinkelacker et al. 1993). The wrinkled flame structures can clearly be identified. Importantly, one can see the overshoot of OH at the flame interface indicating that a single progress variable may describe all scalars in the flow.

With the flamelet submodel assumed, a model is required to describe the transport and evolution of c. With c as input, the flamelet submodel returns temperature, density, and species concentration that is used by the fluid dynamic submodel (the turbulence model). There are several significant variations as to how the flamelet submodel is coupled to the turbulence model; these variations are reviewed by, e. g., Ashurst (1995), Candel et al. (1994), Pope (1991), Libby and Williams (1994), and Peters (1987).

With the flamelet submodel assumed, a model is required to describe the transport and production of c. The most primitive of these models, the eddy-break-up (EBU) model, will now be discussed as it illustrates how any of the flamelet models are coupled to the turbulence model. The EBU model (introduced in Chapter 12) specifies a product generation rate $\overline{\omega}_c$ in terms of the turbulence frequency (an inverse eddy turnover time) and product fluctuation intensity c'_{rms}. Integration of

$$\overline{\omega}_c = -\frac{\overline{\rho}\, C_c}{\overline{M}} c'_{\mathrm{rms}} \frac{\tilde{\varepsilon}}{\tilde{k}} \qquad (14.4)$$

yields the product mean $\bar{c}$ from which density ρ is computed from the flamelet model, e. g., $c = (\rho - \rho_{\mathrm{u}})/(\rho_{\mathrm{b}} - \rho_{\mathrm{u}})$, where b = burnt and u = unburnt. This density is required by the turbulence model.

This early model has the right property that the reaction rate is zero in the reactants and in the products, but, as shown in Chapter 13, the model overpredict reaction rates, especially when the laminar flame speed is zero, where the reaction should be zero. Furthermore, while the model predicts a mean fuel consumption rate, no information is predicted about pollution emission, which is increasingly demanded from a model.

There are several significantly improved flamelet submodels. The models are reviewed by Ashurst (1995), Candel et al. (1994), Pope (1991), Libby and Williams (1994), and Peters (1987).

14.3 Turbulent Flame Velocity

One goal of a turbulent premixed combustion model is prediction of the mean fuel-consumption rate. This is tantamount to predicting the flame angle in Fig. 14.1 and 14.3. One would expect that sufficient input parameters would be Φ, T, $\overline{v}$, and v'. By analogy to the laminar flame, the propagation of a turbulent premixed flame has routinely been described by a *turbulent flame velocity* v_{T}. Damköhler (1940) proposed a pioneering model for this velocity assuming the turbulent flame to be a wrinkled laminar flame. Using

$$\rho_{\mathrm{u}}\, v_{\mathrm{T}} A_{\mathrm{T}} = \rho_{\mathrm{u}}\, v_{\mathrm{L}} A_{\mathrm{L}}\,, \qquad (14.5)$$

where A_{L} denotes the overall area of the wrinkled laminar flame fronts, A_{T} the area of the mean turbulent flame front, and v_{L} the laminar flame velocity (see Fig. 2.7), one obtains the basic relation for the turbulent flame velocity (Fig. 14.4)

$$v_{\mathrm{T}} = v_{\mathrm{L}} A_{\mathrm{L}} / A_{\mathrm{T}}\,. \qquad (14.6)$$

The ratio between v_{T} and v_{L} is given by the ratio between the laminar and (mean) turbulent flame area. Damköhler used, e. g., $A_{\mathrm{L}} / A_{\mathrm{T}} = 1 + v'/v_{\mathrm{L}}$, where v' denotes the root mean square of the fluctuation of the velocity (see Section 14.1). In this way, one obtains

$$v_T = v_L + v' = v_L(1 + v'/v_L) . \qquad (14.7)$$

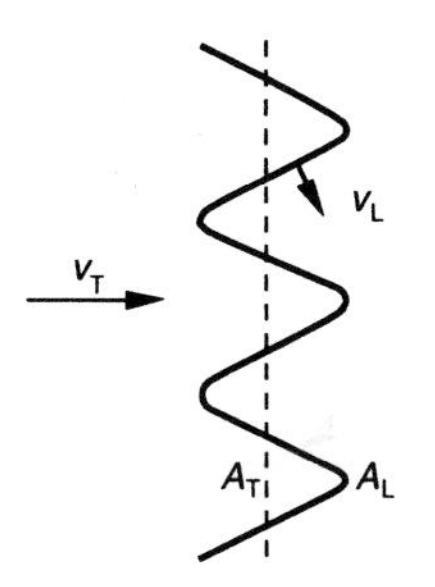

Fig. 14.4. Schematic illustration of the propagation of a premixed flame front into a turbulent mixture.

Like the eddy-break-up model, this model has the correct trend (see Fig. 14.6) as long as the turbulence intensity is not so high that flame extinction takes place (see Section 14.5). In particular, since v' is related to the piston velocity, this model predicts an increase in the piston speed, related to an increase of the speed of an engine (thus v' is proportional to the number of revolutions per minute) increases the turbulent burning velocity. Combustion in piston engines would be restricted to a single rotation speed, if this effect did not exist (Heywood 1988).

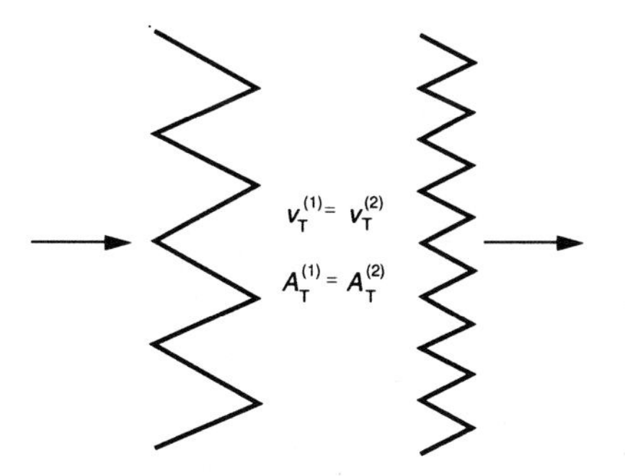

Fig. 14.5. Schematic illustration of two thin flame fronts with different length scales but the same area.

Another phenomenon which is supported by experiments in turbulent premixed flames (Liu and Lenze 1988) is the fact that Eq. (14.5) does not essentially depend on the turbulent length scale (e. g., the integral length scale l_0). This result is consistent with a simple picture given in Fig. 14.5. Although both flame fronts have different length scales, the total area of the laminar flame fronts, and thus the turbulent flame velocity, are the same.

A major failure of the model appears when the premixed mixture is too lean or too rich, to propagate a flame, i. e., beyond the limits of flammability (shown in Fig. 14.7). In this case, v_L is zero so there is no flame, yet the model shamelessly predicts v_T equal to v'. The prediction of the average cone angle in Fig. 14.1 or 14.3 remains as a challenge to models of premixed combustion.

14.4 Flame Extinction

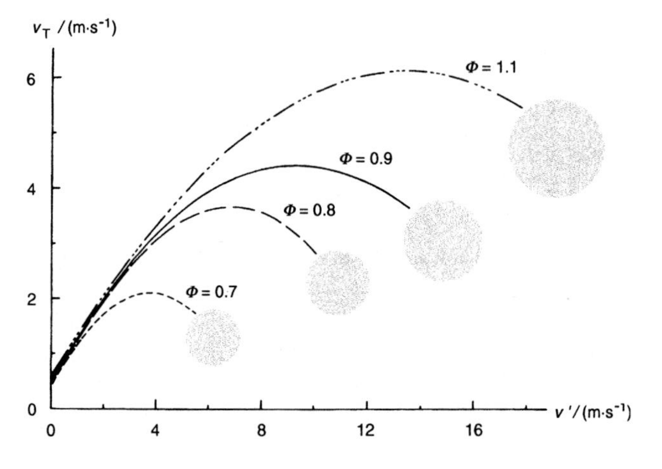

Fig. 14.6. Dependence of the turbulent flame velocity on the turbulence intensity in the combustion of a C_3H_8-air mixture (Abdel-Gayed et al. 1984); hatched: quenching regime

As the turbulence intensity v' increases, a maximum in the turbulent flame velocity v_T is observed. Premixed flame *extinction processes* are being manifested. This has been shown, e. g., by Bradley and coworkers (Abdel-Gayed et al. 1984, Bradley 1993) in a combustion vessel filled with C_3H_8 and air, where several opposed fans generate intense turbulence (Fig. 14.6).

An explanation can be obtained, if the flamelet model is used. Laminar premixed flames extinguish for very high strain rates. Turbulent nonpremixed flames showed the analogous behavior, i. e., one can blow out a match (see Chapter 13).

Figure 14.7 shows the strain, which is necessary to extinguish laminar premixed counterflow flames, as a function of the equivalence ratio. Different reaction mechanisms are used in order to show that the discrepancy between measurements and simulation is not caused by the chemistry. Experience shows that small amounts of heat loss, difficult to quantify in the experiment, can account for the discrepancy (Stahl and Warnatz 1991).

Together with the flamelet model, these measurements and simulations of laminar flames lead to a mechanism of extinction processes in turbulent premixed flames.

From the view fixed with the flame, turbulence convects reactants toward the flame. Higher convection rates lead to steeper gradients at the flame which in turns leads to larger diffusive losses. With increasing turbulence (steeper gradients), the finite rate of the chemical kinetics is unable to generate products as fast as reactants are delivered and products, including enthalpy, are removed. In consequence, the flame temperature decreases, which further lowers the reaction rate.

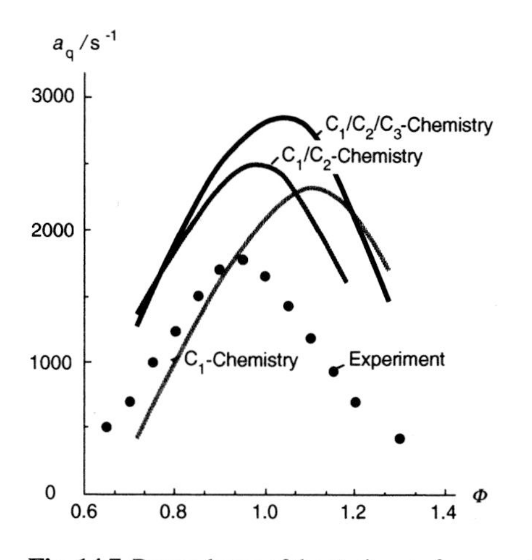

Fig. 14.7. Dependence of the strain rate for quenching a_q on the mixture composition of propane-air flames (Stahl and Warnatz 1991)

The peak temperature would gradually decrease with increasing convection velocity were it not for the fact that at some temperature, near 1700K for hydrocarbon fuels, the net generation of radical species is negative and the flame suddenly "blows out" as shown in Fig. 14.8.

An example of the suddenness of the blowout is presented in Fig. 14.9 where calculations show that the characteristic time for extinction is of the order of less than milliseconds. The sudden contractions of the gas due to extinction (together with the resonance caused by the geometrical configuration) are sources for the roaring noise of turbulent flames (Stahl and Warnatz 1991). As for nonpremixed flames, anemic flames near extinction are sensitive to Lewis number instabilities (Peters and Warnatz 1982).

As Fig. 14.7 shows, lean (or rich) flames extinguish more readily than stoichiometric ones. This is one of the reasons why surprisingly high amounts of hydrocarbon emissions are observed in lean combustion engines, though, naively, one would expect just the contrary, i. e., with excess oxygen, one would think complete combustion of the hydrocarbons is assured.

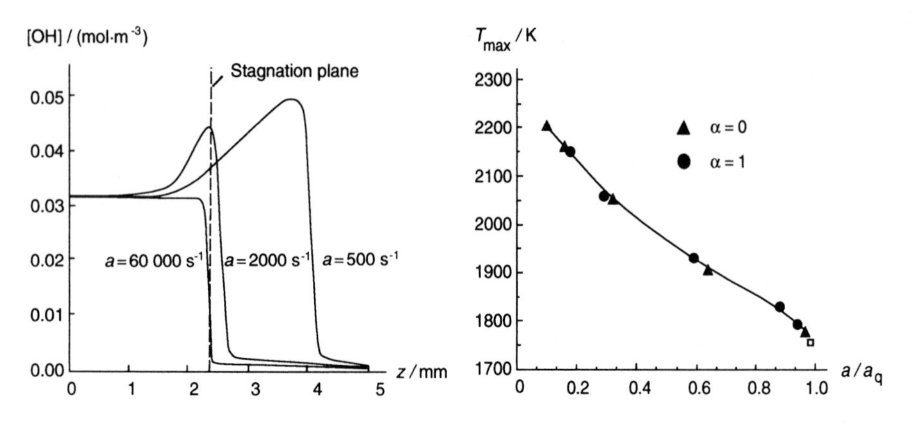

Fig. 14.8. OH-concentration as function of the strain rate (left hand side) for an opposed flow flame of stoichiometric C_3H_8-air against hot combustion products and maximum flame temperature as function of the strain rate a (right hand side) in a stoichiometric CH_4-air flame for planar ($\alpha = 0$) and cylindrical ($\alpha = 1$) counterflow configurations (Stahl and Warnatz 1991)

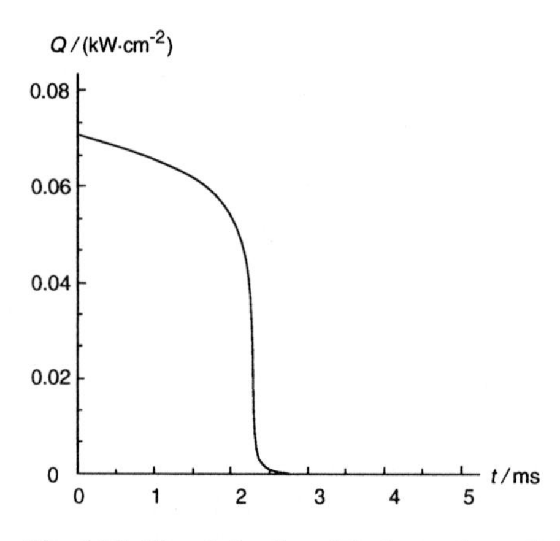

Fig. 14.9. Time behavior of the heat release during extinction of a stoichiometric methane-air premixed counterflow flame (Stahl and Warnatz 1991)

14.5 Other Models of Turbulent Premixed Combustion

Despite of observations that encourage the notion of a turbulent premixed flame being reviewed as an ensemble of strained laminar flames (flamelets), there are extreme mixing circumstances where such notions fail. This extreme zone is above and to the left of the line given by Da = 1 in the Borghi diagram, Fig. 14.2.. Examples include jet-stirred reactors (Malte and Pratt 1974, Glarborg et al. 1986) and cases of very strong turbulence (Roberts et al. 1993).

In such intense mixing case, the reaction zone is broad, if not completely homogenous. The spatial resolution of a numerical code may be relaxed since there are no steep flame fronts. A surprising consequence is reasonable use of *Large Eddy Simulation* (*LES*, Reynolds 1989) where the turbulent flow field is simulated using direct numerical simulation except that grid points are not extended to the smallest scale. Instead, there are fewer grid points so that the large scales are determined with DNS, but the unresolved scales, the *subgrid scales,* are modelled as isotropic turbulence using any of the turbulence models, such as the k-ε model and, more recently, the linear eddy model of Kerstein 1992 (LES-LE). The chemistry submodel can be that of a well stirred reactor. Applications of LES include automotive engine calculations (e. g., with the program KIVA™, Amsden et al. 1989) or weather forecast calculations.

14.6 Exercises

Exercise 14.1. The PDF of the velocity shall be given in a turbulent premixed Bunsen flame by the expression

$$P(u) = 0.0012\,(10\,u^2 - u^3) \qquad \text{for} \qquad 0 \leqq u \leqq 10\ \text{m/s}$$
$$P(u) = 0 \qquad \text{for} \qquad u \geqq 10\ \text{m/s}\ ,$$

where the velocity is inserted dimensionless into the expression for $P(u)$. Calculate the most probable velocity u_{p}, the average $\bar{u}$, and the mean square of the fluctuation $\overline{u'^2}$. What is the mean turbulent flame velocity v_{T}, if the laminar flame velocity is assumed to be v_{L} = 60 cm/s? Specify the ratio between the mean turbulent and the laminar flame area.

Exercise 14.2. The limiting case of turbulent premixed flames is an infinitely fast mixing of reactants and products. Consider a combustion reactor, where fuel and air are added and which is stirred so much, that a homogeneous mixture of fuel, air, and products is obtained. Temperature T_{R} and pressure in the reactor shall be constant, too.

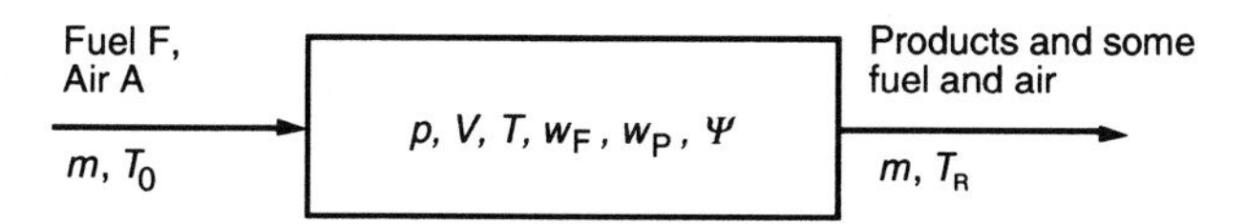

(a) Determine the mass fraction Ψ of unburnt compounds in dependence of the temperature T_{R} of the reactor and the temperature T_{C}, which would correspond to complete combustion. Use a constant mean heat capacity c_p and a specific heat of reaction q.

(b) Derive a formula, which relates the volume based mass flow $\dot{m}/V$ with the temperature of the reactor, the fuel and air mass fractions, and the pressure. What is obtained for $T_R = T_C$? The reaction is assumed to proceed according to

$$\mathrm{F} + \mathrm{L} \rightarrow \text{Products},$$

and the density based rate coefficient shall be given in the form $k = A \cdot \exp(-E/RT)$.

Exercise 14.3. Using Eqs. (14.1) and (14.2), show that the turbulent Reynolds number is given by $R_l = Ka^2\, Da^2$ (Peters, 1987).

Exercise 14.4. According to Peters (1987), the circumferential velocity υ_e of an eddy of size l_e is $\upsilon_e^3 = \varepsilon \cdot l_e$. With this relationship, show that the Kolmogorov time is equal to the diffusion time determined by $l_K = \sqrt{\nu \cdot t}$.

15 Combustion of Liquid and Solid Fuels

Prior to this chapter, this book has focused on combustion between fuel and oxidizer in the gas phase. However, in many practical combustion processes the fuel starts as a liquid, or as a solid, which is then burnt by a gaseous oxidizer. Examples of the combustion of liquids include that in jet aircraft engines, Diesel engines, and oil fired furnaces. Examples of combustion of solids include that of coal, wood (in forest and building fires), plastics, and trash.

Naturally, the additional consideration of a phase change and phase boundary leads to these combustion processes being less well understood than combustion processes in the gas phase. In addition to the processes in the gas phase (i. e., the chemical reaction and molecular transport discussed in the previous 14 chapters), similar processes in the liquid or solid phase and at the interface (energy, mass, and momentum transport) have to be accounted for. Furthermore, the fluid flow is usually turbulent.

As a consequence, the overall model of liquid or solid combustion consists of a wide array of interacting submodels. Anyone of these submodels can be made increasingly accurate, with the attendant (rapid) increase of computational time. Thus the successful (useful) model will allow approximations for certain submodels while retaining some level of detail for other submodels. The choice as to which submodels needs greatest detail is problem-specific, i. e., depends on what question one is investigating. Knowing which submodel to contract and which to expand has evolved into a major study by itself. Such studies are illuminated in the reviews by, e. g., Faeth (1984), Williams (1990), and Sirignano (1992). In this chapter, some salient features of liquid and solid fuel combustion shall be discussed briefly.

15.1 Droplet and Spray Combustion

The combustion of liquids is typically accomplished by injecting a liquid through an orifice into a gas-phase combustion environment. Turbulence inside of the liquid stream (intentionally generated from the high shear inside the injector) causes the emerging liquid stream to break up into a tangle of liquid strands that subsequently evolve into a dense cloud of droplets that ballistically penetrate through the gas into the combustion zone. Heat transfer to the droplet increases vapor pressure, and thus fuel evaporation into the gas phase, until subsequent gas-phase ignition commences. A nonpremixed flame surrounds (at least partly) the droplet or a droplet group. It is important to note that the vapor and not the liquid itself is what ultimately burns. The collection of these concurrent processes is called *spray combustion.* Researchers who investigate models of spray combustion have focused onto two areas:

1. Single Droplet Combustion: If one is allowed to investigate separately the fundamental physicochemical processes that occur in spray combustion, one often focuses on the details of combustion of a single droplet. Here one can employ very detailed models for the description of chemical reaction, evaporation, and molecular transport (of mass and energy) in the gas phase, in the droplet, and at the interface.

2. Spray Combustion: In modelling practical systems (jet engine combustor or Diesel combustion, or direct injected gasoline combustion), submodels for all of the physical processes have to be included. As noted, all of these submodels are approximated to an equal balanced level of detail so that the net model does not require a prohibitive amount of computational work. Thus, jet breakup, droplet dispersion, droplet evaporation, turbulent mixing, gas-phase chemical kinetics, etc., are often described by a simplified model, often called a global or reduced model.

The relationship of single droplet combustion to spray combustion is analogous to the relationship of laminar flame models that become embedded into models of turbulent combustion.

15.1.1 Combustion of Single Droplets

Implicit in the study of single droplet combustion is the assumption that combustion of a dense cloud of many droplets emerging from a fuel spray can be viewed as an ensemble of single droplet combustion. This implicit assumption is analogous to the assumption that turbulent flames can be modelled as an ensemble of laminar flames. Both assumptions have lead to an improved understanding of the combustion process. Thus, a detailed understanding of the processes in singlet droplet combustion is a requirement for the greater understanding of combustion of an ensemble of droplets, i. e., spray combustion.

Great model simplification results when one assumes spherical symmetry, since then the mathematical model becomes one-dimensional. A first step in modeling droplet combustion is to model droplet evaporation. An analytical model of droplet evaporation was developed in Exercises 5.2 and 5.3. These models are easily extended to the droplet combustion by merely adding the additional condition of a spherical nonpremixed flame that surrounds the droplet. These illuminating analytical models are achieved at the expense of several constraining assumptions that include: steady-state (i. e., initial heat and ignition transients are negligible) fast chemical kinetics, transfer of heat and mass is identical (unity Lewis number), and the properties of thermal conductivity λ, specific heat c_p, and the product of ρD are constant, independently of T. The analysis yields for the mass $\dot{m}_\mathrm{f}$ evaporating per unit time

$$\dot{m}_{\text{f}} = \frac{2\pi\lambda_{\text{g}}d}{c_{p,\text{g}}} \cdot \ln[1+B] \quad \text{with} \quad B = \frac{\Delta h_{\text{comb}}/\nu + c_{p,\text{g}}(T_\infty - T_\text{s})}{h_{\text{f,g}}} , \tag{15.1}$$

where d is the droplet diameter, λ_g the thermal conductivity of the gas phase, $c_{p,\text{g}}$ the specific heat of the gas phase, $\Delta h_{\text{comb}}/\nu$ the specific enthalpy of combustion divided by the stoichiometric mass ratio of the oxidizer and the fuel, $T_\infty - T_\text{s}$ the difference between the temperature of the gas phase far away from the droplet and that at the droplet surface, and $h_{\text{f,g}}$ the enthalpy of formation of the gas phase. B is called the *Spalding number*. Noting that the evaporation rate is given by

$$\dot{m}_\text{f} = -\rho_\text{L}\frac{\pi}{2}d^2 \cdot \frac{\text{d}d}{\text{d}t} = -\rho_\text{L}\frac{\pi}{4}d \cdot \frac{\text{d}d^2}{\text{d}t} , \tag{15.2}$$

one obtains

$$\frac{\text{d}d^2}{\text{d}t} = -\frac{8\lambda_\text{g}}{\rho_\text{L}c_{p,\text{g}}} \cdot \ln[1+B] .$$

Integration leads to the well-known d^2-law for droplet lifetime

$$d^2(t) = d_0^2 - Kt \qquad \text{with} \qquad K \equiv \frac{8\lambda_\text{g}}{\rho_\text{L}c_{p,\text{g}}} \cdot \ln[1+B] . \tag{15.3}$$

It can be seen that the mass burning rate is only weakly dependent (logarithmic) on the fuel properties (enthalpy of combustion Δh_{comb}, enthalpy of vaporization $h_{\text{f,g}}$), and is directly related to the properties of the surrounding gas medium and to the initial droplet diameter. A doubling of the initial diameter quadruples the droplet burnout time and, hence, the combustor must be longer. In the limit $\Delta h_{\text{comb}} = 0$, there is pure droplet evaporation.

Such insight is obtained from analytical models. However, with numerical models, all of the above assumptions can be increasingly relaxed. With inclusion of details for all of the submodels, the computation takes on the name of "simulation".

The simulation is performed by solving the conservation equations in the gas phase, the droplet, and at the interface (see, e. g., Cho et al. 1992, Stapf et al. 1991). This system can be realized experimentally, when single droplets are injected into a chamber of hot combustion products. In order to avoid gravitational effects in the experiment, which perturb the spherical symmetry, the combustion gases are convected downwards at the velocity of the droplet, or even better, the combustion chamber is situated in a fall tower where gravity is zero for the short duration of the experiment (see, e. g., Yang and Avedisian 1988). (Note, however, that in the numerical model the effect of gravity is easily included; for the analytic model, this inclusion is much more difficult.)

Three phases of droplet combustion are identified, determined by different physical phenomena:

(I) Heating Phase: Heat from the gas phase causes the droplet surface to heat up. Much of the energy is convected into the droplet until the entire droplet is approaching a boiling temperature; then significant loss of droplet mass commences.

(II) Fuel Evaporation Stage: Fuel evaporates into the gas phase and a combustible mixture is formed; the square of the droplet diameter decreases in time, the d^2-*law*.

(III) Combustion Phase: The mixture ignites and burns as a spherically symmetric, laminar nonpremixed flame; the droplet diameter now decreases in time, again, by the d^2-law with a different K (see Eq. 15.3).

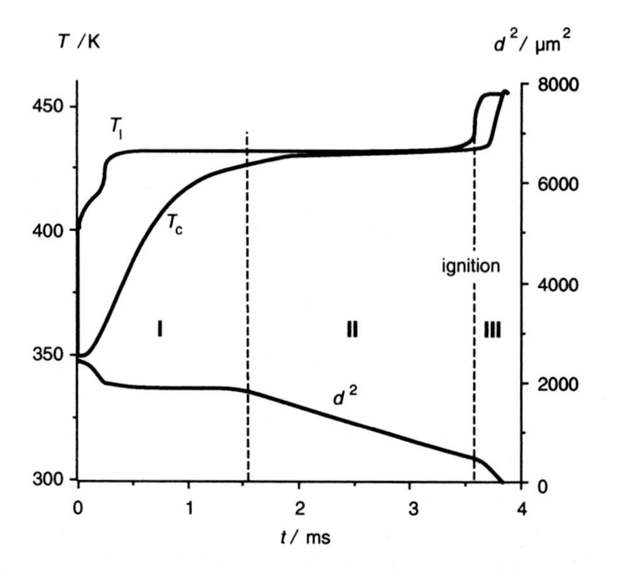

Fig. 15.1. Computational simulation of characteristic parameters during heat-up, vaporization, and combustion of a methanol droplet (T_{drop} = 350 K, d = 50 μm) in air (T = 1100 K, p = 30 bar); plotted are the temperature in the center of the droplet (T_c) and at the interface (T_I), as well as the square of the droplet diameter d; ignition occurring at 3.6 ms (Stapf et al. 1991)

Characteristic parameters during three phases of (I) heat-up, (II) evaporation, and (III) combustion of a methanol droplet surrounded by hot air are shown in Fig. 15.1. As soon as the droplet is exposed to the hot surrounding air, heat transfer occurs from the hot gas phase to the droplet, and the temperature T_I at the droplet surface increases rapidly until the phase equilibrium is reached. Within the droplet, heat conduction to the interior raises the center temperature T_c .

The simulation shows that the assumption of steady-state, needed for the analytical solution, is an over-simplification that results in underpredicting droplet lifetime by nearly 50%; if the analytic model was used as a combustor design criteria, the combustor would be too short. In time, the droplet temperature reaches a steady state where heat conduction to the droplet is balanced by evaporation of liquid at the droplet surface. This temporary balance leads to a temporarily constant evaporation rate. Based on the simplified treatment of the evaporation process described above (see also Exercises 5.2 and 5.3, or Strehlow 1985) it was shown that the square of the droplet diameter decreases linearly with time, $d(d^2)/dt$ = const., where the constant depends on various properties of the droplet and the surrounding gas phase. As the straight lines in Fig 15.1 indicate, a d^2-law is valid for much of the droplet lifetime.

Ignition in the gas phase occurs after an induction time (at $t = 3.5$ ms in Fig. 15.1). The spherical nonpremixed flame surrounding the droplet leads to augmented heating of the droplet and, thus, to an acceleration of the evaporation process as evidenced by the increased negative slope of the d^2 vs. time line at $t = 3.5$ ms in Fig. 15.1.

Due to the wide variety of physicochemical processes involved, some aspects of droplet combustion are influenced by nearly all submodels. For example, a knowledge of ignition delay times is important for practical applications (see also Section 10.4). Auto-ignition occurs when the temperature is high enough and when, at the same place, evaporation has locally sustained a combustible mixture for a time that allows a long sequence of chemical reactions to bring the mixture through the chemical induction phase to ignition. For such predictions, the simulations are essential, the analytical model simply lacks the details for such a prediction.

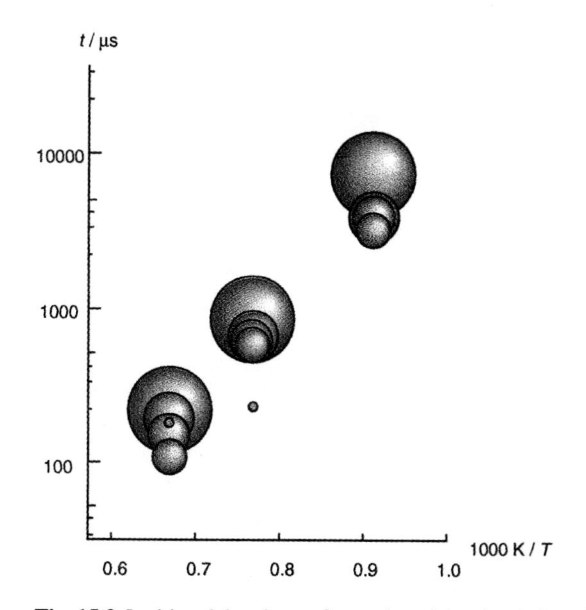

Fig. 15.2. Ignition delay times t for methanol droplets in hot air, dependence on gas temperature and droplet size (diameters from 10 to 100 µm are illustrated by the size of the spheres)

The dependence of ignition delay times on the temperature of the gas phase is illustrated in Fig. 15.2 for various initial droplet diameters (Stapf et al. 1991). Usually the ignition delay times increase with increasing droplet radius. This increase is caused by the fact that heat is withdrawn from the gas phase in order to evaporize the droplet. Deviations from this behavior are observed (see Fig. 15.1) for very small droplet diameters, because the droplet has completely evaporized before ignition takes place.

The examples above have illuminated the combustion of droplets in a quiescent environment. In practical applications, droplets usually move with a certain velocity relative to the surrounding gas. This is caused, e.g., by the injection of the fuel spray or by a turbulent flow field. Thus, it is important to understand the influence of the flow field on the ignition and combustion of droplets. This can be seen from Fig. 15.3, where an ignition process of a methanol droplet in air is shown. The droplet is subject to a flow of air from the left with a velocity of 10 m/s . It can be seen that in this case ignition starts in the downwind region of the droplet. After some time a flame front is formed which is similar to the one observed in nonpremixed counterflow flames (see Fig. 9.1.a).

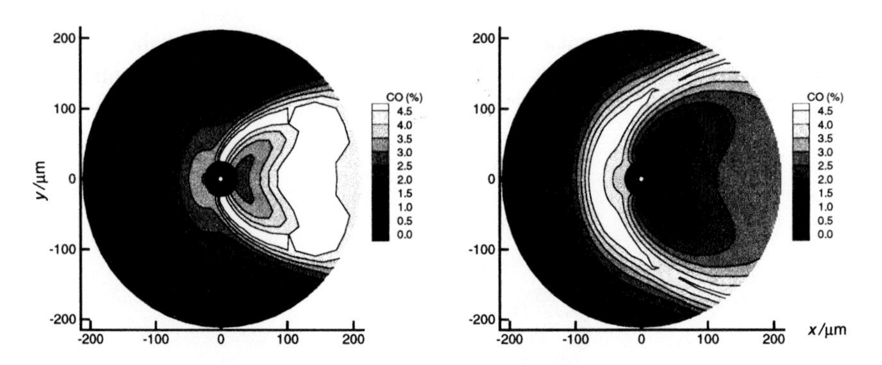

Fig. 15.3. Ignition process of a methanol droplet in hot air which approaches the droplet from the left at a velocity of 10 m/s. Left: mass fraction contours of CO during the ignition process; right: contours after a nonpremixed flame has formed (Aouina et al. 1997)

Most liquid fuels are distilled from petroleum and consist of hundreds of compounds with a range of boiling points. As the liquid drop begins to heat up, the most volatile liquids will evaporate first, followed by liquids of intermediate volatility (and thus higher boiling temperature), and then by the vaporization of remaining low volatility (viscous oils). Fortunately, auto-ignition occurs early in this sequence of events, so the evaporation of the low volatility oils is augmented by the flame that now encircles the remains of the drop.

15.1.2 Combustion of Sprays

As noted above, a first step towards modelling of spray combustion is to assume that the burning spray is merely an ensemble of single, non-interacting, burning droplets. The droplets emerge from the jet as a dense cloud of drops with a wide range of diameters. However, it is not known how these different size drops interact with each other and with the surrounding turbulent gaseous flow field (see, e. g., Williams 1990). These questions have been addressed by dividing the overall process into the formation of the spray, the motion of the droplets, and evaporation, followed by combustion.

The spray is formed, as a fuel jet (coming, e. g., from a nozzle) is shattered by shear forces during injection. This process is similar to the formation of turbulent structures in shear layers (Clift et al. 1978). The liquid fragments, which are not yet spheres, are launched into the (usually turbulent and recirculating) flow fields of oxidizer and combustion products. The distribution of droplet diameters in a spray is not uniform. It has been profitable to characterize the distribution of diameters by various size distribution functions with associated moments that evolve in time. The distribution function and its subsequent evolution is largely determined by the nature of the injection and the flow in the combustor (i. e., the boundary conditions).

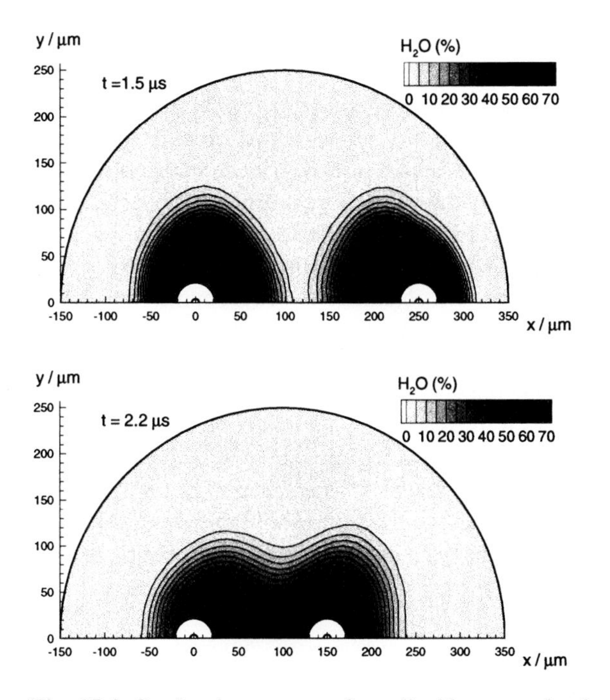

Fig. 15.4. Combustion process of two liquid oxygen droplets in a hydrogen environment; contours denote the mass fractions of water (Aouina et al. 1997).

Evaporation of the droplets and diffusion of the fuel into the gas phase leads to the formation of a combustible mixture, which ignites at sufficiently high temperatures. If *dilute sprays* are considered (distance between the droplets is large), the processes during ignition and combustion can be understood by an isolated examination of the droplets. In *dense sprays,* however, the droplets are too close for their interaction to be neglected. This can be seen from Fig. 15.4 where a combustion process of two droplets of liquid oxygen in a hot hydrogen atmosphere is depicted, a combustion system which arises in cryogenic rocket engines. The economic success of these rocket engines is essential to the vast communication-via-satellite industry. Reliable modelling helps these economics by reducing the amount of (very expensive) experiments at these cryogenic conditions typical of rocket engines, e. g., ~100 K and 400 bar.

The liquid oxygen droplets with an initial diameter of 50 μm and an initial temperature of 85 K are situated in a laminar uniform flow of hot hydrogen at 1500 K and 10 bar, and the velocity of the gas is 25 m/s. In the upper figure the distance between the droplets is 250 μm (5 times the droplet diameter) and in the lower figure the distance is 150 μm (3 times the diameter). Only the upper half of the configuration is shown in the Figures. If the distance between the two droplets is sufficiently small, the two droplets are surrounded by one common non premixed flame zone. Otherwise the two droplets form two distinct flame zones. These numerical experiments showed quantitatively the increase in droplet burnout time and, thus, can aid the rocket motor design process.

Noting that the combustion of sprays is a complex interaction of many different processes such as droplet heating, evaporation, ignition, combustion, interaction of different droplets, interaction of the droplet combustion with the turbulent flow field, etc., it is evident that modelling of spray combustion is a challenging task. Like in the case of turbulent single-phase combustion, a calculation from first principles (i. e., using direct numerical simulations (see Chapter 12) is not possible, but like in gas-phase combustion processes the information obtained from direct numerical simulations of the combustion of single droplets and droplet arrays can be refined and then incorporated into spray models.

This approach is equivalent to the flamelet approach with the difference that the laminar flamelets stem from flames formed around the droplets. Recalling the basic principles of the flamelet approach we note that the laminar nonpremixed flamelets were characterized by the mixture fraction and the strain rate. The strain rate is a measure for the counterflow velocity. As an analogy, droplet flamelets can be characterized by the mixture fraction and the velocity of the droplet relative to the surrounding gases, see Fig. 15.5.

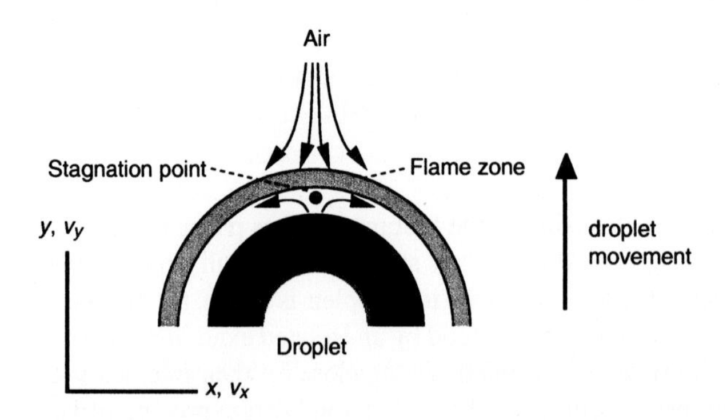

Fig. 15.5. Schematic illustration of counterflow droplet combustion, see Fig. 9.1

For spray flames the flamelet concept is, however, complicated by the fact that additional parameters have to be taken into account, namely varying droplet sizes, varying values of the temperature and composition of the surrounding gases (a droplet may move into regions of hot combustion products produced by the combustion of other droplets), and instationary processes caused by the complete evaporation of the droplet (the supply of fuel and thus the combustion ends).

Although this approach is quite comprehensive, the computation of the large flamelet libraries is computationally very expensive. Furthermore, the approach is complicated by the fact that in spray combustion several domains are identified where the number density of the droplets is so high that, due to a lack of oxidizer, single droplet combustion cannot occur. This is illustrated in Fig. 15.6.

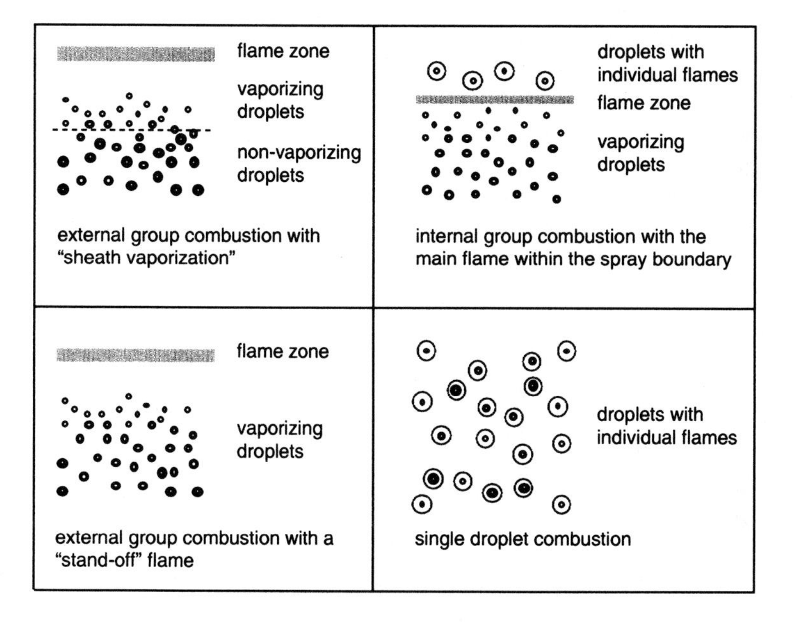

Fig. 15.6. Combustion modes of a droplet cloud (adopted from Chiu et al. 1982)

If the spray is very dense, the density of the fuel in the inner core of the spray is so high that a saturation occurs and the droplets do not evaporate. Close to the surrounding air, evaporation takes place and a cloud of fuel vapor is formed which then diffuses into the oxidizer, forming a nonpremixed flame. This mode of combustion is referred to as "external group combustion with sheath vaporization" (Chiu et al. 1982). At lower densities of the spray all the droplets evaporate, but still the flame zone is at the boundary between the fuel vapor and the surrounding oxidizer ("external group combustion with a standoff flame"). If the spray gets more dilute some of the droplets are in an environment with an excess of oxidizer. These droplets will be surrounded by individual flames, but nevertheless there will still be a flame zone at the boundary of the dense droplet cloud and the oxidizer ("internal group combustion" with the main flame within the spray boundary). In dilute sprays, finally single droplet combustion occurs. Of course, these four different modes are only a crude characterization, and there are many modes in between these limiting cases. In the transition from external to internal group combustion, e. g., flame zones around small groups of droplets (see Fig. 15.4) can be observed.

As with gas phase combustion, insight to the various droplet-flame interactions are being illuminated by investigations done with droplets in opposed flow flames. The aforementioned one-dimensional simplicity of this configuration allows for inclusion of detailed chemical kinetics and fluid transport. Figure 15.7 shows a simulation where the momentum of the droplet carried it through the flame front and through the stagnation plane, where drag from the opposing velocity finally reverses the direction of the droplet and sends the droplet back through the flame front (see exercises for an exploration of droplet drag).

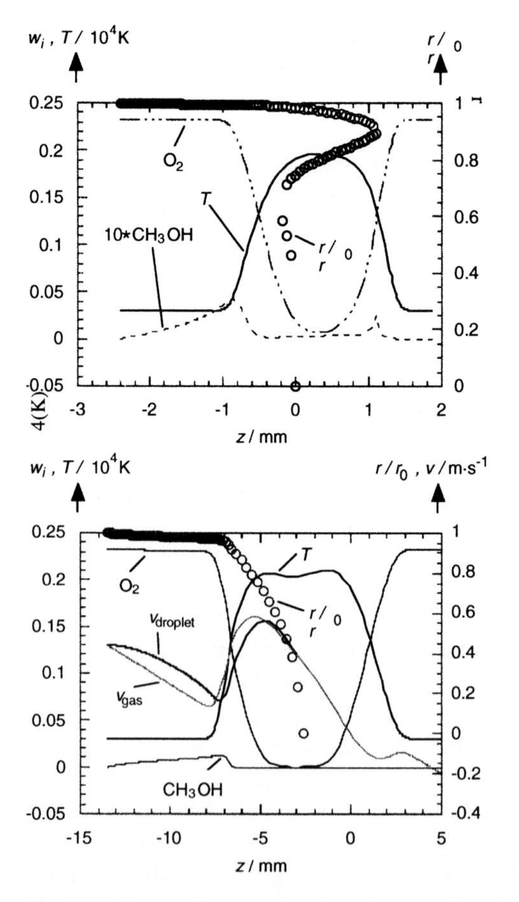

Fig. 15.7. Counterflow combustion of a monodisperse spray of methanol with air at (top) high strain and (bottom) low strain (Gutheil and Sirignano 1998)

Because spray combustion, like droplet combustion, is a typical nonpremixed combustion process, the use of flamelet models is motivated. These flamelet models can be combined with models for the droplet evaporation. Such a model has, e.g., been used to simulate a stratified-charge engine fueled with n-octane (Gill et al. 1994).

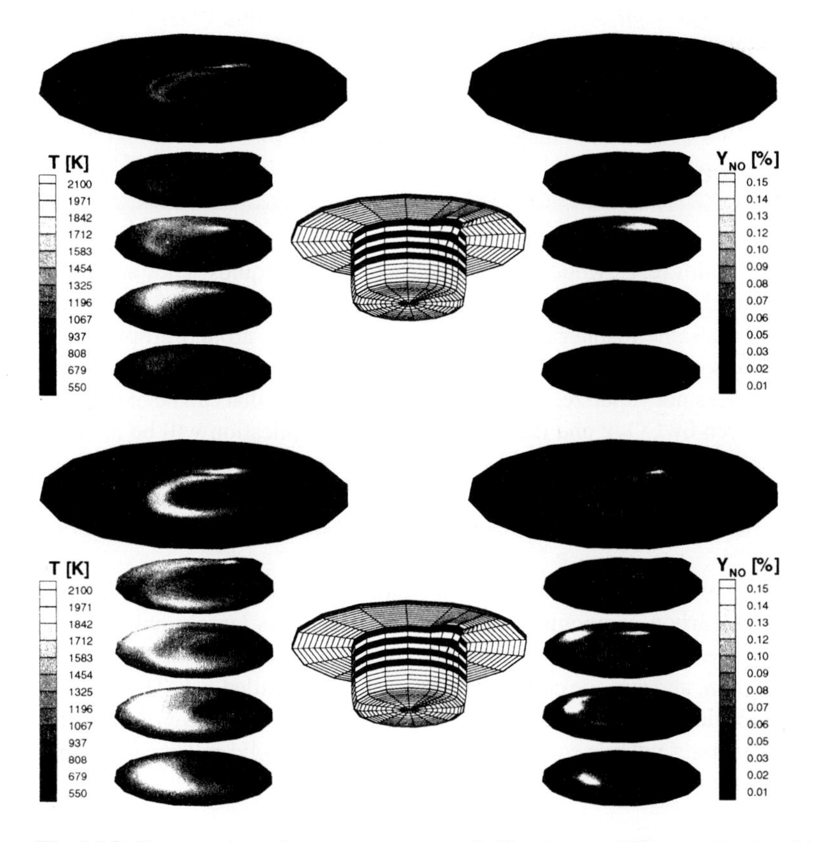

Fig. 15.8. Contour plots of mean temperature (left) and mean NO mass fraction (right) in a stratified charge engine at 5^0 crank angle before top dead-center for partial load (upper drawing) and for full load (lower drawing); color maps are given at five different locations (dark positions marked in the plot of the numerical grid) in the combustion chamber (Gill et al. 1994).

Stratified-charge engines (also called Direct Injected Gasoline (DIG)) combine the advantages of Diesel and Otto engines (Takagi 1998). The fuel is injected into the cylinder during the compression phase. Ignition is initiated by a spark plug close to the mixing zone of fuel and air. In the ignition zone the mixture is rich which supports the ignition process. On the other hand the overall global mixture composition is lean which reduces the formation of nitric oxides (see Chapter 17). Resulting temperatures and nitric oxide mass fractions are shown in Fig. 15.8 for two characteristic cases (partial and full load at 5° before top dead center). These simulations show great progress. In spite of the simplifications to the many submodels (including spray fluid dynamics, droplet evaporation, spark ignition, flame propagation and quenching, to name a few), these simulation retains enough essential ingredients to be an invaluable aid in synthesis of new combustion engines such as the direct-injection gasoline (DIG) engine and the even newer concept of managed engine knock in a homogeneous charge compression ignited (HCCI) engine (Thring 1989, Christensen et al. 1998, Flowers et al. 2000).

15.2 Coal Combustion

At first glance, solids (e. g., wood or plastic) burn much like liquids. They are heated to a point where they generate significant gas-phase volatilizes which subsequently burn in much the same manner as in gas-phase combustion around droplets. The difference occurs when some residual fraction of the solid (and for some liquids, e. g., heavy oils) does not vaporize. This solid may have carbon that can be oxidized to CO by O_2 or even by CO_2); and thus a new level of complication will be included into the model for solid combustion. As with liquids combustion, a brief section on solid combustion will be given with a focus on the combustion of coal. A more detailed discussion is presented by Hobbs et al. (1993), Smoot (1993), Speight (1994) and Turns (1996).

Coal is not a homogeneous chemical compound, but instead is a mixture of a wide variety of hydrocarbon compounds whose structure remains largely unknown. In addition to volatile and nonvolatile combustible compounds, coal contains compounds which do not burn, and survive combustion in a solid form called *ash*. Three different processes which interact are distinguished in coal combustion: *pyrolysis* of the coal (which generates volatile compounds and a carbon rich solid called *coke*), burning of the *volatile compounds,* and burning of the coke.

Pyrolysis of the Coal: The pyrolysis (*thermal decomposition* and *degasification*) of the coal occurs at temperatures higher than 600 K. A separation into coke, tar, and volatile compounds is observed. The pyrolysis is governed by many physicochemical processes, such as shrinking or swelling of the coal particles, the structure of the coal (e. g., pore size), transport processes in the pores and at the surface of the coal particles, the temperature during the pyrolysis, and secondary reactions of the pyrolysis products.

Since the chemical makeup of the coal is not known, the mechanism describing coal pyrolysis can only be, and should only be, approximate (see, e. g., Solomon et al. 1987). Volatile compounds are formed by the separation of functional groups that then form CH_4, H_2, CO, HCN, etc.. This breaking of chemical bonds in the coal leads to fragments, which can rearrange and react, forming tar. The chemical processes are followed by diffusion of the volatile compounds to the surface of the coal particles, where they evaporate and subsequently burn.

The understanding of this pyrolysis is poor compared to that of gas-phase combustion. Accordingly, one resorts to coarse models to describe the pyrolysis, such as a constant rate of pyrolysis that is independent of anything, or a global rate with non-integer exponents. However, analogous to non-elementary gas-phase reactions (see Chapter 6), these empirical models have the disadvantage of being reliable in a small range of conditions; extrapolation to other conditions usually results in unrealistic predictions.

Burning of the Volatile Compounds: The volatile compounds, which are formed during pyrolysis, are burnt in the gas phase. The governing processes (evaporation, diffusion into the gas phase, and then combustion) are similar to those of droplet combustion. However, the volatile compounds are a mixture of unknown composition; as such, a detailed description based on elementary kinetics is not possible and is unwarranted.

Burning of the Coke: Burning of the solid coke adds a new dimension to the combustion process. Coke is largely carbon; it has a low vapor pressure, so that evaporation followed by gas-phase oxidation is not a major reaction path. Instead, the surface carbon is oxidized to CO by gas phase CO_2 (and O_2), that collide with, and/or stick to, the surface. This reaction of CO_2 is $C(s) + CO_2(g) = 2\,CO(s)$. The carbon is now strongly attached to the oxygen and weakly attached to the surface. Thus, the nascent CO does have a high vapor pressure and leaves the surface to the gas phase where CO can be further oxidized to CO_2.

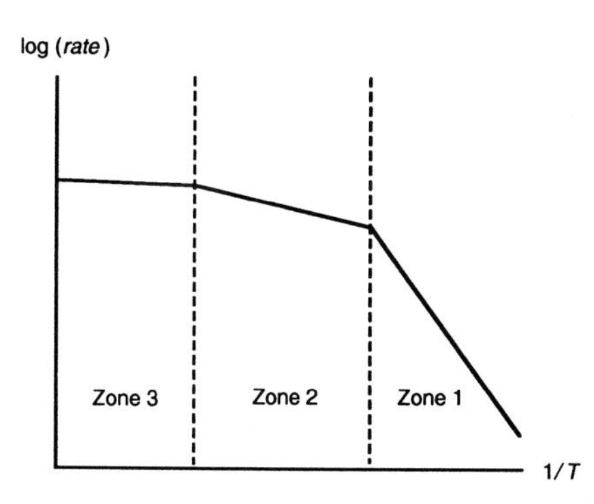

Fig. 15.9. Consequences of reaction limitation (Zone 1) and transport limitations (Zones 2 and 3) on the global activation energy of the rate of coke combustion (Walker jr. et al. 1959)

The modelling of this heterogenous process brings into the combustion process the established formalisms of catalytic chemistry (see Section 6.7). The formalism includes adsorption of molecules on the surface, surface reactions, and desorption of the products, diffusion through the pores, and diffusion at the particle surfaces. As with catalytic combustion, a decrease in the rate of any one of the above sequential processes will limit the rate of the entire process. Thus, as with catalytic combustion, the exponential temperature dependence of surface reaction rates cause them to be the limiting step at low temperatures (called "kinetically controlled"; Zone 1 in Fig. 15.9 with high activation energy) and to become so fast at high temperatures that pore diffusion (Zone 2 in Fig. 15.9) and gas diffusion (Zone 3 in Fig. 15.9) become the rate-limiting processes ("transport controlled process" with relatively low activation energy). In contrast to combustion processes in the gas phase, the heterogeneous reactions involved in burning of coke are relatively unexplored (Lee et al. 1995).

15.3 Exercises

Exercise 15.1. Laboratory tests at $p = 1$ bar show that a droplet of methanol (40 micron diameter) burns out in $\tau_{\text{burnout}} = 1$ ms. Estimate how long it will take in a gas turbine engine to burn out the droplet at the same temperature, but at $p = 40$ bar. What is the τ_{burnout} at 40 bar?

16 Low Temperature Oxidation, Engine Knock

A detailed knowledge of the combustion processes in internal combustion engines is required if one wishes to further improve on the remarkable development of engine technology and related fuel technology. Such improvements aim at the efficient use of fuels with a minimum amount of pollutant emissions. In the specific case of the spark ignited engine, i. e., the Otto engine, a thermodynamic analysis of the engine cycle shows that overall efficiency η will increase with increasing compression ratio ε ($\eta \approx 1 - 1/\varepsilon^{\kappa-1}$; $\kappa = C_p/C_V$). Furthermore, the absolute power output increases when more mass is inducted on each intake stroke. Unfortunately, as the compression ratio is increased, the onset of a phenomenon called engine knock occurs, being ruinous to the engine. Understanding the chemical basis of knock is the subject of this chapter.

16.1 Fundamental Phenomena

The unburnt gases compressed by the piston are additionally compressed by the burnt gases that expand behind the spark-ignited flame front. The last remaining unburnt gas is called the *end gas*. With higher compression ratios, the end-gas temperature increases until spontaneous ignition occurs (Jost 1939). This sudden ignition leads to the formation of pressure peaks in the cylinder that cause the audible knocking noise. Knock is to be avoided as these pressure peaks damage the piston and engine. For a definitive text on piston engines, see Heywood (1988).

Fuels differ very much in their tendency to produce knock. In order to allow a direct comparison, the *Cooperative Fuel Research Committee* (CFR, ca. 1930) defined a scale that determines a fuel's *octane number* (ON); see Table 16.1. For this scale, the knock tendency of a fuel is compared with that of a mixture of n-heptane/iso-octane (2,2,4-trimethyl-pentane) burned in a standardized single cylinder engine. Iso-octane (with a low knock tendency) is defined to have an octane number of 100 while n-heptane, which has a high knock tendency, is defined to have an octane number of 0. Thus, a fuel with ON = 80 has the same knocking tendency as a mixture of 80% iso-octane and 20% n-heptane.

Tab. 16.1. Experimental research octane numbers (RON) of some selected fuels (Lovell 1948)

Formula	Name	RON
CH_4	Methane	120
C_2H_6	Ethane	115
C_3H_8	Propane	112
C_4H_{10}	n-Butane	94
C_4H_{10}	iso-Butane (2-methylpropane)	102
C_5H_{12}	n-Pentane	62
C_5H_{12}	iso-Pentane (2-methylbutane)	93
C_6H_{14}	n-Hexane	25
C_6H_{14}	iso-Hexane (3-methylpentane)	104
C_7H_{16}	n-Heptane	0
C_7H_{16}	Triptane	112
C_8H_{18}	n-Octane	-20
C_8H_{18}	iso-Octane	100
	—	
C_4H_8	Methylcyclopropane	102
C_6H_{12}	Cyclohexane	84

Formula	Name	RON
C_6H_{12}	1,1,2-Trimethyl-cyclopropane	111
C_7H_{14}	Cycloheptane	39
C_8H_{16}	Cyclooctane	71
	—	
C_6H_6	Benzene	103
C_7H_8	Toluene (Methylbenzene)	120
C_8H_{10}	Xylene-m (1,3-Dimethylbenzene)	118
	—	
C_3H_6	Propene	102
C_4H_8	1-Butene	99
C_5H_{10}	1-Pentene	91
C_6H_{12}	1-Hexene	76
C_5H_8	Cyclopentene	93
	—	
CH_3OH	Methanol	106
C_2H_5OH	Ethanol	107

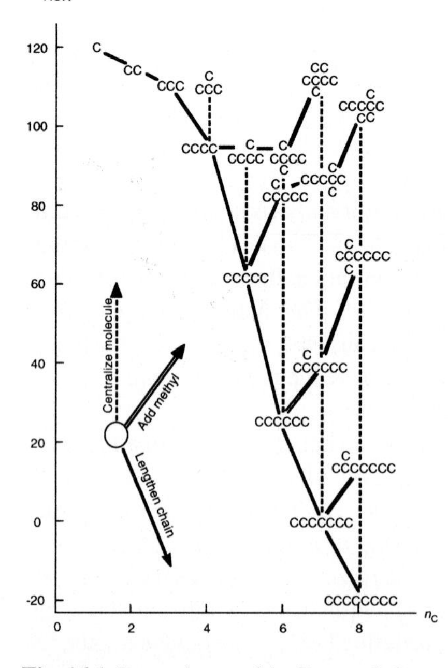

Fig. 16.1. Dependency of the Research Octane Number (RON) on molecular structure changes (Lovell 1948, Morley 1987)

A comparison of octane numbers shows that iso-alkanes tend to knock much less than n-alkanes (Jost 1939, Heywood 1988). The changes of knock tendency due to modifications of the molecular structure (like chain lengthening or methyl addition) are shown in Fig. 16.1 (Lovell 1948, Morley 1987).

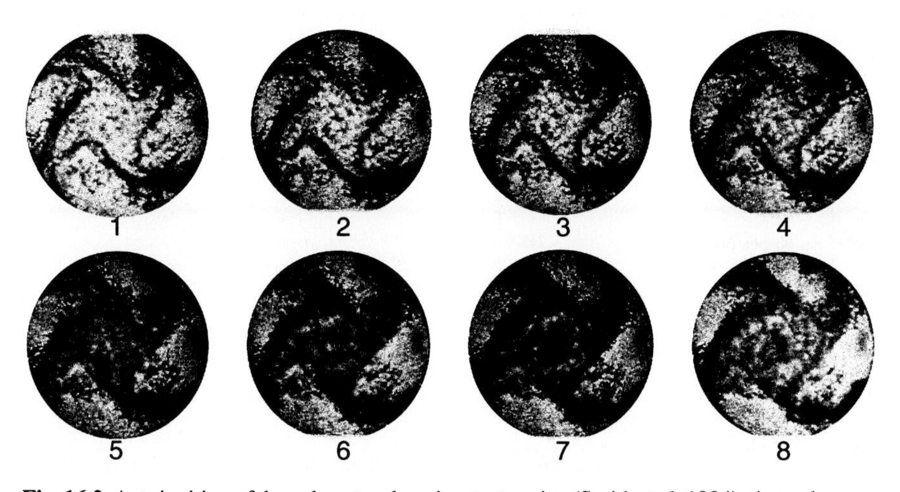

Fig. 16.2. Autoignition of the unburnt end gas in a test engine (Smith et al. 1984); time advances in 25 μs steps from upper left to lower right. Autoignition is occurring between frames 4 and 5

Photographs in a test engine with optical access through the top of the cylinder head (Smith et al. 1984) show the autoignition of the unburnt end gas before arrival of the flame front (Fig. 16.2). The pictures were taken at intervals of 28.6 μs with an exposure time of 1.5 μs. In order to avoid heat losses to the wall, the end gas is compressed and confined by four converging flame fronts ignited by four spark plugs. Assuming adiabatic compression, the time behavior of the end gas temperature is inferred from that of the pressure. Autoignition occurs at between the fourth and fifth photo. The unburnt end gas autoignites almost simultaneously and homogeneously.

The onset of autoignition is almost exclusively governed by the chemical kinetics. Temperature- and pressure-history in the end gas determine the ignition delay time (see Chapter 10). The end gas is compressed (and thus heated) by the piston and by the advancing flame front. If end-gas temperature and pressure are not too high, the flame front will consume the end gas before knock can occur. Otherwise the end gas will autoignite in front of the advancing flame front.

Due to the high sensitivity of the ignition delay time with respect to the temperature, the end gas ignites first at locations with locally increased temperature (*hot spots*), which are caused by the fact that (although the end gas is nearly homogeneous) small temperature- and pressure-fluctuations are present. The reason for these fluctuations is not yet completely known. Ignition of the hot spots leads to pressure-induced flame propagation or the formation of detonation waves, either of which results in a very fast ignition of all the end gas (see Section 10.6 for details).

16.2 High-Temperature Oxidation

Figure 16.3 shows CARS- (*Coherent Anti-Stokes Raman Scattering*) and SRS-spectroscopic (*Spontaneous Raman Scattering*) nonintrusive temperature measurements in the end gas of the engine described in the previous section (Smith et al. 1984).

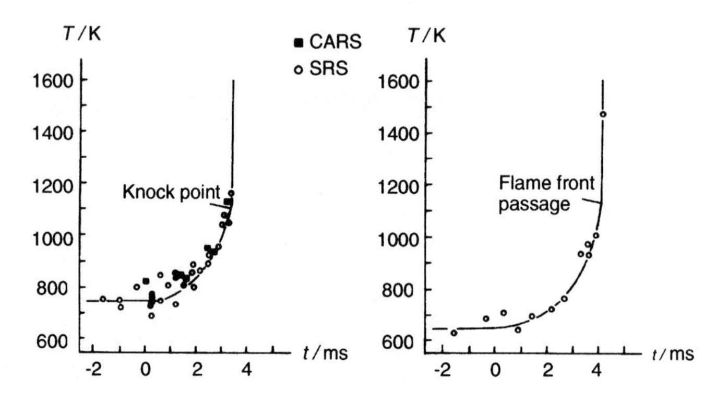

Fig. 16.3. Temperature of the unburnt end gas in a test engine. On the right, the flame front consumes the end gas before the onset of knock (measurements for different cycles)

With this engine and this fuel, knock occurs at about 1100 K. Reaction mechanisms have to be found, which can explain autoignition at these temperatures. The chain branching reaction

$$H\bullet + O_2 \rightarrow O\bullet + OH\bullet,$$

which dominates combustion at high temperatures, is (due to the large activation energy) too slow to explain autoignition at temperatures below 1200 K. Sensitivity analyses and analyses of reaction paths indicate (Esser et al. 1985) that the chain branching, which is responsible for the autoignition, after an initiation reaction like $RH + O_2 \rightarrow R\bullet + HO_2\bullet$ is given by ($R\bullet$ = hydrocarbon radical)

$$HO_2\bullet + RH \rightarrow H_2O_2 + R\bullet$$
$$H_2O_2 + M \rightarrow OH\bullet + OH\bullet + M.$$

The OH-radicals can reproduce the HO_2, e. g., by

$$OH\bullet + H_2 \rightarrow H_2O + H\bullet$$
$$H\bullet + O_2 + M \rightarrow HO_2\bullet + M .$$

Indeed, this branching via the HO_2-radical can explain the knock process in the test engine at temperatures of about 1100 K.

Figure 16.4 shows simulations and experimental results for a knocking and a non-knocking engine cycle (Esser et al. 1985). The ignition delay time is calculated based on the measured pressure histories, which enables the calculation of the temperature histories. In the knocking case, the calculated ignition time corresponds very well to the time when knocking is observed in the experiment. In the non-knocking case, the autoignition time is greater than the flame arrival time, thus regular combustion is completed before autoignition can take place.

However, this result is not general. Current automotive engines knock at much lower temperatures, and the chemistry of engine knock demands additional mechanisms as will be discussed in the next section.

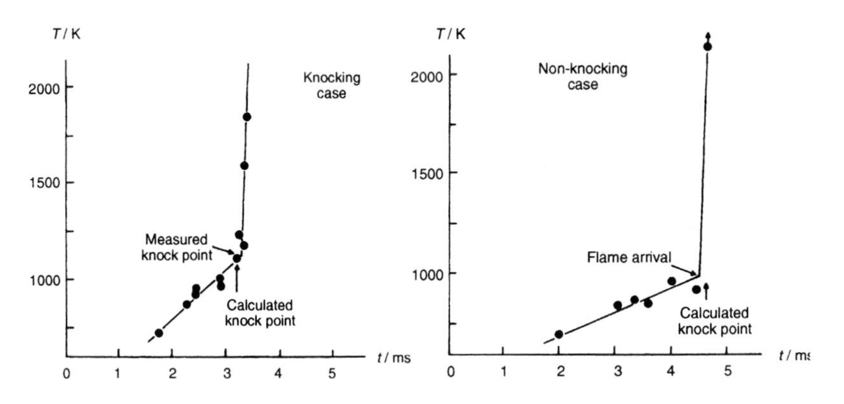

Fig. 16.4. Autoignition in a knocking (upper) and a non-knocking (lower) engine cycle (n-butane used as fuel). Experiment: Smith et al. 1984, simulation: Esser et al. 1985

16.3 Low-Temperature Oxidation

Usually the end gas heat losses in production engines are higher than in the test engine described above. Thus, autoignition, when it happens, occurs at lower temperatures (800 K - 900 K, see example below). The decomposition of H_2O_2 is quite slow at these temperatures, and other (fuel-specific, and, thus, more complicated) chain-branching mechanisms govern the ignition process (Pitz et al. 1989):

R•	+ O_2	$\rightleftarrows$	RO_2•		(first O_2-addition)
RO_2•	+ RH	$\rightarrow$	ROOH	+ R•	(external H-atom abstraction)
ROOH		$\rightarrow$	RO•	+ HO•	(chain branching)
RO_2•		$\rightarrow$	HOOR'•		(internal H-atom abstraction)
HOOR'•		$\rightarrow$	R'O	+ HO•	(chain propagation)

In a first step, hydrocarbon radicals (R•) react with oxygen and form peroxy radicals (RO_2•). These can abstract hydrogen atoms forming hydroperoxy compounds (ROOH). After an external hydrogen atom abstraction (reaction with another molecule), the hydroperoxy compound decomposes into an oxy radical (RO•) and OH. Alternatively, an internal hydrogen-atom abstraction can take place (i. e., the abstraction of a hydrogen atom from the molecule itself); this is probable if R• can form a relatively stable 5-, 6-, or 7-membered ring, where the two oxygen atoms and the H atom transferred also take part in the ring formation (see Fig. 16.5). Subsequently to the internal abstraction, the radical $R'O_2H$• (the free valence is at the position, where the hydrogen atom has been abstracted) reacts to a compound without free valences (aldehyde, ketone) and OH, according to a chain propagation.

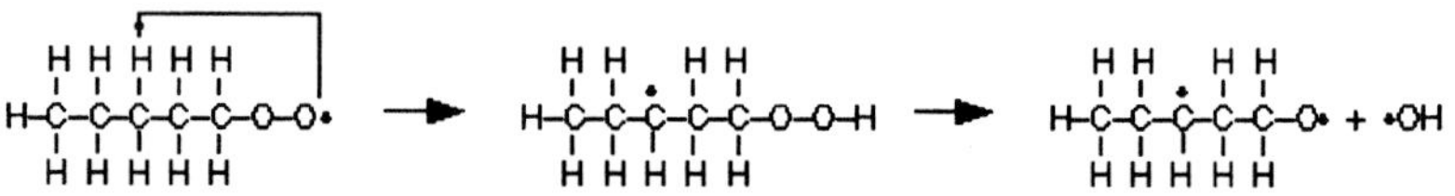

Fig. 16.5. Internal hydrogen atom abstraction in a heptyl peroxi radical via an intermediate 6-membered ring structure. The biradical on the right hand side will immediately isomerize to form, e. g., a stable aldehyde C_4H_9–CHO

However, the external H-atom abstraction is much slower than the internal abstraction, and the mechanism cannot explain an efficient chain branching and thus autoignition. A mechanism which explains the ignition behavior is obtained, if the O_2-addition is repeated at the radical $R'O_2H$•, which has been formed after the first O_2-addition (Chevalier et al 1990a,b):

HO_2R'• + O_2	$\rightleftarrows$	$HO_2R'O_2$•	(second O_2-addition)
$HO_2R'O_2$• + RH	$\rightarrow$	$HO_2R'O_2H$ + R•	(external H-atom abstraction)
$HO_2R'O_2H$	$\rightarrow$	$HO_2R'O$• + HO•	(chain branching)
$HO_2R'O$•	$\rightarrow$	OR'O + HO•	(chain propagation)
$HO_2R'O_2$•	$\rightarrow$	$HO_2\dot{R}''O_2H$	(internal H-atom abstraction)
$HO_2\dot{R}''O_2H$	$\rightarrow$	$HO_2R''O$ + HO•	(chain propagation)
$HO_2R''O$	$\rightarrow$	OR''O• + HO•	(chain branching)

This mechanism can explain the observations of so-called *two-stage ignition* (see Fig. 16.6) and a *negative temperature coefficient* of the ignition delay time (see Fig. 16.7): The precursors of the chain branching, formed by the oxygen addition, decompose back to the reactants at high temperatures due to their instability (*degenerate chain branching*). In two-stage ignitions, a combustible mixture reacts accompanied by a small temperature increase, which (surprisingly) stops the chain branching. After a second, very long, ignition delay time, a second ignition and complete combustion takes place, which is governed by the high-temperature oxidation. The region of negative temperature coefficient is characterized by the fact that a temperature increase causes an increase of the ignition delay time, instead of the usual temperature dependence (see Section 10.4) where delay time would decrease.

Furthermore, multistage ignition in principal can explain the complex structure of *p*-*T* ignition diagrams treated in Section 10.3 (see Bamford and Tipper 1977 for further information).

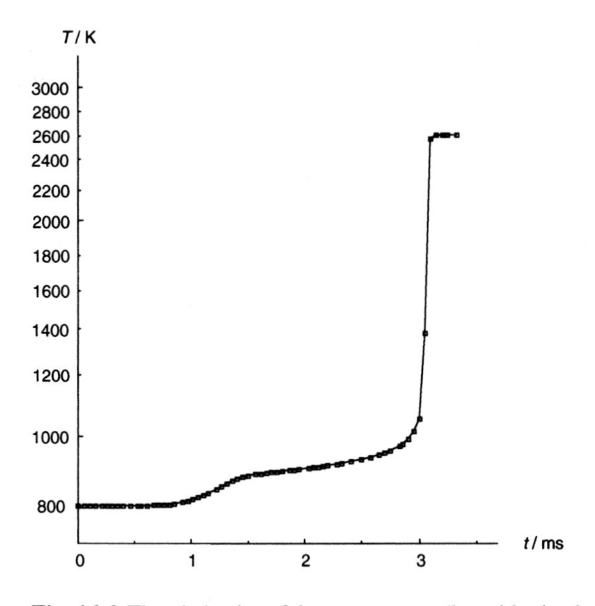

Fig. 16.6. Time behavior of the temperature (logarithmic plot) during two-stage ignition of a stoichiometric n-heptane-air mixture, constant pressure p = 15 bar, T_0 = 800 K, adiabatic conditions (calculated with the mechanism of Chevalier et al 1990a,b)

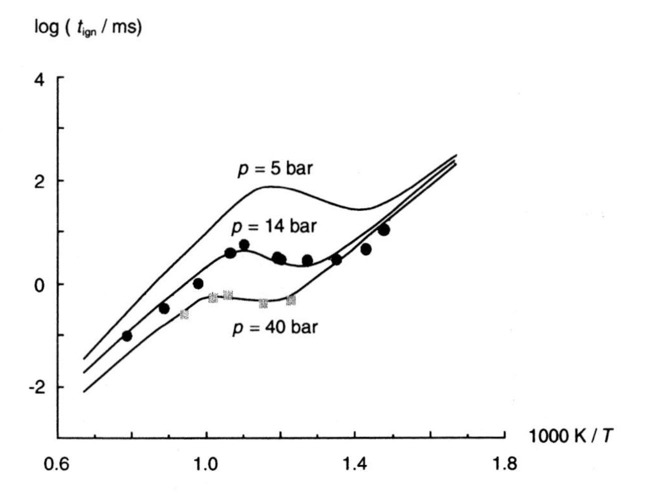

Fig. 16.7. Ignition delay times in stoichiometric n-heptane-air mixtures; negative temperature coefficients occur just below 1000 K (Chevalier et al. 1990a,b)

An example for autoignition in a car engine is shown in Fig. 16.8 (see Warnatz 1991). The fuel is n-octane, which has a high knock tendency. At 900 K knock is observed in the experiment, indicated by pressure oscillations. The simulation is based on temperature- and pressure histories from the experiment; the pressure is measured directly, whereas the temperature is calculated assuming nearly adiabatic compression in the cylinder and a certain heat loss to the cylinder walls. A similar ignition behavior is observed both in the experiment and in the simulation, where the overshoot of the OH-radical and the onset of CO formation are indicating autoignition of the end gas.

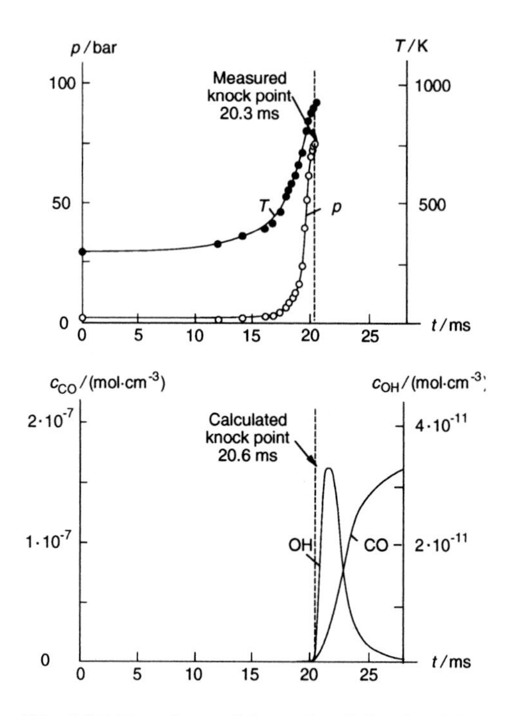

Fig. 16.8. Experimental (upper) and simulated (lower) knocking in an engine (Warnatz 1991)

The low-temperature oxidation leads to very large reaction mechanisms, in part because the many radicals R, R', R",... have a variety of different isomeric structures (~6000 reactions of ~2000 species for n-$C_{16}H_{34}$, cetane, a component of Diesel and jet fuel, see Fig. 16.9) and because the addition of oxygen to these radicals leads to yet more species that need consideration in the mechanism. These reaction mechanisms are best generated automatically by computer programs, as the chance for human error is large (Chevalier et al. 1990b). The reaction types occurring (at high temperature: (1) alkane and alkene decomposition, (2) H atom abstraction from alkanes and alkenes by H, O, OH, HO_2, HC-radicals, (3) β-decomposition of alkyl radicals, (4) isomerization of alkyl radicals; at low temperature: (5) two consecutive O_2 additions to alkyl radicals, (6) isomerization of alkylperoxi, alkylhydroperoxi radicals via cyclic structures, (7) OH elimination after internal rearrangement, and (8) β-decomposition of O=RO$^\bullet$, C=RO$^\bullet$, O=R$^\bullet$, and alkenyl radicals) are written for the actual molecular structures resulting for the fuel considered.

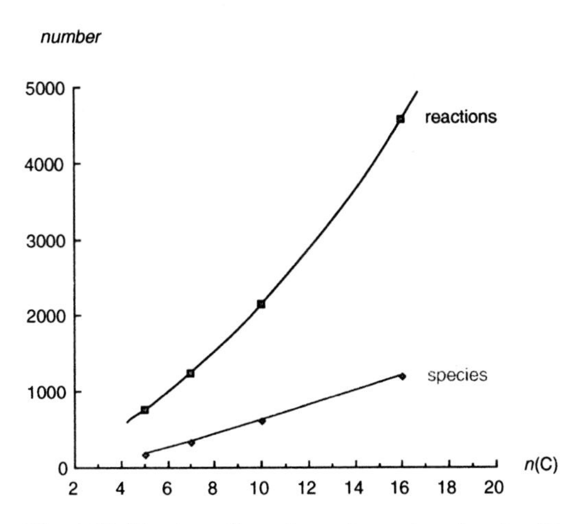

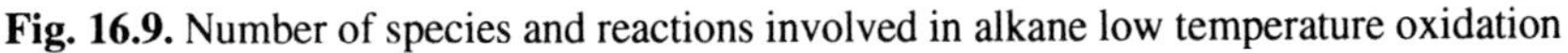

Fig. 16.9. Number of species and reactions involved in alkane low temperature oxidation

16.4 Knock Damages

The ignition of *hot spots* in the unburnt end gas is too fast for the pressure to equilibrate. The hot spots are formed due to non-uniformities in temperature or concentrations. Consequently, pressure waves are generated, which can cause the formation of detonation waves (Lutz et al. 1989, Goyal et al. 1990a,b; see Section 10.6).

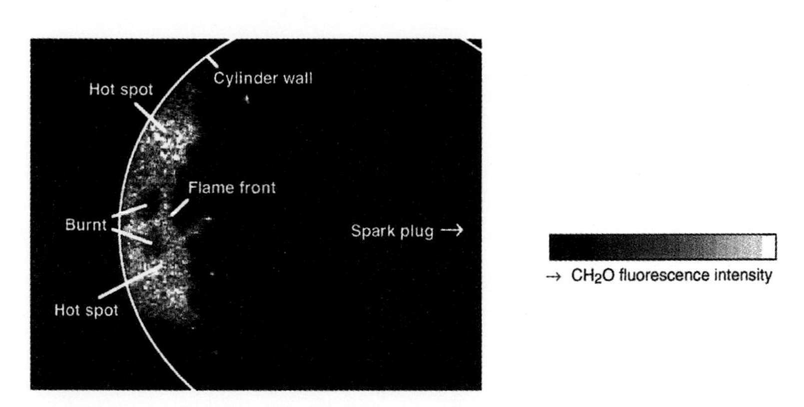

Fig. 16.10. Hot spot formation in the end gas of an Otto engine visualized by 2D-LIF of CH_2O

An experimental detection of hot spots is shown in Fig. 16.10 (Bäuerle et al. 1995). 2D-LIF of formaldehyde (CH_2O) is used for this purpose. CH_2O is built up in the end gas before the flame front, and hot spots can be detected by its higher concentration. Ignition is occurring earlier in the hot spots, leading then to the disappearance of the formaldehyde.

The fast propagation of the detonation waves (velocities higher than 2000 m/s; see Chapter 10) causes an almost simultaneous ignition of the end gas. If the pressure waves (which may superpose or be focussed) hit the cylinder walls, knock damages can result in the form of metal removal from wall or piston.

16.5 Exercises

Exercise 16.1. Derive the reaction rate for the OH-radicals in dependence of R, O_2, and RH, where R and R′ are two different hydrocarbon groups, and the dot denotes a free valence. Assume quasi-steady state for the intermediates. The mechanism is given by the reaction sequence

$$
\begin{array}{llllllll}
\dot{R} & + & O_2 & \rightarrow & \dot{R}O_2 & & & (1) \\
\dot{R}O_2 & & & \rightarrow & \dot{R} & + & O_2 & (-1) \\
\dot{R}O_2 & + & RH & \rightarrow & ROOH & + & \dot{R} & (2) \\
ROOH & & & \rightarrow & \dot{R}O & + & \dot{O}H & (3) \\
\dot{R}O_2 & & & \rightarrow & \dot{R}'OOH & & & (4) \\
\dot{R}'OOH & & & \rightarrow & R'O & + & \dot{O}H & (5)
\end{array}
$$

What happens if Reaction (-1) is much slower than reaction (1)?

17 Formation of Nitric Oxides

With the steady increase in combustion of hydrocarbon fuels, the products of combustion are distinctly identified as a severe source of environmental damage. The major combustion products are carbon dioxide and water. These products were, until recently, considered harmless. Now, even the carbon dioxide is becoming a significant source in the atmospheric balance, and concerns of a global greenhouse effect are being raised.

Less obvious products of combustion are nitric oxides (NO_x). In the last half of the twentieth century, it has become apparent that NO and NO_2, collectively called NO_x, is a major contributor of photochemical smog and ozone in the urban air, more general, the troposphere (Seinfeld 1986). Furthermore, NO_x participates in a chain reaction removing ozone from the stratosphere with the consequence of increased ultraviolet radiation reaching the earth's surface (Johnston 1992). Consequently, minimization of NO_x production has become a most important topic in combustion. This minimization has been and continues to be achieved through increased comprehension of the underlying chemical kinetic mechanisms that generate NO_x and understanding of the interaction of chemical kinetics and fluid dynamics. These models guide investigations toward finding new ways for the minimization of these pollutants.

Four different routes are now identified in the formation of NO_x (Bowman 1993). These are the thermal route, the prompt route, the N_2O route, and the fuel-bound nitrogen route. Each of these shall be discussed in turn. Furthermore, examples will be given of some *primary measures* (modifications of the combustion process itself) used to reduce the amount of NO_x generated and of some *secondary measures* (post-combustion processing) that chemically convert NO_x to harmless products (like H_2O and N_2).

17.1 Thermal NO (Zeldovich-NO)

Thermal NO or *Zeldovich*-NO (after Y. B. Zeldovich, 1946, who postulated the mechanism) is formed by the elementary reactions (Baulch et al. 1994)

$$\mathrm{O + N_2} \xrightarrow{k_1} \mathrm{NO + N} \quad k_1 = 1.8 \cdot 10^{14} \exp(-318\ \mathrm{kJ \cdot mol^{-1}}/(RT))\ \mathrm{cm^3/(mol \cdot s)} \quad (1)$$

$$\mathrm{N + O_2} \xrightarrow{k_2} \mathrm{NO + O} \quad k_2 = 9.0 \cdot 10^{9} \exp(-27\ \mathrm{kJ \cdot mol^{-1}}/(RT))\ \mathrm{cm^3/(mol \cdot s)} \quad (2)$$

$$\mathrm{N + OH} \xrightarrow{k_3} \mathrm{NO + H} \quad k_3 = 2.8 \cdot 10^{13}\ \mathrm{cm^3/(mol \cdot s)} \quad (3)$$

The name "thermal" is used, because the Reaction (1) has a very high activation energy due to the strong triple bond in the N_2-molecule, and is thus sufficiently fast only at high temperatures. Because of its small rate, Reaction (1) is the rate-limiting step of the thermal NO-formation. The temperature dependence of the rate coefficient k_1 is shown in Fig. 17.1.

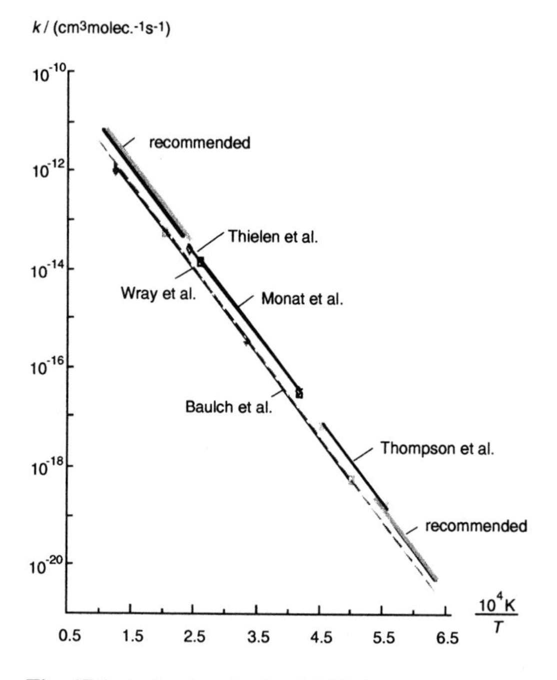

Fig. 17.1. Arrhenius plot $k = k(1/T)$ for the reaction $\mathrm{O + N_2 \rightarrow NO + N}$ (Riedel et al. 1992)

Figure 17.2 shows results of NO-concentration measurements in premixed hydrogen-air flames and compares them with computational results (at a distance $z = 3$ cm behind the flame front), which take into account Reactions 1-3 (Warnatz 1981b). Rather good agreement is obtained, suggesting the rate coefficients k_1, k_2, and k_3 are known quite well (see Fig. 17.1).

While the concentration of, e. g., H_2O or CO_2 is roughly predicted using equilibrium, NO is poorly predicted with equilibrium (note the logarithmic scale in Fig. 17.2). Reaction (1) is so slow that an equilibrium is reached only for times which are much longer than typical residence times in the high temperature range (some ms).

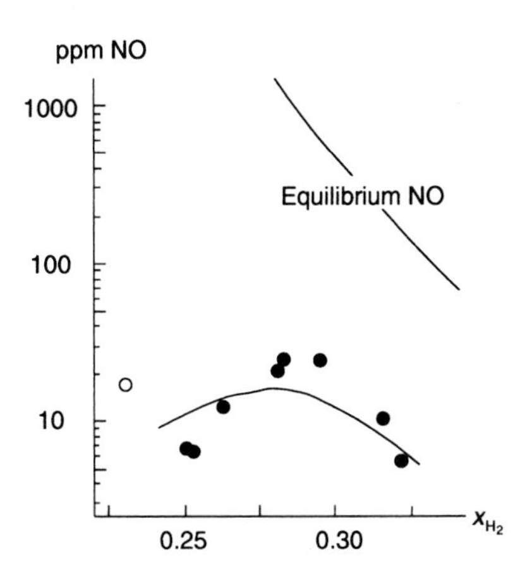

Fig. 17.2. Measured and calculated NO-concentrations in H_2-air flames, dependence on the stoichiometry (Warnatz 1981b)

For the rate of formation of NO, one obtains according to the Reactions (1-3) above

$$\frac{d[NO]}{dt} = k_1[O][N_2] + k_2[N][O_2] + k_3[N][OH] . \tag{17.1}$$

Because

$$\frac{d[N]}{dt} = k_1[O][N_2] - k_2[N][O_2] - k_3[N][OH] \tag{17.2}$$

and the nitrogen atoms can be assumed to be in a quasi-steady state (fast reaction in the steps (2) and (3), see Section 7.1.1 for details), i. e., $d[N]/dt \approx 0$, one obtains for the NO-formation

$$\frac{d[NO]}{dt} = 2k_1[O][N_2] . \tag{17.3}$$

Thus, it can be seen that NO can be minimized by decreasing $[N_2]$, [O], or k_1 (i. e., by decreasing the temperature).

The N_2-concentration can be accurately measured with a probe or estimated rather accurately assuming equilibrium in the burnt gas. For the O-atom concentration, it is tempting to assume the equilibrium, which can be obtained very easily from thermodynamic data. However, equilibrium – particularly at relatively low pressures – underpredicts [O] by a factor of up to 10 (as can be seen from Fig. 17.3). This *super-equilibrium concentration* is generated by kinetics in the flame front.

A better approximation for [O] is the *partial-equilibrium* assumption discussed in Section 7.1.2. There the result was

$$[O] = \frac{k_{H+O_2} \cdot k_{OH+H_2} \cdot [O_2][H_2]}{k_{OH+O} \cdot k_{H+H_2O} \cdot [H_2O]} . \tag{17.4}$$

Thus, the O-atom concentration can be calculated from the concentrations of H_2O,

O_2, and H_2, which can be measured or estimated very easily, because they are stable species. As shown in Section 7.1.2, partial equilibrium is valid only for high temperatures above 1700 K. Here this constraint does not matter very much, because the rate coefficient k_1 is insignificant at temperatures T < 1700 K (Warnatz 1990).

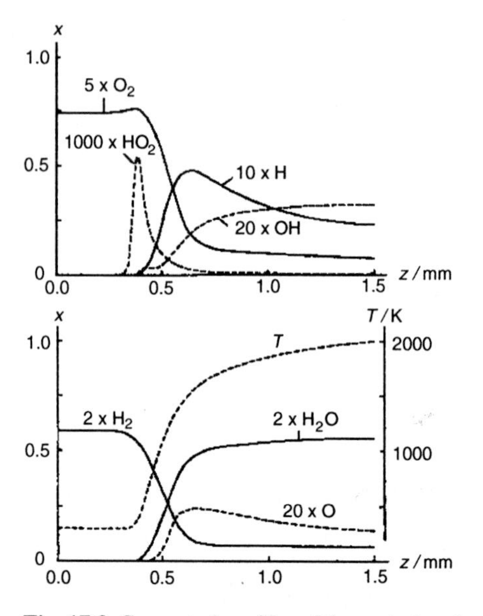

Fig. 17.3. Computed profiles of the mole fractions in a stoichiometric hydrogen-air flame; p = 1 bar, T_u = 298 K (Warnatz 1981b)

17.2 Prompt NO (Fenimore-NO)

The mechanism of *prompt* or *Fenimore* NO was postulated by C. P. Fenimore (1979), who measured [NO] above a hydrocarbon flat flame and noted that the [NO] did not approach zero as the probe approached the flame from the downstream side, as the Zeldovich mechanism predicts. The additional mechanism that is promptly producing NO at the flame front is more complicated than thermal NO, because the prompt NO results from the radical CH, which was previously considered to be an unimportant transient species that is generated through a complex reaction scheme shown in Fig. 17.4. The CH, which is formed as an intermediate at the flame front only (see Fig. 2.8), reacts with the nitrogen of the air, forming hydrocyanic acid (HCN), which reacts further to NO (see Section 17.4 for details),

$$CH + N_2 \rightarrow HCN + N \begin{array}{l} \nearrow \cdots \rightarrow NO \\ \searrow \cdots \rightarrow N_2 \end{array}$$

Precise information about the rate-limiting step $CH + N_2 \rightarrow HCN + N$ is rather rare in the literature, as can be seen from the Arrhenius plot of the rate coefficient in Fig. 17.5. The estimated accuracy is about a factor of 2 at the present time.

Thus, predictions of Fenimore NO are less accurate, as can be seen in Fig. 17.6, which shows mole fraction profiles in a stoichiometric C_3H_8-air low-pressure flame (Bockhorn et al. 1991). Points denote experiments, lines denote simulations.

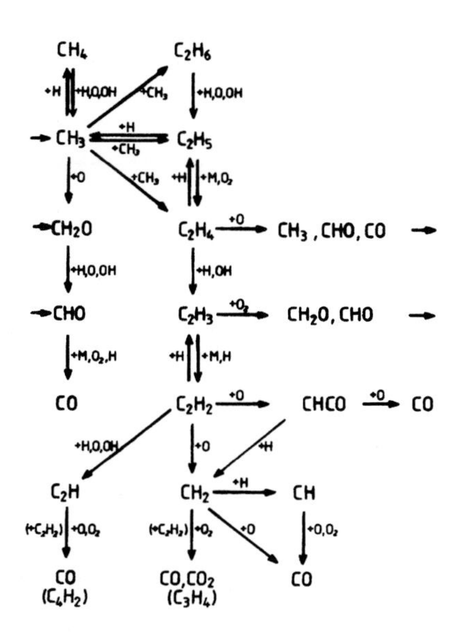

Fig. 17.4. Mechanism of the oxidation of C_1- and C_2-hydrocarbons (Warnatz 1981a, 1993)

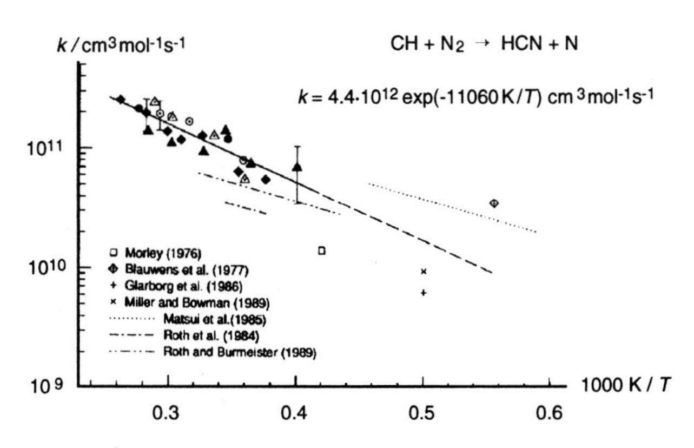

Fig. 17.5. Rate coefficients for the reaction of CH with N_2 (Dean et al. 1990)

Because C_2H_2 as a CH-radical precursor (see Fig. 17.4) is accumulated under fuel-rich conditions (due to CH_3-recombination), prompt NO is favored in rich flames.

The NO production during methane combustion in a *stirred reactor* is shown in Fig. 17.7. Calculations are performed for a purely thermal mechanism and the complete mechanism (Zeldovich- and Fenimore-NO), such that the difference between thermal NO and overall NO may be attributed to prompt NO.

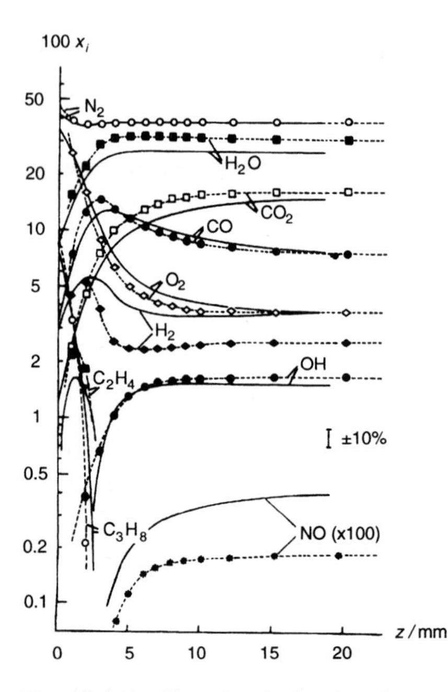

Fig. 17.6. Profiles of mole fractions in a stoichiometric propane-air flame (Bockhorn et al. 1991)

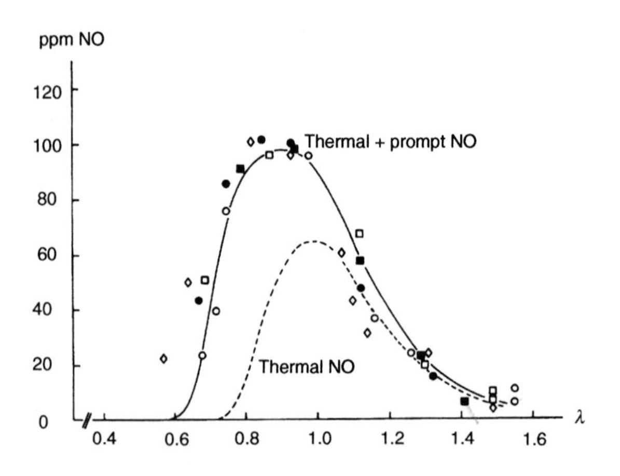

Fig. 17.7. NO production in a stirred reactor; dependence on the air number $\lambda = 1/\Phi$ (Bartok et al. 1972, Glarborg et al. 1986); $\lambda > 1$ characterizes a lean mixture

The activation energy of the reaction $CH + N_2 \rightarrow HCN + N$ is only about 75 kJ/mol ($T_a \approx 9\,000$ K), compared to 318 kJ/mole ($T_a = 38\,200$ K) for the formation of the thermal NO; therefore, in contrast to thermal NO, prompt NO is also produced at relatively low temperatures (about 1000 K).

17.3 NO Generated via Nitrous Oxide

The *nitrous oxide (N_2O) mechanism* is analogous to the thermal mechanism in that O-atom attacks molecular nitrogen. However, with the presence of a third molecule M, the outcome of this reaction is N_2O (postulated first by Wolfrum 1972),

$$N_2 + O + M \rightarrow N_2O + M \ .$$

The N_2O may subsequently react with O atoms to form NO (Malte and Pratt 1974),

$$N_2O + O \rightarrow NO + NO \qquad E_a = 97 \text{ kJ/mol} \ .$$

This reaction has been often overlooked since it usually is a insignificant contributor to the total NO. However, lean conditions can suppress the formation of CH and, hence, lead to less Fenimore NO, and low temperatures can suppress the Zeldovich NO. What remains is NO generated via N_2O, which is promoted at high pressures because of the three-body reaction and, typical for three-body reactions, has a low activation energy so that low temperatures do not penalize this reaction as much as they do the Zeldovich-NO reaction. All of these circumstances lead to the N_2O route being the major source of NO in lean premixed combustion in gas turbine engines (Correa 1992).

17.4 Conversion of Fuel Nitrogen into NO

The conversion of fuel-nitrogen, sometimes called fuel-bound nitrogen (FBN), into NO is mainly observed in coal combustion, because even "clean" coal contains about 1 mass-% chemically bound nitrogen. The nitrogen-containing compounds evaporate during the gasification process and lead to NO formation in the gas phase.

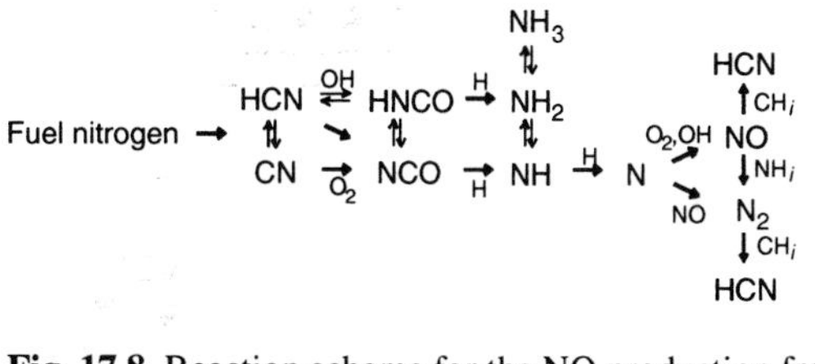

Fig. 17.8. Reaction scheme for the NO production from fuel nitrogen (Glarborg et al. 1986)

The conversion of the nitrogen-containing compounds (see Fig. 17.8) into NH_3 (ammonia) and HCN (hydrocyanic acid) is usually quite fast and, thus, not rate-limiting. The rate-limiting steps are the reactions of the N-atoms (see below).

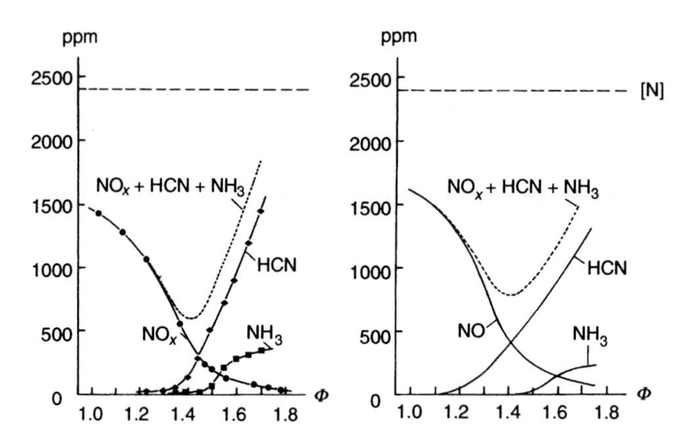

Fig. 17.9 Measurements (left) and simulation (right) of the formation of nitrogen-containing compounds in propane-air flames of different stoichiometry, doped with 2 400 ppm CH_3-NH_2 (Eberius et al. 1987); $\Phi > 1$ characterizes fuel-rich conditions

A propane-air flame, which is doped with 2400 ppm CH_3-NH_2 (methylamine) can be used as a model system for fuel-nitrogen conversion to NO (Fig. 17.9; Eberius et al. 1987). Under oxygen rich conditions ($\Phi < 1.0$) about two thirds of the fuel nitrogen are oxidized to NO. The rest is converted to N_2. Under fuel rich conditions ($\Phi > 1.0$) the amount of NO decreases, but other products like HCN (hydrocyanic acid) and NH_3 (ammonia) are formed, which, again, are oxidized to NO in the atmosphere. The most important fact is, that the sum of the pollutants has a minimum at $\Phi = 1.4$, i. e., the conversion of the fuel-nitrogen into molecular nitrogen (N_2) has a maximum for fuel-rich conditions. The simulations have been performed using a reaction mechanism similar to that in Tab. 17.1 (in addition to the mechanism of propane combustion, Tab. 6.1). This mechanism also accounts for all of the mentioned sources of NO.

Tab. 17.1. Reaction mechanism of NO_x formation and reduction (Klaus and Warnatz 1995)

Reaction				*A*[cm,mol,s]	*b*	*E*/kJ·mol^{-1}
-----	30. - 40. Reactions of H-N-O-Species					
-----	30. Consumption of NH_3					
NH_3	+H	=NH_2	+H_2	$6.36 \cdot 10^{05}$	2.4	42.6
NH_3	+O	=NH_2	+OH	$1.10 \cdot 10^{06}$	2.1	21.8
NH_3	+OH	=NH_2	+H_2O	$2.04 \cdot 10^{06}$	2.0	2.37

NH_3	+M*		$=NH_2$	+H	+M*	$1.40 \cdot 10^{16}$	.06	379.
----	31. Consumption of NH_2							
NH_2	+H		=NH	$+H_2$		$6.00 \cdot 10^{12}$	0.0	0.00
NH_2	+O		=NH	+OH		$7.00 \cdot 10^{12}$	0.0	0.00
NH_2	+O		=HNO	+H		$4.50 \cdot 10^{13}$	0.0	0.00
NH_2	+O		=NO	$+H_2$		$5.00 \cdot 10^{12}$	0.0	0.00
NH_2	+N		$=N_2$	+H	+H	$7.20 \cdot 10^{13}$	0.0	0.00
NH_2	$+O_2$		=HNO	+OH		$4.50 \cdot 10^{12}$	0.0	105.
NH_2	$+O_2$		=NH	$+HO_2$		$1.00 \cdot 10^{14}$	0.0	209.
NH_2	+OH		=NH	$+H_2O$		$9.00 \cdot 10^{07}$	1.5	-1.91
NH_2	$+HO_2$		$=NH_3$	$+O_2$		$4.50 \cdot 10^{13}$	0.0	0.00
NH_2	$+NH_2$		$=NH_3$	+NH		$6.30 \cdot 10^{12}$	0.0	41.8
----	32. Consumption of NH							
NH	+H		=N	$+H_2$		$1.00 \cdot 10^{13}$	0.0	0.00
NH	+O		=NO	+H		$7.00 \cdot 10^{13}$	0.0	0.00
NH	+OH		=NO	$+H_2$		$2.40 \cdot 10^{13}$	0.0	0.00
NH	+OH		=N	$+H_2O$		$2.00 \cdot 10^{09}$	1.2	0.02
NH	+OH		=HNO	+H		$4.00 \cdot 10^{13}$	0.0	0.00
NH	$+O_2$		=NO	+OH		$1.00 \cdot 10^{13}$	-0.2	20.8
NH	$+O_2$		=HNO	+O		$4.60 \cdot 10^{05}$	2.0	27.2
NH	+NH		$=N_2$	+H	+H	$2.54 \cdot 10^{13}$	0.0	0.40
----	33. Consumption of N							
N	+OH		=NO	+H		$3.80 \cdot 10^{13}$	0.0	0.00
N	$+O_2$		=NO	+O		$6.40 \cdot 10^{09}$	1.0	26.1
N	$+CO_2$		=NO	+CO		$1.90 \cdot 10^{11}$	0.0	14.2
N	+NO		$=N_2$	+O		$3.27 \cdot 10^{12}$	0.3	0.00
N	+N	+M*	$=N_2$	+M*		$2.26 \cdot 10^{17}$	0.0	32.3
N	+NH		$=N_2$	+H		$3.00 \cdot 10^{13}$	0.0	0.00
N	+CH		=CN	+H		$1.30 \cdot 10^{12}$	0.0	0.00
N	$+^3CH_2$		=HCN	+H		$5.00 \cdot 10^{13}$	0.0	0.00
N	$+CH_3$		$=H_2CN$	+H		$7.10 \cdot 10^{13}$	0.0	0.00
N	+HCCO		=HCN	+CO		$5.00 \cdot 10^{13}$	0.0	0.00
N	$+C_2H_2$		=HCN	+CH		$1.04 \cdot 10^{15}$	-0.5	0.00
N	$+C_2H_3$		=HCN	$+^3CH_2$		$2.00 \cdot 10^{13}$	0.0	0.00
----	34. Consumption of N_2H							
N_2H	+O		$=N_2O$	+H		$1.00 \cdot 10^{14}$	0.0	0.00
N_2H	+O		=NO	+NH		$1.00 \cdot 10^{13}$	0.0	0.00
N_2H	+OH		$=N_2$	$+H_2O$		$3.00 \cdot 10^{13}$	0.0	0.00
N_2H	+M*		$=N_2$	+H	+M*	$1.70 \cdot 10^{12}$	0.0	59.9
N_2H	+NO		$=N_2$	+HNO		$5.00 \cdot 10^{13}$	0.0	0.00
----	35. Consumption of N_2							
N_2	+CH		=HCN	+N	++)	$1.56 \cdot 10^{11}$	0.0	75.1
N_2	$+^3CH_2$		=HCN	+NH		$4.28 \cdot 10^{12}$	0.0	150.
----	36. Consumption of NO							
NO	+OH	+M*	$=HNO_2$	+M*		$5.08 \cdot 10^{12}$	-2.5	0.28

NO	$+HO_2$	$=NO_2$	$+OH$		$2.10\cdot10^{12}$	0.0	-2.01
NO	$+NH$	$=N_2$	$+OH$		$2.16\cdot10^{13}$	-.23	0.00
NO	$+NH$	$=N_2O$	$+H$	+)	$2.94\cdot10^{14}$	-0.4	0.00
NO	$+NH$	$=N_2O$	$+H$	+)	$-2.16\cdot10^{13}$	-.23	0.00
NO	$+NH_2$	$=N_2$	$+H_2O$		$2.00\cdot10^{20}$	-2.6	3.87
NO	$+NH_2$	$=N_2$	$+H$	$+OH$	$4.76\cdot10^{15}$	-1.1	0.81
NO	$+NH_2$	$=N_2H$	$+OH$		$3.97\cdot10^{11}$	0.0	-1.63
NO	$+CH$	$=HCN$	$+O$		$1.20\cdot10^{14}$	0.0	0.00
NO	$+{}^1CH_2$	$=HCN$	$+OH$		$2.00\cdot10^{13}$	0.0	0.00
NO	$+{}^3CH_2$	$=HCNO$	$+H$		$2.59\cdot10^{12}$	0.0	25.0
NO	$+{}^3CH_2$	$=HCN$	$+OH$		$5.01\cdot10^{11}$	0.0	12.0
NO	$+CH_3$	$=HCN$	$+H_2O$		$1.50\cdot10^{12}$	0.0	91.0
NO	$+CH_3$	$=H_2CN$	$+OH$		$1.00\cdot10^{12}$	0.0	91.0
NO	$+CHO$	$=CO$	$+HNO$		$7.20\cdot10^{12}$	0.0	0.00
NO	$+C_2H$	$=HCN$	$+CO$		$2.11\cdot10^{13}$	0.0	0.00
NO	$+HCCO$	$=HCNO$	$+CO$		$1.30\cdot10^{13}$	0.0	0.00

---- 37. Consumption of N_2O

N_2O	$+H$	$=OH$	$+N_2$		$9.64\cdot10^{13}$	0.0	63.1
N_2O	$+O$	$=NO$	$+NO$		$6.60\cdot10^{13}$	0.0	111.
N_2O	$+O$	$=N_2$	$+O_2$		$1.02\cdot10^{14}$	0.0	117.
N_2O	$+OH$	$=HO_2$	$+N_2$		$2.00\cdot10^{12}$	0.0	41.8
N_2O	$+CO$	$=N_2$	$+CO_2$		$1.25\cdot10^{12}$	0.0	72.3
N_2O	$+CH_3$	$=CH_3O$	$+N_2$		$1.00\cdot10^{15}$	0.0	119.
N_2O	$+M^*$	$=O$	$+N_2$	$+M^*$	$7.23\cdot10^{17}$	-.73	263.

---- 38. Consumption of NO_2

NO_2	$+O$	$=NO$	$+O_2$		$1.00\cdot10^{13}$	0.0	2.51
NO_2	$+H$	$=NO$	$+OH$		$1.00\cdot10^{14}$	0.0	6.27
NO_2	$+N$	$=N_2$	$+O_2$		$1.18\cdot10^{12}$	0.0	0.00
NO_2	$+CO$	$=NO$	$+CO_2$		$1.20\cdot10^{14}$	0.0	132.
NO_2	$+CH$	$=CHO$	$+NO$		$5.90\cdot10^{13}$	0.0	0.00
NO_2	$+{}^3CH_2$	$=CH_2O$	$+NO$		$5.90\cdot10^{13}$	0.0	0.00
NO_2	$+CH_3$	$=CH_3O$	$+NO$		$1.30\cdot10^{13}$	0.0	0.00
NO_2	$+CHO$	$=CO_2$	$+H$	$+NO$	$8.40\cdot10^{15}$	-.75	8.07
NO_2	$+CHO$	$=CO$	$+HNO$		$2.10\cdot10^{00}$	3.3	9.82
NO_2	$+HCCO$	$=NCO$	$+CO$	$+OH$	$5.00\cdot10^{12}$	0.0	0.00
NO_2	$+HCCO$	$=HNCO$	$+CO_2$		$5.00\cdot10^{12}$	0.0	0.00
NO_2	$+HCCO$	$=HCN$	$+CO_2$	$+O$	$5.00\cdot10^{12}$	0.0	0.00
NO_2	$+M^*$	$=NO$	$+O$	$+M^*$	$1.10\cdot10^{16}$	0.0	276.
NO_2	$+NO_2$	$=NO$	$+NO$	$+O_2$	$1.60\cdot10^{12}$	0.0	109.

---- 39. Consumption of HNO

HNO	$+H$	$=NO$	$+H_2$		$1.81\cdot10^{13}$	1.9	4.16
HNO	$+OH$	$=NO$	$+H_2O$		$1.32\cdot10^{07}$	1.9	-4.00
HNO	$+N$	$=NO$	$+NH$		$1.00\cdot10^{13}$	0.0	8.30
HNO	$+O_2$	$=NO$	$+HO_2$		$3.16\cdot10^{12}$	0.0	12.5
HNO	$+NH_2$	$=NO$	$+NH_3$		$5.00\cdot10^{13}$	0.0	4.20
HNO	$+HNO$	$=N_2O$	$+OH$		$3.90\cdot10^{12}$	0.0	209.

HNO	+NO	$=N_2O$	$+H_2O$		$2.00\cdot10^{12}$	0.0	109.
HNO	$+NO_2$	$=HNO_2$	+NO		$6.02\cdot10^{11}$	0.0	8.31
HNO	+M*	=NO	+H	+M*	$1.50\cdot10^{16}$	0.0	203.

---- 40. Consumption of HNO_2

HNO_2	+H	$=NO_2$	$+H_2$		$1.20\cdot10^{13}$	0.0	30.7
HNO_2	+O	$=NO_2$	+OH		$1.20\cdot10^{13}$	0.0	25.1
HNO_2	+OH	$=NO_2$	$+H_2O$		$1.30\cdot10^{10}$	1.0	0.56

----- 50. - 55. Reactions of C-H-N-O-Species

---- 50. Consumption of HCN

HCN	+O	=NCO	+H		$1.11\cdot10^{06}$	2.1	25.6
HCN	+O	=NH	+CO		$2.77\cdot10^{05}$	2.1	25.6
HCN	+OH	=HNCO	+H		$4.77\cdot10^{11}$	0.0	91.4
HCN	+CN	$=C_2N_2$	+H		$2.00\cdot10^{13}$	0.0	0.00

---- 51. Consumption of CN/C_2N_2

CN	+O	=CO	+N		$1.00\cdot10^{13}$	0.0	0.00
CN	+OH	=NCO	+H		$6.00\cdot10^{13}$	0.0	0.00
CN	$+O_2$	=NCO	+O		$6.60\cdot10^{12}$	0.0	-1.70
CN	$+H_2$	=HCN	+H		$3.10\cdot10^{05}$	2.4	9.30
CN	$+H_2O$	=HCN	+OH		$7.83\cdot10^{12}$	0.0	31.1
CN	+N	$=N_2$	+C		$1.04\cdot10^{15}$	-0.5	0.00
CN	+NO	$=N_2$	+CO		$1.07\cdot10^{14}$	0.0	33.4
CN	+NO	=NCO	+N		$9.64\cdot10^{13}$	0.0	176.
CN	$+N_2O$	=NCO	$+N_2$		$1.00\cdot10^{13}$	0.0	0.00
CN	$+NO_2$	=NCO	+NO		$3.00\cdot10^{13}$	0.0	0.00
CN	+CH4	=HCN	$+CH_3$		$9.03\cdot10^{12}$	0.0	7.82
C_2N_2	+O	=NCO	+CN		$4.57\cdot10^{12}$	0.0	37.1

---- 52. Consumption of HNCO/HCNO

HCNO	+H	=HCN	+OH		$1.00\cdot10^{14}$	0.0	0.00
HCNO	+H	=HNCO	+H		$1.00\cdot10^{11}$	0.0	0.00
HNCO	+H	$=NH_2$	+CO		$2.25\cdot10^{07}$	1.7	15.9
HNCO	+O	=NH	$+CO_2$		$9.60\cdot10^{07}$	1.4	35.6
HNCO	+O	=NCO	+OH		$2.20\cdot10^{06}$	2.1	47.8
HNCO	+O	=HNO	+CO		$1.50\cdot10^{08}$	1.6	184.
HNCO	+OH	=NCO	$+H_2O$		$6.40\cdot10^{05}$	2.0	10.7
HNCO	$+O_2$	=HNO	$+CO_2$		$1.00\cdot10^{12}$	0.0	146.
HNCO	$+HO_2$	=NCO	$+H_2O_2$		$3.00\cdot10^{11}$	0.0	121.
HNCO	+M*	=NH	+CO	+M*	$1.10\cdot10^{16}$	0.0	359.
HNCO	+NH	=NCO	$+NH_2$		$3.03\cdot10^{13}$	0.0	99.1
HNCO	$+NH_2$	=NCO	$+NH_3$		$5.00\cdot10^{12}$	0.0	25.9

---- 53. Consumption of NCO

NCO	+O	=NO	+CO		$4.20\cdot10^{13}$	0.0	0.00
NCO	+H	=NH	+CO		$5.20\cdot10^{13}$	0.0	0.00
NCO	+OH	=CHO	+NO		$5.00\cdot10^{12}$	0.0	62.7

NCO	$+H_2$	$=HNCO$	$+H$		$7.60\cdot10^{02}$	3.0	16.7
NCO	$+N$	$=N_2$	$+CO$		$2.00\cdot10^{13}$	0.0	0.00
NCO	$+O_2$	$=NO$	$+CO_2$		$2.00\cdot10^{12}$	0.0	83.6
NCO	$+M^*$	$=N$	$+CO$	$+M^*$	$1.00\cdot10^{15}$	0.0	195.
NCO	$+NO$	$=N_2O$	$+CO$		$6.20\cdot10^{17}$	-1.7	3.19
NCO	$+NO$	$=N_2$	$+CO_2$		$7.80\cdot10^{17}$	-1.7	3.19
NCO	$+NCO$	$=N_2$	$+CO$	$+CO$	$1.80\cdot10^{13}$	0.0	0.00
NCO	$+NO_2$	$=CO$	$+NO$	$+NO$	$1.30\cdot10^{13}$	0.0	0.00
NCO	$+NO_2$	$=CO_2$	$+N_2O$		$5.40\cdot10^{12}$	0.0	0.00
NCO	$+HNO$	$=HNCO$	$+NO$		$1.80\cdot10^{13}$	0.0	0.00
NCO	$+HNO_2$	$=HNCO$	$+NO_2$		$3.60\cdot10^{12}$	0.0	0.00
NCO	$+CHO$	$=HNCO$	$+CO$		$3.60\cdot10^{13}$	0.0	0.00
---- 54. Consumption of C							
CH	$+H$	$=C$	$+H_2$		$1.50\cdot10^{14}$	0.0	0.00
C	$+O_2$	$=CO$	$+O$		$5.00\cdot10^{13}$	0.0	0.00
C	$+NO$	$=CN$	$+O$		$6.60\cdot10^{13}$	0.0	0.00
---- 55. Consumption of H_2CN							
H_2CN	$+N$	$=N_2$	$+^3CH_2$		$2.00\cdot10^{13}$	0.0	0.00
H_2CN	$+M^*$	$=HCN$	$+H$	$+M^*$	$3.00\cdot10^{14}$	0.0	92.0

$k=A\cdot T^b\cdot\exp(-E/RT)$, $[M^*]=[H_2]+6.5\cdot[H_2O]+0.4\cdot[O_2]+0.4\cdot[N_2]+0.75\cdot[CO]+1.5\cdot[CO_2]+3.0\cdot[CH_4]$
$\rightarrow$: only forward reaction is considered; =: the reverse reaction to be calculated with (6.9); $^+$) The rate coefficients of these two (identical) reactions have to be added up; $^{++}$) rate coefficient slightly different from the recommendation given in Fig. 17.5

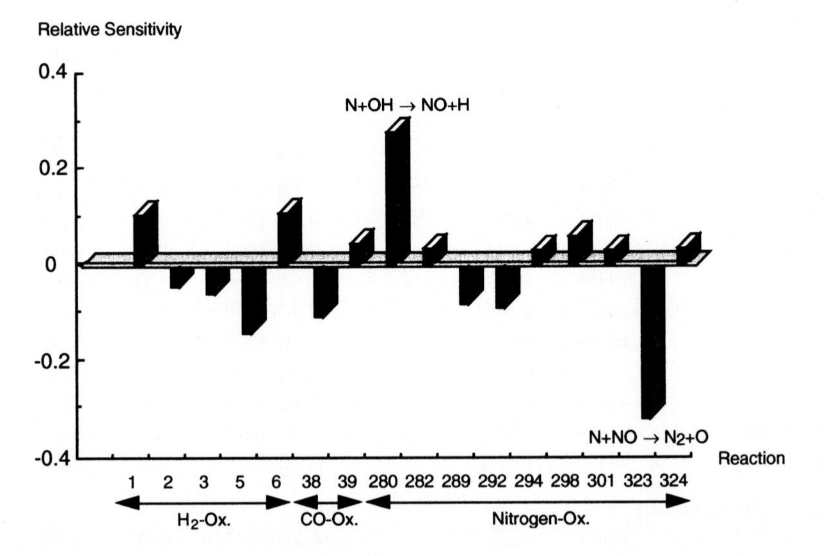

Fig. 17.10. Sensitivity analysis of the NO formation with respect to a reaction mechanism similar to that in Tab. 17.1 (Bockhorn et al. 1991)

A sensitivity analysis (Fig. 17.10) shows, that the rate-limiting steps for the NO formation from fuel nitrogen are the reactions

$$\begin{aligned} N + OH &\rightarrow NO + H \\ N + NO &\rightarrow N_2 + O\,, \end{aligned}$$

which compete for the N-atoms. The data for these reactions are (due to the simplicity of the reactions) rather reliable, so that fuel-NO conversion can be understood in a quantitative way.

17.5 NO Reduction by Combustion Modifications

Using the basic research results given above as a guide, engineers have devised modifications to combustion devices that minimize the amount of NO generated. These combustion modifications are often called *primary measures*. Naturally it is hoped that combustion modifications are without great cost and that they do not need any addition of other compounds. On the other hand, primary methods typically have special new geometrical requirements with respect to the combustion device, and it is very difficult to change old combustion devices to meet those requirements. Thus, primary methods are normally used in new combustion devices only. In old combustion devices there is the possibility to use *secondary methods*, which will be described in a later section.

The results described in Section 17.4 are used technologically in *staged combustion*. In the first stage, fuel rich conditions are chosen for the combustion (about $\Phi =$ 1.4), in order to produce a minimum of compounds NO_x+HCN+NH_3. Then oxygen rich conditions are used in order to obtain stoichiometric combustion in the overall process. The N_2, which is formed in the first stage, is not converted to thermal NO, because the combustion temperature is steadily reduced due to radiation and convective heat transfer. A further reduction of NO can be obtained, if an excess of air is used in the second stage. Then a third stage can be used to add further fuel (*reburn*) and reduce NO by reactions NO + $CH_i \rightarrow$ products (Kolb et al. 1988).

Because of the high activation energy ($T_a \approx$ 38 200 K) of the thermal NO mechanism any scheme that suppresses peak temperatures will lower the NO output. In nonpremixed jet flames, the radiation from the flame, which lowers peak temperatures, has a significant effect on the NO generated. It would be beneficial to inject an "inert" diluent gas such as nitrogen or water, whose additional heat capacity lowers the peak temperature. For this purpose, exhaust gases are reasonably inert. When this effective process is done in piston engines, it is called *exhaust-gas recirculation* (*EGR*), and when done on atmospheric boiler flame, *flue-gas recirculation* (*FGR*). In spite of the success of EGR, the high temperatures and pressures inside of Diesel and Otto engines promotes NO formation as (17.3)-(17.4) indicate. For this reason, devices that burn at lower temperature and pressure are receiving increased attention. Such devices include steam power (Rankine and Kalina cycles), gas turbines, Stirling engines, and fuel cells.

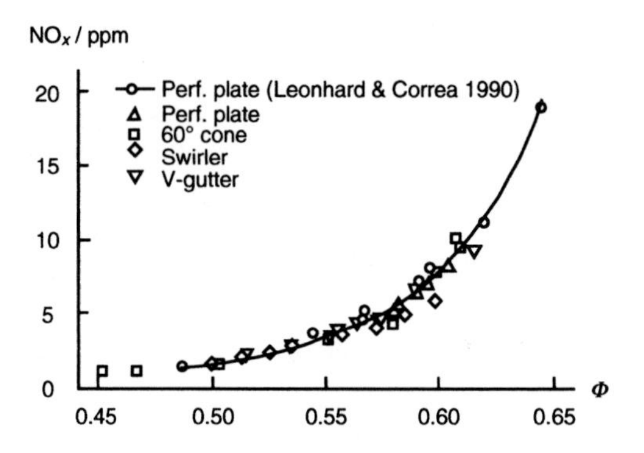

Fig. 17.11. NO_x emissions as function of the equivalence ratio Φ for various types of flame holders ($T = 615$ K, $p = 10.2$ bar)

NO_x formation in gas turbines is reviewed by Correa (1992). Water injection steadily lowers NO_x output until the mass flow of water is about equal to the fuel mass flow rate, at which point the amount of CO and unburnt hydrocarbons increases rapidly to unacceptable levels. Further improvements have been made by operating premixed flames at low temperatures afforded by lean operation. The potential for low NO production under lean conditions could have been anticipated from Fig. 17.7 and is verified by the lean-premix findings presented in Fig 17.11 (Lovett and Abuaf 1992). Note that the NO generated is largely independent of flame holder, consistent with the notion that most of the NO is generated from CH (via the Fenimore mechanism) and super-equilibrium O-atoms (via the N_2O mechanism) that occur in the flame front (see Figs. 2.8 and 17.3 respectively).

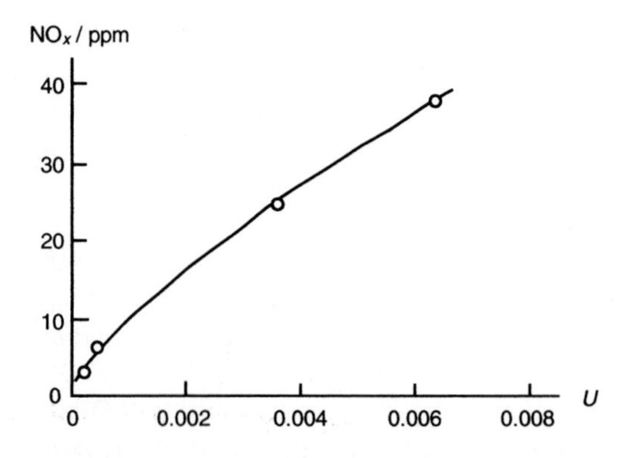

Fig. 17.12. Qualitative relationship between NO_x formation and unpremixedness parameter U

Of course, in any combustion system, the longer the high temperature residence time, the closer to equilibrium NO the system becomes. Thus an optimum time is desired such that all of the fuel is oxidized, as well as intermediates such as CO, and the formation of NO is terminated by rapid cooling (Takeno et al. 1993).

For lean premixed combustion, it is important that the fuel is well mixed. The results presented in Fig. 17.12 (Fric 1993, Mongia et al. 1996) show for a given mean fuel concentration $\bar{c}$, the lowest NO_x is generated when $\overline{c'^2}$ is zero. The extent of non-premixedness U is $\overline{c'^2}$ normalized by its maximum possible value for a given $\bar{c}$, which is evaluated by the *Housdorf relation* $\overline{c'^2}_{max} = \bar{c} \cdot (1 - \bar{c})$; $\overline{c'^2}_{max}$ approaches zero as $\bar{c}$ approaches zero or unity (Dimotakis and Miller 1990). When U is zero, the scalar is completely mixed and homogeneous; when U is unity, no mixing has occurred, even if $\overline{c'^2}$ is low.

As Fig 17.11 indicates, low NO_x is achievable by increasingly lean combustion. In practice, two barriers are impeding the achievement of this goal. First, as the system is increasingly lean, the final flame temperature is lower and, thus, the NO_x is lower. At the same time, the rate with that CO is converted to CO_2 is reduced. Thus, a lower bound to Φ occurs when the CO emission becomes unacceptably high. A closer inspection of this phenomenon shows that the reaction $CO + OH \rightarrow CO_2 + H$ has a surprisingly low activation energy (see Table 6.1). For flame temperatures of 1500 K, the Arrhenius term $\exp(-E_{act}/RT)$ is essentially 1 and, thus, the rate coefficient is effectively temperature-independent. The strong temperature dependence of the CO removal is due to the strong temperature dependence of OH concentration, as shown in Fig 7.3. Thus, the poor CO oxidation rate is due to a decreasing OH concentration and not a decreasing rate coefficient, in contrast to the rate of NO_x production.

The second barrier to lean low NO_x combustion in gas turbines is the onset of large pressure fluctuations in the combustor, called *pressure dynamics*. For a gas turbine combustor at 15 bar, the often observed pressure dynamics of ± 2 bar will in time damage the engine. As Φ decreases, the flame speed decreases, as illustrated in Fig 8.10 and 8.12. This lowered chemical reaction rate is increasingly sensitive to changes in temperature and concentration such as occurs when there are pressure oscillations in the combustor. Pressure oscillations in the combustor, typically at frequencies determined by acoustic modes of the combustor (e. g. $\nu = 100$ Hz to 1000 Hz), cause the slow reaction rate and, hence, the heat release rate to be modulated at the same frequency, potentially strongly amplifying the pressure waves. Pressure waves from the combustor can cause modulation of the input fuel or air flow. These modulations can amplify the combustor pressure oscillations. Such modulations have been measured (Mongia et al. 1998) and modelled (Lieuwen and Zinn 1998).

17.6 Catalytic Combustion

Both of the aforementioned barriers to lean premixed combustion are alleviated with the use of a catalytic combustor. The fuel (and CO) are oxidized on the surface through a sequence of low activation energy reactions, thus oxidation reactions are vigorous at a lower Φ (lower temperature) than for gas phase reactions (see Section 6.7). Furthermore, the surface reactions do not produce NO; the NO_x emissions are below 1 ppm. As for acoustic dynamics, the large surface area of the catalyst supplies viscous drag that dampens pressure pulses and furthermore, the chemical reaction rate on the surface is less closely coupled to the pressure modulations.

However, adoption of a catalyst has been impeded by several hurdles. First, the active surface is usually from platinum (Pt) or palladium (Pd). These noble metals oxidize and vaporize at temperatures above ~1500 K. Consequently, sustained operation of a noble metal catalyst above $T \approx 1300$ K results in an unacceptably high rate of catalyst loss.

The temperature desired by the gas turbine is approaching (1800 K), the strategy mostly has been to burn about half the fuel, the *primary fuel*, in a catalyst section followed by homogeneous gas-phase auto-ignition and combustion of the remaining *secondary fuel*. The homogeneous combustion can be very lean due to extended ignition limits at high temperature; it produces gas-phase NO_x and needs sufficient temperature and time to oxidize CO to CO_2.

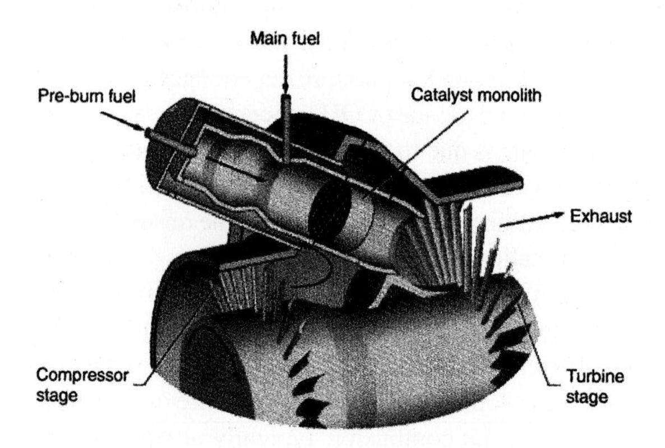

Fig. 17.13. Catalytic combustion of primary fuel and gas-phase combustion of secondary fuel in a turbine

For the turbine combustor shown in the Fig. 17.13, a residence time of ~20 ms is required (Beebe et al. 1995, Dalla Betta et al. 1996, Schlatter et al. 1997, Raja et al. 2000). Injection of the secondary fuel is done in several ways. In one route, the catalyst is active in only one-half of the channels so that the secondary fuel passes through the catalyst and then auto-ignites after contact with the primary combustion products behind the catalyst.

In other devices, the secondary fuel is premixed with air and injected after the catalyst. (Fujii et al. 1996, Smith et al. 1997). In either case, the secondary fuel-in-air mixture mixes with the catalytic combustion products from the primary stage and then auto-ignites.

17.7 NO Reduction by Post-Combustion Processes

If combustion modifications are not efficient enough or not possible at all, post-combustion processes (or *secondary measures*) are necessary to remove pollutants like NO. Probably the most well known of NO removal tool is the catalytic converter that is in the exhaust system of many automobiles (Heywood 1988). The catalyst is a remarkable combination of noble metals that oxidize CO to CO_2 and simultaneously reduce NO to N_2. Key to the success of the catalytic convertor is the *λ-sensor* that detects if there is O_2 in the exhaust. When O_2 is detected, electronic feedback control to the engine directs a slow increase in the fuel flow-rate, and when no O_2 is detected, the fuel rate is slowly decreased. Thus, on average the engine operates at stoichiometric conditions ($\Phi = \lambda = 1$) and likewise, the catalyst operates in a low O_2 and low fuel environment. Unfortunately, the λ-sensor is much less sensitive in the oxygen-rich environments of furnace exhaust (~3 % O_2), Diesel exhaust (~10 % O_2), or gas turbine exhaust (~15 % O_2).

For stationary power plants (including furnaces), gas turbines, and some Diesel engines, a catalyst is often used with the addition of ammonia which combines with NO on the catalyst to produce N_2 and water. The *selective catalytic reduction* (*SCR*) catalyst is active over a wide range of temperatures but is sensitive to fouling from particulates and sulfur in the exhaust (Bowman 1993).

A different use of catalysis is to change the fuel before it is burned. In time, further NO reduction and increase in thermal efficiency (with attendant reduction of CO_2 emissions, now not regulated) may be achieved by *reforming* the fuel first via (global) reactions like $CH_4 + H_2O = CO + 3\ H_2$, which is endothermic and thus capable of recovering waste heat in the exhaust. Combustion of the reformed fuel gives even lower NO due to lack of CH and the leaner flammability limit of H_2-containing fuels.

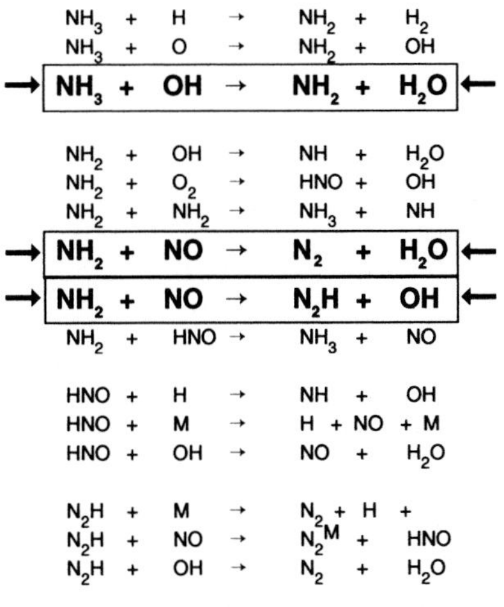

$$\mathrm{NH_3 + H \rightarrow NH_2 + H_2}$$
$$\mathrm{NH_3 + O \rightarrow NH_2 + OH}$$
$$\boxed{\mathrm{NH_3 + OH \rightarrow NH_2 + H_2O}}$$

$$\mathrm{NH_2 + OH \rightarrow NH + H_2O}$$
$$\mathrm{NH_2 + O_2 \rightarrow HNO + OH}$$
$$\mathrm{NH_2 + NH_2 \rightarrow NH_3 + NH}$$
$$\boxed{\mathrm{NH_2 + NO \rightarrow N_2 + H_2O}}$$
$$\boxed{\mathrm{NH_2 + NO \rightarrow N_2H + OH}}$$
$$\mathrm{NH_2 + HNO \rightarrow NH_3 + NO}$$

$$\mathrm{HNO + H \rightarrow NH + OH}$$
$$\mathrm{HNO + M \rightarrow H + NO + M}$$
$$\mathrm{HNO + OH \rightarrow NO + H_2O}$$

$$\mathrm{N_2H + M \rightarrow N_2 + H + M}$$
$$\mathrm{N_2H + NO \rightarrow N_2 + HNO}$$
$$\mathrm{N_2H + OH \rightarrow N_2 + H_2O}$$

Fig. 17.14. Key reactions for the NO reduction by secondary methods (Glarborg et al. 1986)

In contrast to the ammonia with catalyst (SCR) approach above, the *selective non-catalytic reduction* of NO (*SNCR, thermal DeNOx*™) is widely used when possible. In this case NH_3 (ammonia) is added to the exhaust stream, which at sufficiently high temperatures reacts with OH to NH_2. NO reacts with NH_2 and forms water and N_2 (or N_2H, which also leads to N_2) (Lyon 1974, Gehring et al. 1973). The most important elementary reactions of this process are shown in Fig. 17.15.

If the temperature is too high, NH_2 is oxidized to form NO. Thus, the selective reduction of NO is possible only within a rather small temperature window. Figure 17.14 shows the ratio between the initial NO and the NO after the reduction. It can be seen that an efficient reduction is optimal in the region around ~1300 K; this temperature can be shifted down to as low as 1000 K by concurrent addition of H_2. In any event, this narrow window severely limits the applications to which SNCR can be applied. More details are given, e. g., by Miller (1996)

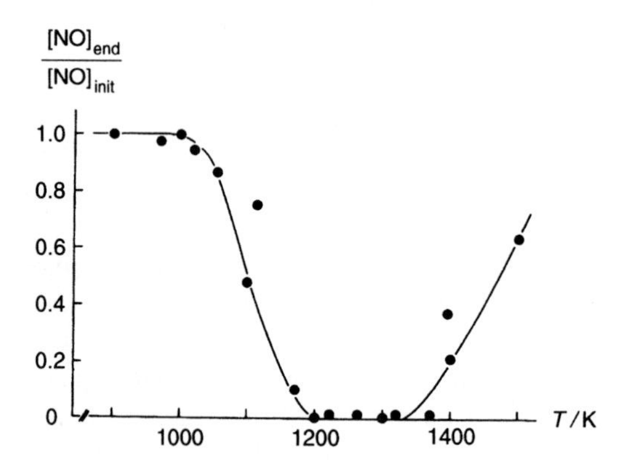

Fig. 17.15. Temperature window for the NO reduction by thermal DeNOx ; Points: measurements (Lyon 1974), line: calculations (Warnatz 1987); 4.6% O_2 with 0.074% NO and 0.85% NH_3 in N_2, residence time 0.15 s

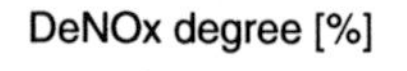

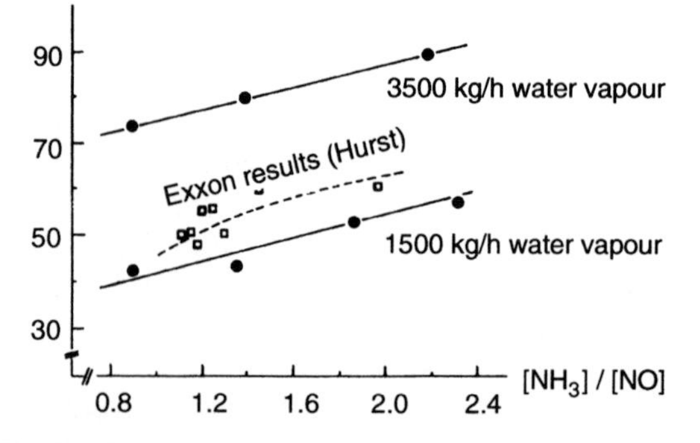

Fig. 17.16. Measured efficiencies of NO reduction by thermal DeNOx (see text); mixing done with steam injection

The excess of ammonia must not be too large ($[NH_3]/[NO] < 1.5$), because the NH_3, which is brought into the atmosphere would again lead to NO_x. Furthermore, a very good mixing of the components is required, given the ppm concentration of all reactants; this amount of mixing is a challenge. Effects of mixing are illustrated in Fig. 17.16, where results of measurements of the degree of NO-reduction are shown (Mittelbach and Voje 1986). Together with high-pressure steam, the same amount of ammonia is sprayed into the exhaust ducts using different jet velocities. Different degrees of NO reduction are the result. These results are based on a careful optimization of the NO reduction, which has been tested in a power plant. Some older experimental results from a power plant in Long Beach, CA (Hurst 1984) are shown for comparison (dotted line).

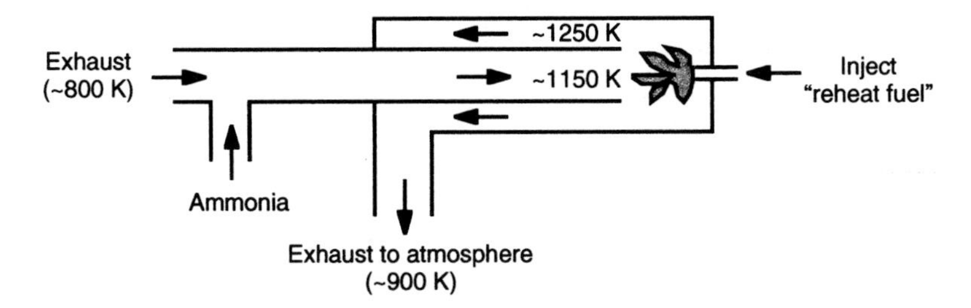

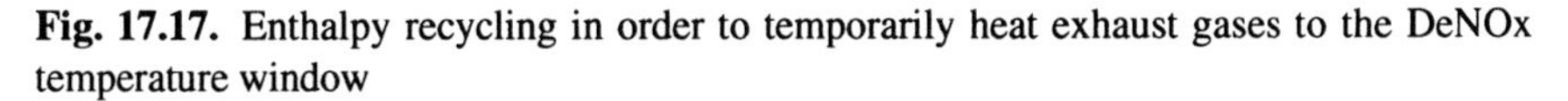

Fig. 17.17. Enthalpy recycling in order to temporarily heat exhaust gases to the DeNOx temperature window

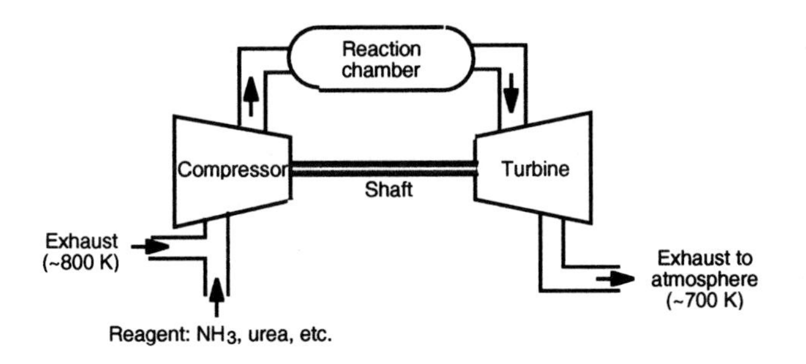

Fig. 17.18. Compression to bring exhaust to the DeNOx temperature window

Rather than inject the NH_3 into the exhaust stream, where the exhaust has the right temperature window, some schemes take the cool exhaust and reheat by burning more fuel, or by passing the exhaust into a counter current heat exchanger in which the hot gases from the end are used to preheat the incoming gases. A wide variety of such devices can be imagined. One example, of many, is that enthalpy is recycled in order to temporarily heat exhaust gases to the temperature window where ammonia reactions remove NO (see Fig. 17.17). A more detailed discussion of such devices is published by Weinberg (1975, 1986).

More subtle is the device that uses a turbine to compress the exhaust to the desired temperature and then uses an expander to recover the energy used for compression. This device would look like a turbocharger on an automobile engine and the process is called *TurboNOx*. This process uses compression to bring exhaust gas to the temperature window needed for ammonia removal of NO from exhaust streams (see Fig. 17.18). Much of the work of compression will be recovered by the turbine (Edgar and Dibble 1996)

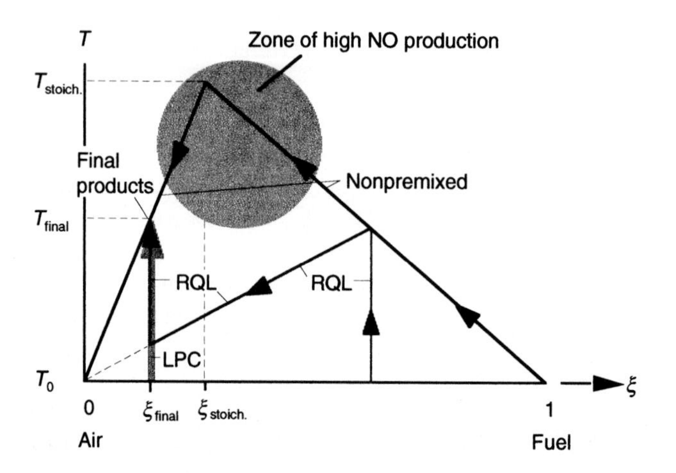

Fig. 17.19. Schematic depicting of mixing and reaction; LPC = lean premixed combustion, RQL = rich-quick-lean approach

Figure 17.19 is a graphical construct that visualizes various combustion modification schemes. Fuel mixing with air is movement from right to left, while chemical reaction is upward movement. In nonpremixed flames, the fuel evolves from the right ($\xi = 1$, $T = T_0$) to maximum temperature at $\xi = \xi_{\text{stoich}}$, $T = T_{\text{stoich}}$ to, e. g., the final composition $\xi = \xi_{\text{final}}$ and temperature $T = T_{\text{final}}$. The zone near the maximum temperature T_{stoich} is where NO generation is largest and, thus, is to be avoided.

The strategy of FGR and EGR is to dilute the air with inert products with the result that the maximum temperature T_{stoich} is lower; hence, the NO generation rate is reduced.

With *lean premixed combustion*, the NO production zone is avoided by mixing fuel ($\xi = 1$, $T = T_0$) with air ($\xi = 0$, $T = T_0$), without reaction, followed by reaction (vertical movement) to low NO products at $\xi = \xi_{\text{final}}$, $T = T_{\text{final}}$. These hot products can then serve to ignite an injected secondary fuel-rich mixture depicted as ($\xi = \xi_2$, $T = T_0$).

The *Rich-Quick-Lean* approach is to rapidly mix rich products with air along the mixing line connecting with $\xi = 0$, $T = T_0$ with the goal of reacting to $\xi = \xi_{\text{final}}$, $T = T_{\text{final}}$. In practice, in spite of the fast mixing, a fraction of the system evolves along the nonpremixed line, and some NO is generated.

18 Formation of Hydrocarbons and Soot

Hydrocarbons and *soot* predate NO_x (Chapter 17) as pollutants from combustion. In earlier times, smoke from the factory smokestack was a sign of prosperity. In time it became a nuisance and finally a health concern. The remedy to the appearance of soot and smoke are the three "t's" of combustion: time, temperature, and turbulence (Babcock and Wilcox, 1972). By allowing for more time at high temperatures with good mixing, one is usually assured of oxidizing soot and other hydrocarbons. However, these conditions also lead to a larger production of NO.

Thus, there is an urgent need to understand the subtle processes of soot formation and subsequent oxidation. Furthermore, there is the question how any hydrocarbon survives the conditions in the flame. The present level of knowledge on the formation of these pollutants is disappointingly low; much understanding is still missing. However, some models exist for specific problems and aspects (see Bockhorn 1994).

18.1 Unburnt Hydrocarbons

In the case of unburnt hydrocarbons, the terminology "pollutant formation" may or may not be correct depending on whether the unburnt hydrocarbon is directly from the fuel or whether the hydrocarbon is generated during the combustion process. For example, large-bore stationary piston engines (say 1 MW = 1 300 hp) burning natural gas will, not surprisingly, have small amounts of methane in the exhaust. Surprisingly, these engines also emit formaldehyde and aromatics such as benzene, toluene, and xylene; all at ppm-levels.

As there are none of these compounds in the fuel, it is interesting to learn how they were generated and why they were not consumed by the flame. This question is even more urgent as combustion is trusted to destroy toxic compounds, medical wastes, and municipal solid waste.

In general, unburnt hydrocarbons are a consequence of local flame extinction. There are two effects: flame extinction (or leakage) by strain (which has been discussed in detail earlier) and flame extinction at walls and in gaps.

18.1.1 Flame Extinction Due to Strain

Flame extinction due to strain is a phenomenon, which depends only on the processes in the gas mixture. High strain of flame fronts (caused, e. g., by intense turbulence) leads to local flame extinction (see Chapters 13 and 14). If the mixture does not reignite, the fuel leaves the reaction zone without being burnt.

The effect of flame extinction due to strain is increasingly important for rich or lean mixtures (see Chapter 14), where temperatures are lower and thus reaction times may become larger than the mixing times. This is, e. g., the reason for the high emission of hydrocarbons from lean combustion engines and from excessive steam injection into nonpremixed gas turbine combustors to suppress NO_x (Bowman 1993). Importantly, low reaction rates are the reason why water can be used to extinguish a fire.

18.1.2 Flame Extinction at Walls and in Gaps

Flame extinctions at walls and in gaps are caused by the interaction of the flame with the walls of the reaction system. Heat transfer (cooling of the reaction zone), as well as the removal of reactive intermediates (e. g., radicals) by surface reactions, are the reason for extinction. From a geometrical point of view, extinction of flame fronts parallel to a wall, of flame fronts perpendicular to a wall, and flame extinction in gaps can be distinguished.

Extinction of a flame front parallel to the wall: Flame fronts cannot exist near cold walls. The *quenching distance* is of the order of the flame thickness (Williams 1984). The movement of a flame front towards a wall (shown schematically in Fig. 18.1) can be treated as a time-dependent one-dimensional problem; thus, the time-dependent laminar conservation equations have to be solved (see Chapter 10). Results are available, e. g., for methanol combustion at high pressures (Westbrook and Dryer 1981).

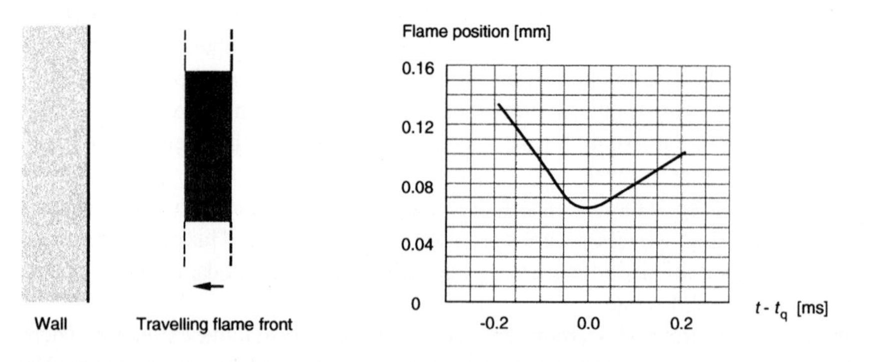

Fig. 18.1. Extinction of a flame front parallel to a wall; CH_3OH-air flame, $p = 10$ bar, $\Phi = 1$, $T_w = 300$ K (t_q is the time of nearest approach of the flame to the wall)

Figure 18.1 shows the time behavior of the flame position (t_q is the time when the flame front is closest to the wall). The minimum distance is only 70 μm. After the flame has reached this position, heat conduction and diffusion broaden the flame front, and the flame position changes again.

Some years ago, it was still assumed that a high amount of the unburnt hydrocarbons in Otto engines stems from flame extinction at walls. However, credible numerical simulations with complete transport like that mentioned above challenged these assumptions by showing that the unburnt hydrocarbons do not stay in the extinction zone, but instead diffuse into the extinguishing flame, which has a surprisingly long life time. In this way the hydrocarbons are consumed, leaving only a few ppm.

This modelling prompted a series of carefully contrived combustion experiments where flames were ignited at the center of a highly polished sphere. The result is that few hydrocarbons survived the flame impingement onto smooth walls (Bergner et al. 1983). Thus, the flame extinction at walls contributes much less to the emission of unburnt hydrocarbons from Otto engines than previously expected. (Here is a good example where modelling and experiment jointly improved understanding.)

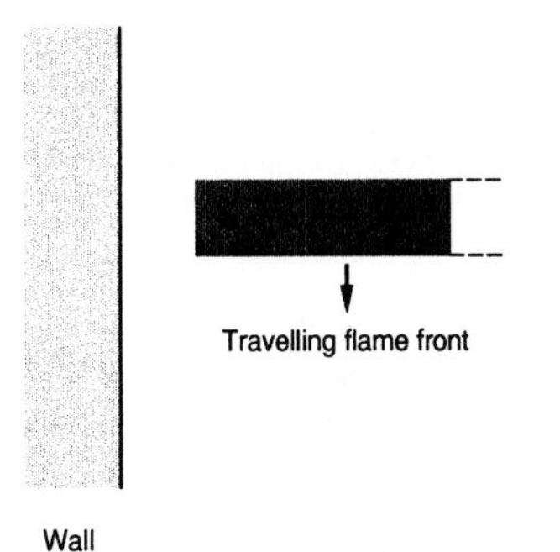

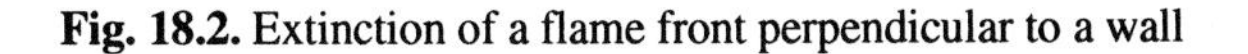
Fig. 18.2. Extinction of a flame front perpendicular to a wall

Extinction of a flame front perpendicular to a wall: The extinction of a flame front perpendicular to a wall is shown in Fig. 18.2. This is a more probable occurrence than extinction of a flame parallel to a wall. Here the flame has greater speed, and therefore the time for hydrocarbons to diffuse into the reaction zone is less. The expectation is that more hydrocarbons will survive in this configuration. However, the quantitative treatment requires the solution of the two-dimensional conservation equations with complex chemistry (at least 100 reactions of about 20 species). Based upon the simulation of smaller 2D systems (see, e. g., Behrendt et al. 1992, Braack 1998), it can be estimated that such computations are now possible and that they will improve the understanding of the flame-wall process.

Extinction of a flame front in a gap: If a flame front enters a gap (e. g., between the cylinder and piston, above the piston rings), extinction is observed (see Fig. 18.3).

There are systematic experiments on the influence of gaps and roughness of walls on the emission of unburnt hydrocarbons (see, e. g., Bergner et al. 1983). This problem, also, is at a frontier where convincing solutions can be expected soon.

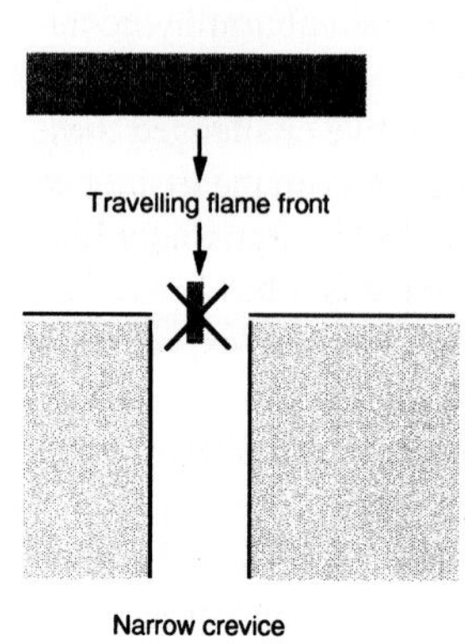

Fig. 18.3. Extinction of a flame front in a gap

18.2 Formation of Polycyclic Aromatic Hydrocarbons (PAH)

If there is no extinction, the fuel seems to be completely broken down to C_1- and C_2-hydrocarbons (Warnatz 1981a). Higher hydrocarbons which are formed after this breakdown, thus, have to be formed from these smaller hydrocarbon fragments.

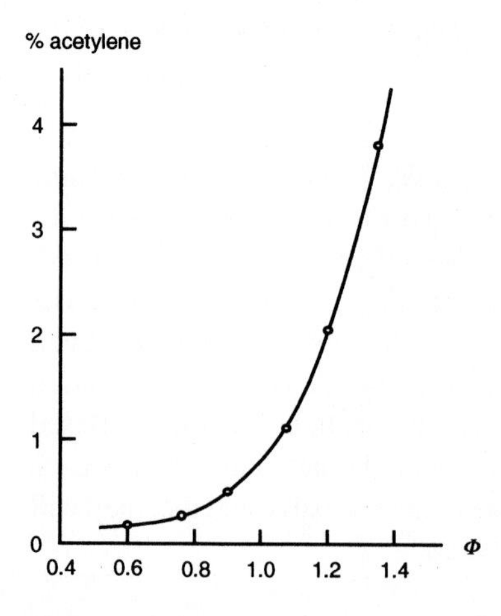

Fig. 18.4. Dependence of acetylene formation in CH_4-O_2 flames on the equivalence ratio Φ (Wagner 1979)

An important class of higher hydrocarbons are the *polycyclic aromatic hydrocarbons (PAH)*. These compounds are usually formed under fuel-rich conditions (present in rich premixed and always in nonpremixed flames) and can be carcinogenic (e. g., benzpyrene). They are important precursors in soot formation (see next section). Acetylene formed in high amounts under rich conditions (see Fig. 18.4 and reaction scheme Fig. 17.4) is the most important precursor of PAH's.

PAH formation is started by C_3H_4 decomposition or reaction of CH or CH_2 with C_2H_2 to C_3H_3, which can form the first ring (benzene C_6H_6) after recombination to an aliphatic C_6H_6 and rearrangement (Alkemade and Homann 1989, Stein et al. 1991, Melius et al. 1992). The reason is that the competing oxidation reactions of C_3H_3 are very slow,

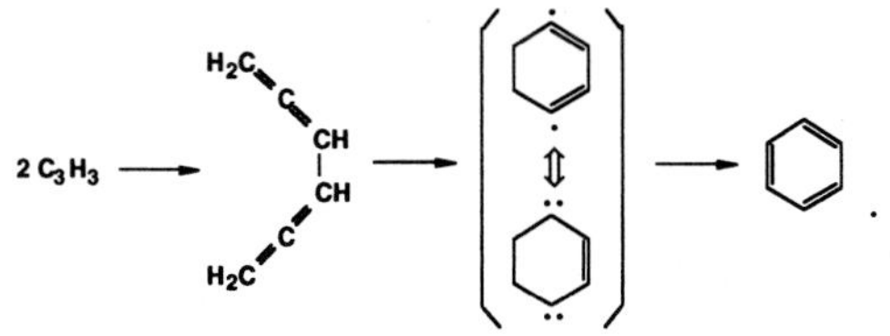

Bittner and Howard (1981) proposed the first elementary reaction mechanism for growth of PAH from acetylene as surface growth species. It starts with the addition of C_2H_2 to phenyl radicals forming styryl radical. A second C_2H_2 adds to the styryl radical, and ring closure follows forming naphthalene. Frenklach and Wang (1991) and Bockhorn and Schäfer (1994) have proposed similar ring growth elementary mechanisms. Further addition of acetylene C_2H_2 to the ring leads to the growth of the molecule (see below); furthermore, it is suggested that PAH growth partly is caused by aromatic structures as well (McKinnon 1989, Böhm et al. 1998). A typical phenomenon of such *condensation processes* is the fact that the larger the number of steps needed for growth, the larger is the dependence on the equivalence ratio Φ.

An example for the growth of the molecules by alternating radical formation by H-atom attack and acetylene addition is given in Fig. 18.5 (Frenklach and Clary 1983, Frenklach and Wang 1991).

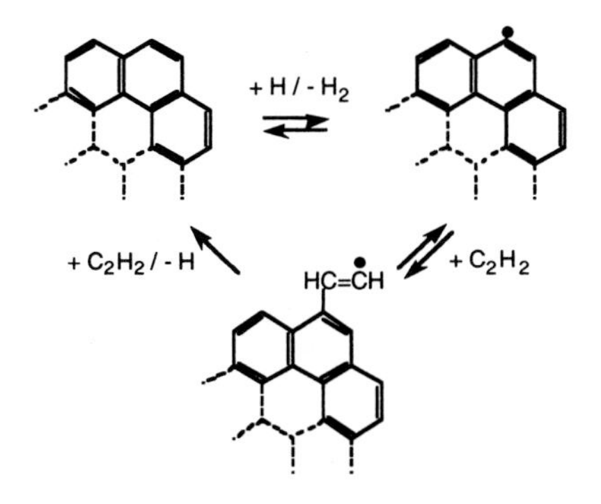

Fig. 18.5. Ring growth by radicalic acetylene addition in PAH formation (Frenklach and Clary 1983, Frenklach and Wang 1991)

A comparison of experimental results (Bockhorn et al. 1983) and simulations (Frenklach and Warnatz 1987) for PAH-formation is shown in Fig. 18.6. The maxima of the concentration are reproduced quite well by the calculations, but the rate of the PAH oxidation is over-predicted.

It is apparent that there is much room for improvement in the underlying models for aromatization and PAH formation.

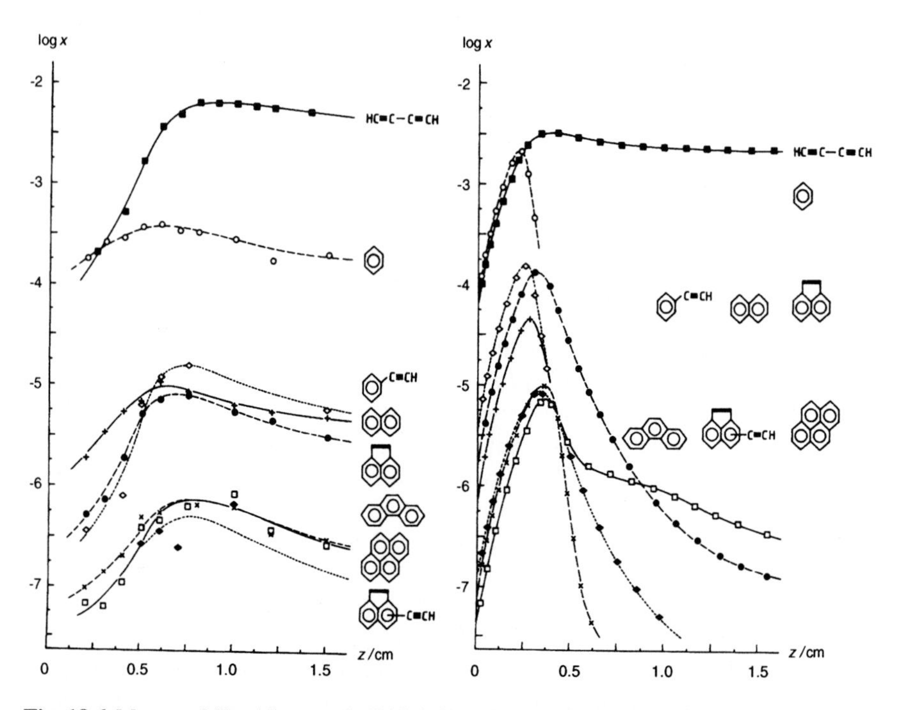

Fig. 18.6. Measured (Bockhorn et al. 1983, left) and calculated (Frenklach and Warnatz 1987, right) profiles of PAH in a laminar premixed acetylene-oxygen-argon flame at low pressure and fuel rich conditions

18.3 The Phenomenology of Soot Formation

It is widely accepted that further growth of the PAH (see Section 18.2) leads to soot (see, e. g., Wagner 1979, Haynes and Wagner 1981, Homann 1984, Bockhorn 1994). The first step here is formation of particle-like structures by conglomeration of molecules (Bockhorn 1994). This *particle inception* takes place at molecular masses between 500 and 2000 a.m.u. Subsequently (see Fig. 18.7), the particles grow by *surface growth* by addition of mainly acetylene, and by *coagulation*. Mainly in nonpremixed systems, *soot oxidation* is taking place after mixing with oxygen-containing gas.

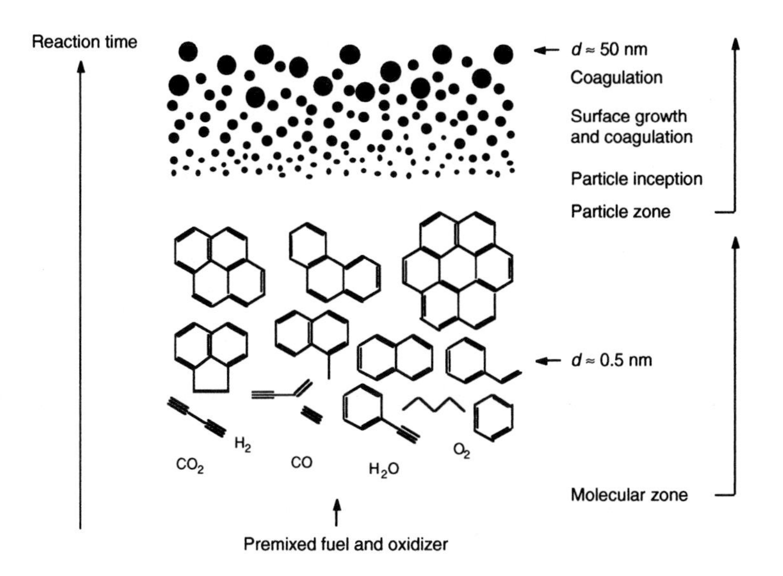

Fig. 18.7. Schematic reaction path leading to soot formation in homogeneous mixtures or premixed flames (Bockhorn 1994); ≡ is an acetylenic triple bond (≡ = acetylene, e. g.)

Combustion of hydrocarbons in fuel-rich mixtures (with CO and H_2 as main products at these conditions) can be described by the formal reaction

$$C_nH_m \;+\; k\,O_2 \;\rightarrow\; 2k\,CO \;+\; m/2\,H_2 \;+\; (n-2k)\,C_s \;,$$

where C_s is solid carbon. If the soot formation would be thermodynamically controlled, solid carbon should appear for $n > 2k$ or a C/O-ratio larger than 1. This is not the case, as can be seen from Table 18.1. Thus, it must be concluded that soot formation is kinetically controlled.

Tab. 18.1. Measured soot limits in terms of C/O ratios at 1800 K (Haynes and Wagner 1981); —— = not available

Combustion system	Bunsen burner	Well stirred reactor	Flat flame 1 bar	Flat flame 26 mbar
CH_4 -O_2	——	——	0.45	——
C_2H_6 -air	0.48	——	0.47	——
C_3H_8 -air	0.47	——	0.53	——
C_2H_4 -air	0.61	0.70	0.60	——
C_2H_4 -O_2	——	——	0.71	——
C_4H_8 -air	0.52	0.68	——	——
C_2H_2 -air	0.83	——	——	——
C_2H_2 -O_2 ($T \approx 3000$ K)	0.95	——	——	0.95
C_6H_6 -air	0.57	0.57	0.65	——
C_6H_6 -O_2	——	——	——	0.74
$C_{11}H_{10}$-air	0.42	0.50	——	——

Soot formation usually is described in terms of the *soot volume fraction* f_V, which is the ratio of the soot volume and the total volume V_{total},

$$f_V = V_{\text{soot}} / V_{\text{total}} , \qquad (18.1)$$

and the *soot particle number density* $[n]_{\text{soot}}$, which is the ratio of the soot particle number n_{soot} and the total volume and can be connected with the soot concentration c_{soot} via the Avogadro constant N_{A},

$$[n]_{\text{soot}} = n_{\text{soot}} / V_{\text{total}} = N_A \cdot c_{\text{soot}} . \qquad (18.2)$$

If the soot particles are assumed to be monodisperse, the resulting *soot particle diameter* is given by

$$d_{\text{soot}} = \sqrt[3]{\frac{6 f_V}{\pi c_{\text{soot}} N_{\text{A}}}} . \qquad (18.3)$$

A variety of measurements of two of the properties defined above (soot volume fraction, soot concentration, and particle size) can be found in the literature. However, the problem with these measurements consists in the fact that they are unrelated and cannot lead to a systematic understanding of soot formation. Table 18.2 presents a (necessarily not complete) selection of experimental results presented in the literature.

Tab. 18.2. Experiments on soot volume fraction, soot concentration, and particle size in sooting combustion systems

Configuration	System	Authors	Year
Homogeneous mixture (shock tube)	$CH_4/C_2H_2/C_2H_4/C_3H_8/C_7H_{16}$-air	Kellerer et al.	1996
Laminar premixed flame	CH_4-O_2	d'Alessio et al.	1975
	C_2H_4-Ar-O_2	Harris et al.	1986a,b,1988
	C_2H_4-Ar-O_2	Wieschnowski et al.	1988
	CH_4-O_2	d'Anna et al.	1994
	C_2H_2-Ar-O_2	Mauss et al.	1994a
Laminar nonpremixed counterflow	C_2H_4/C_3H_8-N_2-O_2	Vandsburger et al.	1984
	C_2H_2-air	Mauss et al.	1994b
Laminar nonpremixed coflow	C_2H_4-air	Santoro et al.	1987
	C_2H_4-N_2-air	Moss et al.	1995
Turbulent nonpremixed coflow	Kerosene-N_2-O_2	Moss	1994
	C_2H_2-N_2-O_2	Geitlinger et al.	1998

A global result of these measurements is that the soot volume fraction f_V is increasing with increasing pressure p, increasing C/O ratio, and that the T dependency is described by a bell-shaped curve (see Figs. 18.8) which is caused by two facts: Soot formation is needing radicalic precursors (like the C_3H_3 mentioned in Section 18.2) and, thus, does not occur at low temperatures. Furthermore, soot precursors are pyrolyzed and oxidized at elevated temperatures so that the soot formation is limited to a temperature range between 1000 K and 2000 K.

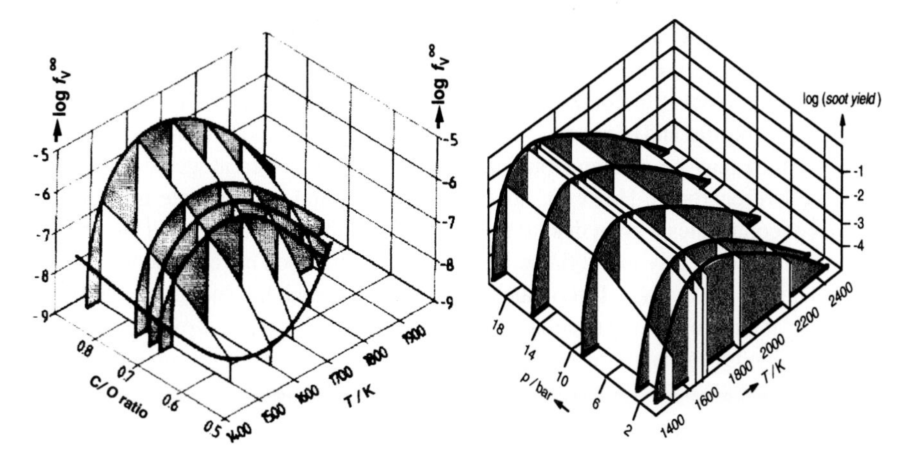

Fig. 18.8. Experimental temperature, pressure, and C/O dependence of the final soot volume fraction f_V^∞ and soot yield = fraction of C appearing as soot (Böhm et al. 1989, Jander 1995)

A variety of products can be formed, differing in their carbon- and hydrogen-content (see, e. g., Wagner 1979, Homann 1984, Bockhorn 1994). In spite of its variety, soot often can be characterized by a (usually log-normal, see Fig. 18.16) distribution of the molar masses of the particles formed. The structure of soot is difficult to characterize, because there is not a clear transition from gas to liquid, or liquid to solid phase. Freshly formed soot seems to consist of polycycles with side chains, having a molar H/C ratio of about 1. Aging, caused by heating, leads to compounds with a higher carbon content, similar to graphite. Physically, soot appears as grape-like clusters of small spheres (*spherules*); see Fig. 18.9. The single spherules typically reach 20-50 nm in diameter (Palmer and Cullis 1965), depending on the combustion conditions.

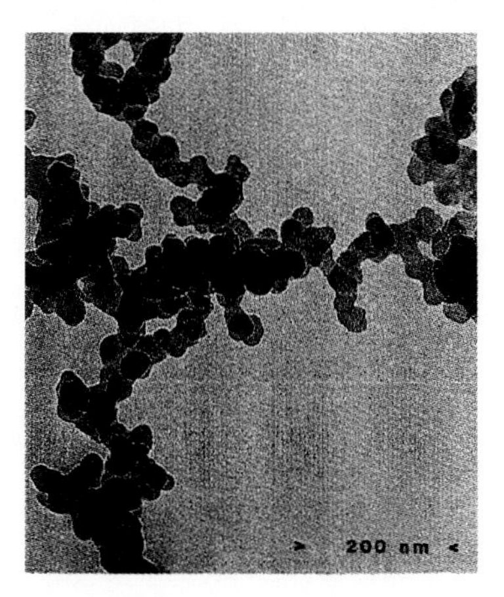

Fig. 18.9. Photograph of a soot particle from a nonpremixed flame (Dobbins and Subramaniasivam 1994)

Soot is used in many industrial processes (often called *carbon black*), like in the production of printing ink or as a filling material in "rubber" tires (60% of the mass of tires is carbon black). In combustion processes soot is an undesirable end-product. Higher temperatures and pressures, such as in Diesel engines, lead to increased amounts of soot which may be carcinogenic itself, or absorbs other carcinogenic polycyclic aromatic hydrocarbons. However, soot is a much desired intermediate in furnaces, because it contributes a great deal to the heat transfer through radiation. The strategy is to generate soot early in the flame, have it radiate, and then have the soot oxidized before leaving the furnace. If too much radiation is allowed, the soot is too cool ($T <$ 1 500 K) for rapid oxidation and thus appears in the exhaust. (It is this overcooling that is responsible for the smoking of kerosene lamps if the wick is turned up too high.)

18.4 Modelling and Simulation of Soot Formation

Soot Inception: The first step here is formation of particle-like structures by coagulation of PAH molecules (Bockhorn 1994) or PAH molecule growth into the third dimension via addition of 5-membered ring structures, as known from the C_{60} molecule (Zhang et al. 1986). Measurements on the rate of this *particle inception* and its delay time τ (see Fig. 18.10) are available (see, e. g., Harris et al. 1986a, Mauss et al. 1994a). Particle inception seems to take place at molecular masses between 500 a.m.u. (Pfefferle et al. 1994), 300-700 a.m.u. (Frenklach and Ebert 1988), 1600 a.m.u. (Miller 1990) and 2000 a.m.u. (Löffler et al. 1994, d'Anna et al. 1994), where molecules are large enough to be held together by van-der-Waals forces. A theoretical approach (Reh 1991) leads to a van-der-Waals potential well depth of ~6 kJ/mol per C atom. According to this picture, coagulation with a bond strength corresponding to chemical bonds would start at 300-500 a.m.u.

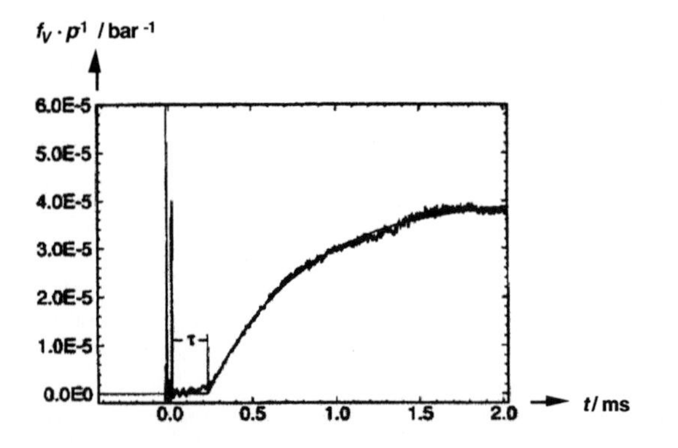

Fig. 18.10. Experimental increase of the (reduced) soot volume fraction by surface growth; C_3H_8, $\Phi = 5$, diluted in 99% Ar, $p = 60$ bar, $T = 1940$ K (Kellerer et al. 1996)

Soot Surface Growth: The overwhelming part of the soot (> 95 %) is formed by surface growth rather than soot inception (see, e. g., Harris et al. 1986a, Mauss et al. 1994a).

Mostly it is assumed that particle growth is similar to formation of PAH, i. e., addition of acetylene and, probably, aromatics. The problem in this connection is that surface growth is not a gas-phase reaction of small molecules, but a heterogeneous process (see Section 6.7), where adsorption and desorption processes at the surface have to be considered as well.

Because of shortage of precise data, phenomenological approaches are used in the literature. Mass growth of soot in premixed flames typically rises to an asymptotic value even though C_2H_2 is present and temperatures are high in the region of no mass growth. This can be described through the first order differential equation (Wagner 1981)

$$\frac{\mathrm{d}f_V}{\mathrm{d}t} = k_{\mathrm{sg}}(f_V^{\infty} - f_V) , \tag{18.4}$$

where k_{sg} is a fitted surface growth rate coefficient and f_V^{∞} is a fitted parameter which represents the ultimate volume fraction of soot formed. The temperature effects for both these parameters (k_{sg} and f_V^{∞}) have been empirically determined. The rate coefficient k_{sg} fits well to an Arrhenius form with an activation energy of 180 kJ/mol and a preexponential factor of ~1.5 x 10^7 s^{-1} (derived from Baumgärtner et al. 1985). The ultimate soot volume fraction f_V^{∞} has a maximum at a temperature around 1600 K (Böhm et al. 1989).

However, these growth rate parameters are empirical fits and do not predict mass growth at conditions not studied, and their simple structure does not reveal the underlying mechanisms. Surface growth by acetylene as deposition species would naively (if the heterogeneous character of the process is not studied in detail; see Section 6.7) lead to the rate expression (Dasch 1985, Harris and Weiner 1990)

$$\frac{\mathrm{d}f_V}{\mathrm{d}t} = k_{\mathrm{C_2H_2}} \cdot p_{\mathrm{C_2H_2}} \cdot S , \tag{18.5}$$

where S = soot surface area density (in. e. g., $\mathrm{m}^2/\mathrm{m}^3$) and $p_{\mathrm{C_2H_2}}$ = partial pressure of the gas-phase acetylene. If constant properties in the sooting zone of the flame are assumed (i. e., $p_{\mathrm{C_2H_2}}$ = const, S = const), a comparison of Eq. 18.4, predicting an exponential approach to f_V^{∞} (see Fig. 18.10), and of Eq. 18.5 leads to the consequence that $k_{\mathrm{C_2H_2}}$ has to decay exponentially, $k_{\mathrm{C_2H_2}} = k_{(t=0)} \cdot \exp(-\delta t)$, where $k_{(t=0)}$ is the surface growth rate coefficient at the begin of the soot zone and δ is the characteristic time of some "deactivation" of the soot surface.

Interpretations of this deactivation process have been given in the literature (Woods and Haynes 1994), but a really convincing physical explanation is not yet available and urgently needed.

Soot Coagulation: Coagulation takes place only for relatively small particles, which are characterized by high rates of growth (up to a diameter of ~10 nm in low pressure premixed systems; Homann 1967, Howard et al. 1973).

Fig. 18.11. Simple model of soot particle coagulation (Ishiguro et al. 1997)

The coagulation process seems to consist of sticking of two particles, which subsequently are "glued" together by a common outer shell generated by deposition analogously to surface growth (see Fig. 18.11, where this process can be recognized twice).

Coagulation essentially is a sticking process, the rate of which can be calculated with relatively little difficulty (Smoluchowski 1917). Assumptions characteristic of such calculations are: (1) The soot particles are small in comparison to the gas mean free path, (2) each collision of two particles results in coagulation, (3) all particles are spherical. The first assumption is valid for small particles or low densities such that the particle diameter is much less than the mean free path. This is satisfied in low pressure flames (Homann 1967, Howard et al. 1973) for the entire coagulation process, but can fail at high pressures (Kellerer et al. 1996). The second and third assumptions seem to be well justified for coagulating soot particles (Homann 1967, Wersborg et al. 1973, Delfau et al. 1979). An alternative treatment for continuum systems is of similar complexity and is described, e. g., by Friedlander (1977).

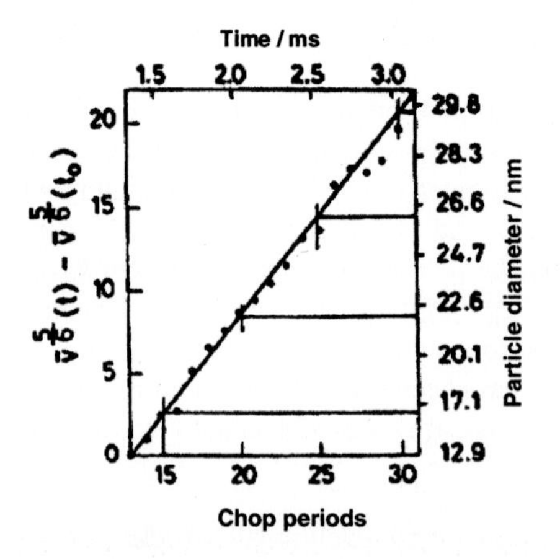

Fig. 18.12. Variation of the mean particle volume $\overline{\upsilon}$ with time for soot formed by shock-heating ethylbenzene in argon at 1750 K (Graham et al. 1975a,b)

Graham (1976), in a study of soot coagulation in shock heated hydrocarbon/argon mixtures, has shown that the previous assumptions apply, resulting in a coagulation rate (expressed in terms of the rate of decrease of the particle number density $[n]$) of the form

$$-\mathrm{d}[n]/\mathrm{d}t = (5/6)\, k_{\mathrm{theory}}\, f_V^{1/6}\, [n]^{11/6} \quad \text{with } k_{\mathrm{theory}} = (5/12)\,(3/4\pi)^{1/6}\,(6k_B T/\rho_{\mathrm{soot}})^{1/2} \cdot G \cdot \alpha\,.$$

Here, f_V is the soot volume fraction, k_B is the Boltzmann constant, ρ is the condensed particle density, α is a factor related to the polydisperse nature of the system, and G is a factor accounting for the increase in collision cross-section over the hard sphere value due to electronic and dispersion forces. Graham's results indicate that for spherical particles $G \sim 2$ and for a self-preserving size distribution $\alpha = 6.55$.

For constant soot volume fraction f_V (coagulation alone determines particle growth), the mean particle volume $\bar{\upsilon} = f_V/[n]$ is given by

$$\frac{\mathrm{d}\bar{\upsilon}}{\mathrm{d}t} = \frac{6}{5} k_{\mathrm{theory}} \cdot f_V \cdot \bar{\upsilon}^{1/6} \qquad \text{or} \qquad \bar{\upsilon}^{5/6} - \bar{\upsilon}_0^{5/6} = k_{\mathrm{theory}} \cdot f_V \cdot t\;.$$

An experimental verification of this relationship is given in Fig. 18.12, supporting the coagulation theory described above.

Soot Oxidation: Soot particles can be oxidized by O atoms, OH radicals, and O_2. These surface reactions in principal should be treated by the catalytic combustion formalism described in Section 6.7. However, due to lack of data and urgent need to simplify, a one-step treatment is often used, assuming a rate law for the CO formed given as

$$\frac{\mathrm{d[CO]}}{\mathrm{d}t} = \gamma_i \cdot Z_i \cdot a_S \quad ; \quad i = \mathrm{O, OH, O_2}\;,$$

where γ_i = reaction probability when molecule i hits the soot surface, Z_i = collision number of molecule i per unit time and area (see Section 6.7), and a_S = soot surface per unit volume.

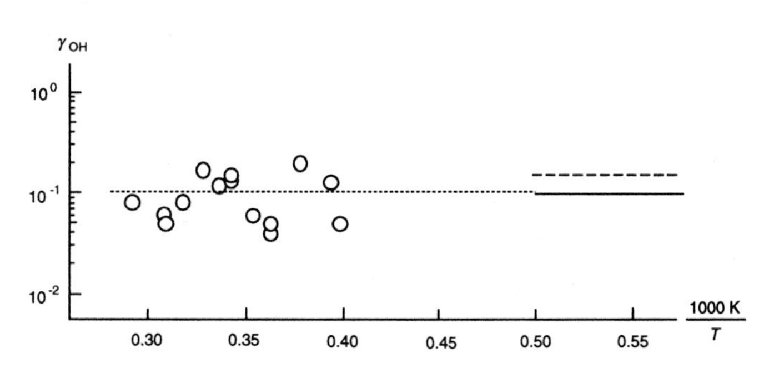

Fig. 18.13. Temperature dependence of the probability γ of the reaction per collision of OH radicals with soot; O von Gersum and Roth (1992), —— Fenimore and Jones (1967), Garo et al. (1990), – – – Neoh et al. (1974), ----- mean value

Because of the low concentration of O in sooting flames ($[O] = 0.01 \cdot [OH]$; El-Gamal 1995) and its limited reaction probability on soot surfaces of $\gamma_O \approx 0.5$ at flame temperatures (see for reference von Gersum and Roth 1992, Roth and von Gersum 1993), the oxidation of soot is, by process of elimination, considered to be primarily due to OH and O_2; reaction probabilities are given in Figs. 18.13 and 18.14.

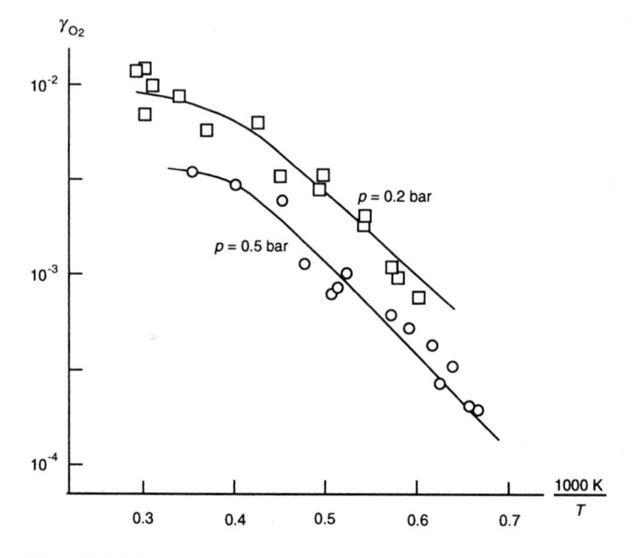

Fig. 18.14. Temperature dependence of the probability γ of the reaction per collision of O_2 molecules with soot at different pressures (Roth et al. 1990, von Gersum and Roth 1992)

Soot Agglomeration: *Soot agglomeration* is taking place in the late phase of soot formation when, due to lack of surface growth, coagulation as sketched in Fig. 18.11 is no longer possible. Consequently, open structured aggregates are formed (see Fig. 18.9), containing 30 - 1800 primary particles (spherules) and characterized by a log-normal size distribution (Köylü and Faeth 1992). A relationship between the number N of primary particles and the maximum length L of the aggregates can be derived to be $N = k_f \cdot (L/3d_p)^{D_f}$, where k_f is a constant fractal prefactor, d_p the primary particle diameter, and D_f a fractal dimension around 1.8 (Köylü et al. 1995).

Detailed Simulation of soot formation and oxidation: *Lumping* is a means to describe formation, e. g., of polymer formation, not by a huge number of single species and an untreatable reaction mechanism, but by use of a distribution function for the degree of polymerization and a repeating reaction cycle for particle growth. The discrete Galerkin method (Deuflhard and Wulkow 1989, Ackermann and Wulkow 1990) offers an elegant way to overcome disadvantages of moment methods (Frenklach 1985), which need to prescribe a certain form of the particle size distribution function. The method is based on the fact that a distribution function can be expressed as

$$P_s^{(\infty)}(t) = \Psi(s;\rho)\cdot\sum_{k=0}^{\infty} a_k(t;\rho)\cdot l_k(s;\rho) \ ,$$

where $\Psi(s;\rho)$ is a weight function (here: $\Psi(s;\rho) = (1-\rho)\rho^{s-1}$ with $0 \le \rho \le 1$, s = 1,2,3,...; i. e. the Schulz-Flory distribution), $l_k(s;\rho)$ are orthogonal polynomials associated with the weight functions (here: Laguerre-type polynomials), $a_k(t;\rho)$ are the time-dependent coefficients (to be obtained from the theory of orthogonal polynomials), and ρ is a parameter in order to optimize the representation. For application, the Galerkin approach is good even for finite summation. For implementation of this scheme, the truncated distribution function $P_{s,\rho}^{(n)}(t)$ is "preprocessed". For this purpose, the function is differentiated with respect to time *t*, and then the chemical rate equations are inserted. This leads to a set of ordinary differential equations for the coefficients $a_k(t;\rho)$ and the parameter ρ of the Schulz-Flory distribution.

As a test for the Galerkin method, the soot yield in a homogeneous system with various initial temperatures can be calculated (for comparison with measurements by Frenklach et al. 1984). The gas-phase mechanism used is similar to that displayed in Table 6.1. Cyclization to benzene P(1) is occurring via C_3H_3 combination (Alkemade and Homann 1989, Miller and Melius 1991, d'Anna and Violi 1998). The rate coefficients used in the Galerkin approximation scheme (P(*N*) is a stable polyaromatic hydrocarbon, A(*N*) and B(*N*) are intermediates during ring growth)

C_3H_3	+ C_3H_3	=	P(1)			Formation of the first ring
P(*N*)	+ H	=	A(*N*)	+ H_2	}	
A(*N*)	+ C_2H_2	=	B(*N*)		}	PAH growth
B(*N*)	+ C_2H_2	=	P(*N*+1)	+ H	}	
P(*N*)	+ P(*M*)	⇒	P(*N*+*M*)			Coagulation

are the same as proposed by Frenklach et al. (1985, 1986); the coagulation rate coefficient is $k = 6\cdot10^{15}\text{cm}^3\cdot\text{mol}^{-1}\cdot\text{s}$.

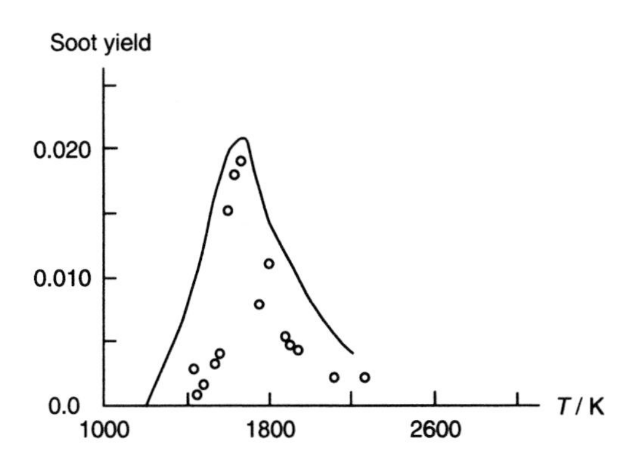

Fig. 18.15. Soot yield in a spatially homogeneous C_2H_2-O_2-Ar mixture at a C/O ratio of 3.1 and p = 1 bar as function of temperature; points: (uncalibrated) measurements (Frenklach et al. 1984), line: simulation (El-Gamal 1995)

A comparison with corresponding experiments is shown in Fig. 18.15. Comments: Soot inception is not correctly described in this simple example (see section on soot inception above); soot oxidation can be neglected in a homogeneous system. Similar results can be produced in premixed flame computations (see Fig. 18.16; El-Gamal and Warnatz 1995).

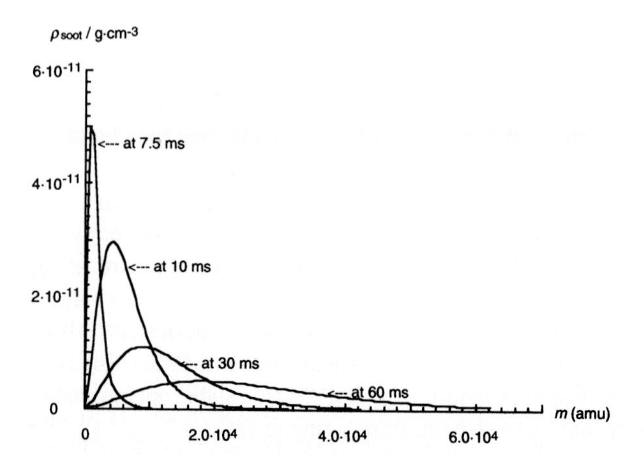

Fig. 18.16. Computed size distribution function for PAH at different times (corresponding to different heights above the burner) in a fuel-rich premixed flat C_2H_2-O_2-Ar flame at $p = 120$ mbar

19 References

Abdel-Gayed RG, Bradley D, Hamid NM, Lawes M (1984) Lewis number effects on turbulent burning velocity. Proc Comb Inst 20:505

Ackermann J, Wulkow M (1990) MACRON - A program package for macromolecular kinetics. Konrad-Zuse-Zentrum Berlin, Preprint SC-90-14

Alkemade V, Homann KH (1989) Formation of C_6H_6 isomers by recombination of propynyl in the system sodium vapour/propynylhalide. Z Phys Chem NF 161:19

Amsden AA, O'Rourke PJ, Butler TD (1989) KIVA II: A computer program for chemically reactive flows with sprays. LA-11560-MS, Los Alamos National Laboratory, Los Alamos

Aouina Y (1997) Modellierung der Tropfenverbrennung unter Einbeziehung detaillierter Reaktionsmechanismen. Dissertation, Universität Stuttgart

Aris R (1962) Vectors, tensors, and the basic equations of fluid mechanics. Prentice Hall, New York

Arnold A, Becker H, Hemberger R, Hentschel W, Ketterle W, Köllner M, Meienburg W, Monkhouse P, Neckel H, Schäfer M, Schindler KP, Sick V, Suntz R, Wolfrum J (1990a) Laser in situ monitoring of combustion processes. Appl Optics 29:4860

Arnold A, Hemberger R, Herden R, Ketterle W, Wolfrum J (1990b) Laser stimulation and observation of ignition processes in CH_3OH-O_2-mixtures. Proc Comb Inst 23:1783

Arrhenius S (1889) Über die Reaktionsgeschwindigkeit bei der Inversion von Rohrzucker durch Säuren. Z Phys Chem 4:226

Ashurst WT (1995) Modeling turbulent flame propagation. Proc Comb Inst 25:1075

Atkins PW (1996) Physical chemistry, 5th ed. Freeman, New York

Babcock and Wilcox (eds) (1972) Steam – Its generation and use. Babcock and Wilcox, Barberton

Bachalo W (1995) A review of laser scattering in spray. Proc Comb Inst 25:333

Bamford CH, Tipper CFH (eds) (1977) Comprehensive chemical kinetics, Vol 17: Gas phase combustion. Elsevier, Amsterdam/Oxford/New York

Bar M, Nettesheim S, Totermund HH, Eiswirth M, Ertl G (1995) Transition between fronts and spiral waves in a bistable surface reaction. Phys Rev Lett 74:1246

Barlow R (1998) Private communication. Combustion Research Facility, Sandia National Laboratories, Livermore

Bartok W, Engleman VS, Goldstein R, del Valle EG (1972) Basic kinetic studies and modeling of nitrogen oxide formation in combustion processes. AIChE Symp Ser 68(126):30

Bäuerle B, Hoffmann F, Behrendt F, Warnatz J (1995) Detection of hot spots in the end gas of an IC engine using two-dimensional LIF of formaldehyde. Proc Comb Inst 25:135

Baulch DL, Cox AM, Just T, Kerr JA, Pilling M, Troe J, Walker RW, Warnatz J (1991) Compilation of rate data on C_1/C_2 Species Oxidation. J Phys Chem Ref Data 21:3

Baulch DL, Cobos CJ, Cox AM, Frank P, Hayman G, Just T, Kerr JA, Murrels T, Pilling MJ, Troe J, Walker RW, Warnatz J (1994) Compilation of rate data for combustion modelling Supplement I. J Phys Chem Ref Data 23:847

Baumgärtner L, Hess D, Jander H, Wagner HG (1985) Rate of soot growth in atmospheric premixed laminar flames. Proc Comb Inst 20:959

Bazil R, Stepowski (1995) Measurement of vaporized and liquid fuel concentration fields in a burning spray jet of acetone using planar laser induced fluorescence. Exp Fluids 20:1

Becker H, Monkhouse PB, Wolfrum J, Cant RS, Bray KNC, Maly R, Pfister W, Stahl G, Warnatz J (1991) Investigation of extinction in unsteady flames in turbulent combustion by 2D-LIF of OH radicals and flamelet analysis. Proc Comb Inst 23:817

Beebe KW, Cutrone MB, Matthews R, Dalla Betta RA, Schlatter JC, Furuse Y, Tsuchiya T (1995) Design and test of a catalytic combustor for a heavy duty industrial gas turbine. ASME paper no 95-GT-137

Behrendt F, Bockhorn H, Rogg B, Warnatz J (1987) Modelling of turbulent diffusion flames with detailed chemistry. In: Warnatz J, Jäger W (eds.) Complex chemical reaction systems: Mathematical modelling and simulation, Springer, Heidelberg, p. 376

Behrendt F, Goyal G, Maas U, Warnatz J (1992) Numerical study of 2-dimensional effects associated with laser-induced ignition in hydrogen-oxygen mixtures. Proc Comb Inst 24:83

Behrendt F, Deutschmann O, Maas U, Warnatz J (1995) Simulation and sensitivity analysis of the heterogeneous oxidation of methane on a platinum foil. J Vac Sci Technol A13:1373

Behrendt F, Deutschmann O, Schmidt R, Warnatz J (1996) Simulation and sensitivity analysis of the heterogeneous oxidation of methane on a platinum foil. In: Warren BK, Oyama ST (eds) Heterogeneous hydrocarbon oxidation, ACS symposium series 638, p 48

Bergner P, Eberius H, Just T, Pokorny H (1983) Untersuchung zur Kohlenwasserstoff-Emission eingeschlossener Flammen im Hinblick auf die motorische Verbrennung. VDI-Berichte 498:233

Bertagnolli KE, Lucht RP (1996) Temperature profile measurements in stagnation-flow diamond-forming flames using hydrogen CARS spectroscopy. Proc Comb Inst 26:1825

Bilger RW (1976) The structure of diffusion flames. Comb Sci Technol 13:155

Bilger RW (1980) Turbulent flows with nonpremixed reactants. In: Libby PA, Williams FA (eds) Turbulent reactive flows. Springer, New York

Bird RB, Stewart WE, Lightfoot EN (1960) Transport phenomena. J. Wiley & Sons, New York

Bish ES, Dahm WJA (1995) Strained dissipation and reaction layer analysis of nonequilibrium chemistry in turbulent reacting flows. Comb Flame 100:457

Bittner JD, Howard JB (1981) Pre-particle chemistry in soot formation. In: Siegla DC, Smith GW (eds) Particulate carbon formation during combustion. Plenum Press, New York .

Bockhorn H (ed) (1994) Soot formation in combustion. Springer, Berlin/Heidelberg

Bockhorn H, Schäfer T (1994) Growth of soot particles in premixed flames by surface reactions. In: Bockhorn H (ed), Soot formation in combustion. Springer, Berlin/Heidelberg

Bockhorn H, Fetting F, Wenz HW (1983) Investigation of the formation of high molecular hydrocarbons and soot in premixed hydrocarbon-oxygen flames. Ber Bunsenges Phys Chem 87:1067

Bockhorn H, Chevalier C, Warnatz J, Weyrauch V (1990) Bildung von promptem NO in Kohlenwasserstoff-Luft-Flammen. 6. TECFLAM-Seminar, TECFLAM, DLR Stuttgart

Bockhorn H, Chevalier C, Warnatz J, Weyrauch V (1991) Experimental Investigation and modeling of prompt NO formation in hydrocarbon flames. In: Santoro RJ, Felske JD (eds) HTD-Vol 166, Heat transfer in fire and combustion systems, Book No G00629-1991

Boddington T, Gray P, Kordylewski W, Scott SK (1983) Thermal explosions with extensive reactant consumption: A new criterion for criticality. Proc R Soc London, Ser A, 390:13
Bodenstein M, Lind SC (1906) Geschwindigkeit der Bildung des Bromwasserstoffs aus seinen Elementen. Z Phys Chem 57:168
Böhm H, Hesse D, Jander H, Lüers B, Pietscher J, Wagner HG, Weiss M (1989) The influence of pressure and temperature on soot formation in premixed flames. Proc Comb Inst 22:403
Böhm H, Jander H, Tanke D (1998) PAH growth and soot formation in the pyrolysis of acetylene and benzene at high temperatures and pressures. Proc Comb Inst 27:1605
Bond GC (1990) Heterogeneous catalysis: Principles and applications, 2nd ed. Oxford Press, Oxford
Borghi R (1984) In: Bruno C, Casci C (eds) Recent advances in aeronautical science. Pergamon, London
Boudart M, Djega-Mariadassou G (1984) Kinetics of heterogeneous catalytic reactions. Princeton University Press, Princeton
Bowman CT (1993) Control of combustion-generated nitrogen oxide emissions: Technology driven by regulation. Proc Comb Inst 24:859
Braack M (1998) An adaptive finite element method for reactive-flow problems. Dissertation, Universität Heidelberg
Bradley D (1993) How fast can we burn? Proc Comb Inst 24:247
Braun M (1988) Differentialgleichungen und ihre Anwendungen. Springer, Berlin/Heidelberg/ New York/London/Paris/Tokyo, p 521
Bray KNC (1980) Turbulent flows with premixed reactants. In: Libby PA, Williams FA (eds) Turbulent reacting flows. Springer, New York
Bray KNC, Libby PA (1976) Interaction effects in turbulent premixed flames. Phys Fluids 19: 1687
Bray KNC, Moss JB (1977) A unified statistical model of the premixed turbulent flame. Acta Astron 4:291
Brena de la Rosa A, Sankar SV, Wang G, Balchalo WD (1992) Particle diagnostics and turbulence measurements in a confined isothermal liquid spray. ASME paper no 92-GT-113
Brown GM, Kent JC (1985) In: Yang WC (ed) Flow visualization III. Hemisphere, London, p 118
Buch KA, Dahm WJA (1996) Fine scale structure of conserved scalar mixing in turbulent flows Part I: $Sc \gg 1$. J Fluid Mech 317:21
Buch KA, Dahm WJA (1998) Fine scale structure of conserved scalar mixing in turbulent flows Part II: $Sc \approx 1$, J Fluid Mech 364:1
Burcat A (1984) Thermochemical data for combustion. In: Gardiner WC (ed) Combustion chemistry. Springer, New York
Burke SP, Schumann TEW (1928) Diffusion flames. Ind Eng Chem 20:998
Candel S, Veynante D, Lacas F, Darabiha N (1994) Current progress and future trends in turbulent combustion. Combust Sci Technol 98:245
Chen JH, Hong GI (1998) Correlation of flame speed with stretch in turbulent premixed methane/air flames. Proc Comb Inst 27:819
Chen JY, Kollmann W, Dibble RW (1989) PDF modeling of turbulent nonpremixed methane jet flames. Comb Sci Technol 64:315
Chevalier C, Louessard P, Müller UC, Warnatz J (1990a) A detailed low-temperature reaction mechanism of n-heptane auto-ignition. Proc. 2nd Int. Symp. on diagnostics and modeling of combustion in reciprocating Engines. The Japanese Society of Mechanical Engineers, Tokyo, p 93
Chevalier C, Warnatz J, Melenk H (1990b) Automatic generation of reaction mechanisms for description of oxidation of higher hydrocarbons. Ber Bunsenges Phys Chem 94:1362

Chiu HH, Kim HY, Croke EJ (1982) Internal group combustion of liquid droplets. Proc Comb Inst 19:971

Cho SY, Yetter RA, Dryer FL (1992) A computer model for one-dimensional mass and energy transport in and around chemically reacting particles, including complex gas-phase chemistry, multicomponent molecular diffusion, surface evaporation, and heterogeneous reaction. J Comp Phys 102:160

Christensen M, Johansson B, Ammneus P, Mauss F (1998) Supercharged homogeneous charge compression ignition. SAE paper 980787

Christmann K (1991) Introduction to surface physical chemistry. Springer, Berlin/Heidelberg

Chue RS, Lee JHS, Scarinci T, Papyrin A, Knystautas R (1993) Transition from fast deflagration to detonation under the influence of wall obstacles. In: Kuhl AL, Leyer JC, Borisov AA, Sirignano WA (eds), Dynamic aspects of detonation and explosion phenomena. Progress in Astronautics and Aeronautics 153:270

Clift R, Grace JR, Weber ME (1978) Bubbles, drops, and particles. Academic Press, New York

Coltrin ME, Kee RJ, Rupley FM (1993) Surface Chemkin: A general formalism and software for analyzing heterogeneous chemical kinetics at a gas-surface interface. Intl J Chem Kin 23:1111

Correa SM (1992) A review of NO_x formation under gas-turbine combustion conditions. Comb Sci Technol 87:329

Curtiss CF, Hirschfelder JO (1959) Transport properties of multicomponent gas mixtures. J Chem Phys 17:550

Dahm WJA, Bish ES (1993) High resolution measurements of molecular transport and reaction processes in turbulent combustion. In: Takeno T (ed), Turbulence and molecular processes in combustion, p 287. Elsevier, New York

Dahm WJA, Tryggvason G, Zhuang MM (1995) Integral method solution of time-dependent strained diffusion-reaction equations with multi-step kinetics, to appear in SIAM Journal of Applied Mathematics

Dalla Betta RA, Schlatter JC, Nickolas SG, Cutrone MB, Beebe KW, Furuse Y, Tsuchiya T (1996) Development of a catalytic combustor for a heavy duty utility gas turbine. ASME paper no 96-GT-485

Damköhler G (1940) Der Einfluss der Turbulenz auf die Flammengeschwindigkeit in Gasgemischen. Z Elektrochem 46:601

D'Alessio A, Lorenzo A, Sarofim AF, Beretta F, Masi S, Venitozzi C (1975) Soot formation in methane-oxygen flames. Proc Comb Inst 15:1427

D'Anna A, D'Alessio A, Minutulo P (1994) Spectroscopic and chemical characterization of soot inception processes in premixed laminar flames at atmospheric pressure. In: Bockhorn H (ed), Soot formation in combustion. Springer, Berlin/Heidelberg

D'Anna A, Violi A (1998) A kinetic model for the formation of aromatic hydrocarbons in premixed laminar flames. Proc Comb Inst 27:425

Dasch JC (1985) Decay of soot surface growth reactivity and its importance in total soot formation. Comb Flame 61:219

Dean AM, Hanson RK, Bowman CT (1990) High temperature shock tube study of reactions of CH and C-atoms with N_2. Proc Comb Inst 23:259

Delfau JL, Michaud P, Barassin A (1979) Formation of small and large positive ions in rich and in sooting low-pressure ethylene and acetylene premixed flames. Comb Sci Tech 20:165

Deuflhard P, Wulkow M (1989) Computational treatment of polyreaction kinetics by orthogonal polynomials of a discrete variable. Impact of Computing in Science and Engineering 1:269

Deutschmann O, Behrendt F, Warnatz J (1994) Modelling and simulation of heterogeneous oxidation of methane on a platinum foil. Catalysis Today 21:461

Deutschmann O, Schmidt R, Behrendt F, Warnatz J (1996) Numerical modelling of catalytic ignition. Proc Comb Inst 26:1747

Dibble RW, Masri AR, Bilger RW (1987) The spontaneous Raman scattering technique applied to non-premixed flames of methane. Comb Flame 67:189

Dimotakis PE, Miller PL (1990) Some consequences of the boundedness of scalar fluctuations. Phys Fluids A2:1919

Dinkelacker F, Buschmann A, Schäfer M, Wolfrum J (1993) Spatially resolved joint measurements of OH- and temperature fields in a large premixed turbulent flame. Proceedings of the Joint Meeting of the British and German Sections of the Combustion Institute, Queens College, Cambridge, p 295

Dixon-Lewis G, Fukutani S, Miller JA, Peters N, Warnatz J et al. (1985) Calculation of the structure and extinction limit of a methane-air counterflow diffusion flame in the forward stagnation region of a porous cylinder. Proc Comb Inst 20:1893

Dobbins AR, Subramaniasivam H (1994) Soot precursor particles in flames. In: Bockhorn H (ed) Soot formation in combustion. Springer, Berlin/Heidelberg

Dopazo C, O'Brien EE (1974) An approach to the description of a turbulent mixture. Acta Astron 1:1239

Dreier T, Lange B, Wolfrum J, Zahn M, Behrendt F, Warnatz J (1987) CARS measurements and computations of the structure of laminar stagnation-point methane-air counterflow diffusion flames. Proc Comb Inst 21:1729

Du DX, Axelbaum RL, Law CK (1989) Experiments on the sooting limits of aerodynamically-strained diffusion flames. Proc Comb Inst 22:387

Eberius H, Just T, Kelm S, Warnatz J, Nowak U (1987) Konversion von brennstoffgebundenem Stickstoff am Beispiel von dotierten Propan-Luft-Flammen. VDI-Berichte 645:626

Eckbreth AC (1996) Laser diagnostics for combustion temperature and species, 2nd edition. In: Sirignano WA (ed), Combustion science and technology book series Vol 3, Gordon and Breach

Edgar B, Dibble RW (1996) Process for removal of oxides of nitrogen. U.S. Patent No 5 547 650

Edwards DH (1969) A survey of recent work on the structure of detonation waves, Proc Comb Inst 12:819

El-Gamal M (1995) Simulation der Rußbildung in vorgemischten Verbrennungssystemen. Dissertation, Universität Stuttgart

El-Gamal M, Warnatz J (1995) Soot formation in combustion processes. In: Der Arbeitsprozess des Verbrennungsmotors, p. 154. Technische Universität Graz

Ern A, Giovangigli V (1996) Optimized Transport Algorithms for Flame Codes, Combust Sci Tech 118:387

Esser C, Maas U, Warnatz J (1985) Chemistry of the combustion of higher hydrocarbons and its relation to engine knock. Proc. 1st Int. Symp. on diagnostics and modeling of combustion in reciprocating Engines. The Japanese Society of Mechanical Engineers, Tokyo, p 335

Faeth GM (1984) Evaporation and combustion of sprays. Prog Energy Combust Sci 9:1

Farrow RL, Mattern PL, Rahn LA (1982) Comparison between CARS and corrected thermocouple temperature measurements in a diffusion flame. Appl Opt 21:3119

Fenimore CP, Jones GW (1967) Oxidation of soot by hydroxyl radicals. J Phys Chem 71:593

Fenimore CP (1979) Studies of fuel-nitrogen in rich flame gases. Proc Comb Inst 17:661

Flower WL, Bowman CT (1986) Soot production in axisymmetric laminar diffusion flames at pressures one to ten atmospheres. Proc Comb Inst 21:1115

Flowers D, Aceves S, Smith R, Torres J, Girard J, Dibble R (2000) HCCI in a CFR engine: Experiments and detailed kinetic modeling. SAE paper 2000-01-0328
Forsythe GE, Wasow WR (1969) Finite-difference methods for partial differential equations. Wiley, New York
Frank-Kamenetskii DA (1955) Diffusion and heat exchange in chemical kinetics. Princeton University Press, Princeton
Frenklach M (1985) Computer modeling of infinite reactions sequences: A chemical lumping. Chem Eng Sci 40:1843
Frenklach M, Clary D (1983) Aspects of autocatalytic reaction kinetics. Ind Eng Chem Fundam 22:433
Frenklach M, Warnatz J (1987) Detailed modeling of PAH profiles in a sooting low pressure acetylen flame. Comb Sci Technol 51:265
Frenklach M, Ebert LB (1988) Comment on the proposed role of spheroidal carbon clusters in soot formation. J Phys Chem 92:561
Frenklach M, Wang H (1991) Detailed modeling of soot particle nucleation and growth. Proc Comb Inst 23:1559
Frenklach M, Ramachandra MK, Matula MA (1984) Soot formation in shock-tube oxidation of hydrocarbons. Proc Comb Inst 20:871
Frenklach M, Clary DW, Gardiner jr WC, Stein SE (1985) Detailed kinetic modeling of soot formation in shock-tube pyrolysis of acetylene. Proc Comb Inst 20:887
Frenklach M, Clary DW, Yuan T, Gardiner jr WC, Stein SE (1986) Mechanism of soot formation in acetylene-oxygen mixtures. Combust. Sci. Tech. 50:79
Fric TF (1993) Effects of fuel-air unpremixedness on NO_x emissions. J Propulsion Power 9:708
Friedlander SK (1977) Smoke, dust and haze. John Wiley and Sons, New York
Fristrom RM, Westenberg AA (1965) Flame structure. McGraw-Hill, New York
Fristrom RM (1995) Flame structure and processes. Oxford University Press, New York/Oxford
Fujii T, Ozawa Y, Kikumoto S, Sato M, Yuasa Y, Inoue H, (1996) High pressure results of a catalytic combustor for gas turbine. ASME paper no 96-GT-382
Garo A, Prado G, Lahaye J (1990) Chemical aspects of soot particles oxidation in a laminar methane-air diffusion flame. Comb. Flame 79, 226 (1990)
Gaydon A, Wolfhard H (1979) Flames, their structure, radiation, and temperature. Chapman and Hall, London
Gehring M, Hoyermann K, Schacke H, Wolfrum J (1973) Direct studies of some elementary steps for the formation and destruction of nitric oxide in the H-N-O system. Proc Comb Inst 14:99
Geitlinger H, Streibel T, Suntz R, Bockhorn H (1998) Two-dimensional imaging of soot volume fractions, particle number densities and particle radii in laminar and turbulent diffusion flames. Proc Comb Inst 27:1613
Gill A, Warnatz J, Gutheil E (1994) Numerical investigation of the turbulent combustion in a direct-injection stratified-charge engine with emphasis on pollutant formation. Proc. COMODIA (1994), p 583. JSME, Yokohama
Glarborg P, Miller JA, Kee RJ (1986) Kinetic modeling and sensitivity analysis of nitrogen oxide formation in well-stirred reactors. Comb Flame 65:177
Golden DM (1994) Gas phase homogeneous kinetics. In: Low-temperature chemistry of the atmosphere (Moortgat GK ed.), pp 69-92, Springer, Berlin/Heidelberg
Gordon S, McBride BJ (1971) Computer program for calculation of complex chemical eqilibrium compositions, rocket performance, incident and reflected shocks and Chapman-Jouguet detonations. NASA SP-273

Goyal G, Warnatz J, Maas U (1990a) Numerical studies of hot spot ignition in H_2-O_2 and CH_4-air mixtures. Proc Comb Inst 23:1767

Goyal G, Maas U, Warnatz J (1990b) Simulation of the transition from deflagration to detonation. SAE 1990 Transactions, Journal of Fuels & Lubricants, Section 4, Vol 99, Society of Automotive Engineers, Inc, Warrendale, PA, p 1

Graham SC (1976) The collisional growth of soot particles at high temperatures. Proc Comb Inst 16:663

Graham SC, Homer JB, Rosenfeld JLJ (1975a) The formation and coagulation of soot aerosols. In: Kamimoto G (ed) Modern developments in shock-tube research: Proceedings of the tenth shock tube symposium, p 621

Graham SC, Homer JB, Rosenfeld JLJ (1975b) The formation and coagulation of soot aerosols generated by the pyrolysis of aromatic hydrocarbons. Proc Roy Soc A 344:259

Grimstead JH, Finkelstein ND, Lempert W, Miles R, Lavid M (1996) Frequency-modulated filtered Rayleigh scattering (FM-FRS): A new technique for real-time velocimetry. AIAA paper no 96-0302

Günther R (1987) 50 Jahre Wissenschaft und Technik der Verbrennung. BWK 39 Nr 9

Gutheil E, Bockhorn H (1987) The effect of multi-dimensional PDF's in turbulent reactive flows at moderate Damköhler number. Physicochemical Hydrodynamics 9:525

Gutheil E, Sirignano WA (1998) Counterflow spray combustion modeling with detailed transport and detailed chemistry. Combustion and Flame 113:92

Hanson RK (1986) Combustion Diagnostics: Planar Imaging Techniques. Proc Comb Inst 21:1677

Hanson RK, Seitzman JM, Paul P (1990) Planar laser-fluorescence imaging of combustion gases. Appl Phys B50:441

Härle H, Lehnert A, Metka U, Volpp HR, Willms L, Wolfrum J (1998) In-situ detection of chemisorbed CO on a polycristalline platinum foil using infrared-visible sum-frequency generation (SFG). Chem Phys Lett 293:26

Harris SJ, Weiner AM (1990) Surface growth and soot particle reactivity. Combust Sci Technol 72:67

Harris SJ, Weiner AM, Ashcraft CC (1986a) Soot particle inception kinetics in a premixed ethylene flame. Comb Flame 64:65

Harris SJ, Weiner AM, Blint RJ, Goldsmith JEM (1986b) A picture of soot particle inception. Proc Comb Inst 22:333

Harris SJ, Weiner AM, Blint RJ (1988) Formation of small aromatic molecules in a sooting ethylene flame. Comb Flame 72:91

Harville T, and Holve D (1997) Method for measuring particle size in presence of multiple scattering. U.S. Patent No 5 619 324

Haynes BS, Wagner HG (1981) Soot formation. Prog Energy Combust Sci 7:229

He LT, Lee JHS (1995) The dynamical limit of one-dimensional detonations. Phys Fluids 7:1151

Heard DE, Jeffries JB, Smith GP, Crosley DR (1992) LIF measurements in methane/air flames of radicals important in prompt-NO formation. Comb Flame 88:137

Heywood JB (1988) Internal combustion engine fundamentals. McGraw-Hill, New York

Hinze J (1972) Turbulence, 2nd ed. McGraw-Hill, New York

Hirschfelder JO (1963) Some remarks on the theory of flame propagation. Proc Comb Inst 9:553

Hirschfelder JO, Curtiss CF (1949) Theory of propagation of flames. Part I: General equations. Proc Comb Inst 3:121

Hirschfelder JO, Curtiss CF, Bird RB (1964) Molecular theory of gases and liquids. Wiley, New York

Hobbs ML, Radulovic PT, Smoot LD (1993) Combustion and gasification of coals in fixed-beds. Progr Energy Comb Sci 19:505

Hodkinson JR (1963) Computational light scattering and extinction by spheres according to diffraction and geometrical optics and some comparison with Mie theory. J Opt Soc Amer 53:577

Homann, KH (1967) Carbon formation in premixed flames. Comb Flame 11:265

Homann KH (1975) Reaktionskinetik. Steinkopff, Darmstadt

Homann KH (1984) Formation of large molecules, particulates, and ions in premixed hydrocarbon flames; progress and unresolved questions. Proc Comb Inst 20:857

Homann K, Solomon WC, Warnatz J, Wagner HGg, Zetzsch C (1970) Eine Methode zur Erzeugung von Fluoratomen in inerter Atmosphäre. Ber Bunsenges Phys Chem 74:585

Holve DJ, Self SA (1979a) Optical particle sizing for in-situ measurement I. Appl. Opt. 18:1632

Holve DJ, Self SA (1979b) Optical particle sizing for in-situ measurement II. Appl. Opt. 18:1646

Hottel HC, Hawthorne WR (1949) Diffusion in laminar flame jets. Proc Comb Inst 3:254

Howard JB, Wersborg BL, Williams GC (1973) Coagulation of carbon particles in premixed flames. Faraday Symp Chem Soc 7:109

Hsu DSY, Hoffbauer MA, Lin MC (1987) Dynamics of OH desorption from Ssingle crystal Pt(111) and polycrystalline Pt foil surfaces. Surface Sci 184:25

Hurst BE (1984) Report 84-42-1, Exxon Research

Ishiguro T, Takatori Y, Akihama K (1997) Microstructure of Diesel soot particles probed by electron microscopy: First observation of inner core and outer shell. Comb Flame 108:231

Jander H (1995) private communication. Universität Göttingen

John F (1981) Partial differential equations. In: Applied mathematical sciences Vol 1. Springer, New York Heidelberg Berlin, p 4

Johnston HS (1992) Atmospheric ozone. Annu Rev Phys Chem 43:1

Jones WP, Whitelaw JH (1985) Modelling and measurement in turbulent combustion. Proc Comb Inst 20:233

Jost W (1939) Explosions und Verbrennungsvorgänge in Gasen. Julius Springer, Berlin

Kauzmann W (1966) Kinetic theory of gases. Benjamin/Cummings, London

Kee RJ, Rupley FM, Miller JA (1987) The CHEMKIN thermodynamic data base. SANDIA Report SAND87-8215, Sandia National Laboratories, Livermore CA

Kee RJ, Miller JA, Evans GH, Dixon-Lewis G (1989b) A computational model of the structure and extinction of strained opposed-flow premixed methane-air flames. Proc Comb Inst 22:1479

Kellerer H, Müller A, Bauer HJ, Wittig S (1996) Soot formation in a shock tube under elevated pressure conditions. Combust Sci Technol 113:67

Kent JH, Bilger RW (1976) The prediction of turbulent diffusion flame fields and nitric oxide formation. Proc Comb Inst 16:1643

Kerstein AR (1992) Linear-eddy modelling of turbulent transport 7. Finite-rate chemistry and multistream mixing. J Fluid Mech 240:289

Kissel-Osterrieder R, Behrendt F, Warnatz J (1998) Detailed modeling of the oxidation of CO on platinum: A Monte-Carlo model. Proc Comb Inst 27:2267

Klaus P, Warnatz J (1995) A contribution towards a complete mechanism for the formation of NO in flames. Joint meeting of the French and German Sections of the Combustion Institute, Mulhouse

Kolb T, Jansohn P, Leuckel W (1988) Reduction of NO_x emission in turbulent combustion by fuel-staging / effects of mixing and stoichiometry in the reduction zone. Proc Comb Inst 22:1193

Kolmogorov AN (1942) The equation of turbulent motion in an incompressible viscous fluid. Izw Akad Nauk SSSR Ser Phys 6:56

Kompa K, Sick V, Wolfrum J (1993) Laser diagnostics for industrial processes. Ber Bunsenges Phys Chem 97:1503

Kordylewski W, Wach J (1982) Criticality for thermal ignition with reactant consumption. Comb Flame 45:219

Köylü ÜÖ, Faeth GM (1992) Structure of overfire soot in buoyant turbulent diffusion flames at long residence times. Comb Flame 89:140

Köylü ÜÖ, Faeth GM, Farias TL, Carvalho MG (1995) Fractal and projected structure properties of soot aggregates. Comb Flame 100:621

Kramer MA, Kee RJ, Rabitz H (1982) CHEMSEN: A computer code for sensitivity analysis of elementary reaction models. SANDIA Report SAND82-8230, Sandia National Laboratories, Livermore CA

Lam SH, Goussis DA (1989) Understanding complex chemical kinetics with computational singular perturbation. Proc Comb Inst 22:931

Lange M, Riedel U, Warnatz J (1998) Parallel DNS of turbulent flames with detailed reaction schemes. AIAA paper no 98-2979

Launder BE, Spalding DB (1972) Mathematical models of turbulence. Academic Press, London/New York

Lauterbach J, Asakura K, Rotermund HH (1995) Subsurface oxygen on Pt(100): kinetics of the transition from chemisorbed to subsurface state and its reaction with CO, H_2, and O_2. Surf Sci 313:52

Law CK (1989) Dynamics of stretched flames. Proc Comb Inst 22:1381

Lee JC, Yetter RA, Dryer FL (1995) Transient numerical modeling of carbon particle ignition and oxidation. Comb. Flame 101:387

Libby PA (1996) Introduction to turbulence. Taylor and Francis, Washington, D.C.

Libby PA, Williams FA (1980) Fundamental aspects of turbulent reacting flows. In: Libby PA, Williams FA (eds) Turbulent reacting flows. Springer, New York

Libby PA, Williams FA (1994) Turbulent reacting flows. Academic Press, New York

Lieuwen T, Zinn BT (1998) The role of equivalence ration oscillations in driving combustion instabilities in low NO_x gas turbines. Proc Comb Inst 27:1809

Liew SK, Bray KNC, Moss JB (1984) A stretched laminar flamelet model of turbulent nonpremixed combustion. Comb Flame 56:199

Liñán A, Williams FA (1993) Fundamental aspects of combustion. Oxford University Press, Oxford

Lindemann FA (1922) Discussion on "The radiation theory of chemical action". Trans Farad Soc 17:599

Liu Y, Lenze B (1988) The influence of turbulence on the burning velocity of premixed CH_4-H_2 flames with different laminar burning velocities. Proc Comb Inst 22:747

Ljungström S, Kasemo B, Rosen A, Wahnström T, Fridell E (1989) An experimental study of the kinetics of OH and H_2O formation on Pt in the $H_2 + O_2$ reaction. Surface Sci. 216:63

Löffler L, Löffler P, Weilmünster P, Homann K-H (1994) Growth of large ionic polycyclic aromatic hydrocarbons in sooting flames. In: Bockhorn H (ed), Soot formation in combustion. Springer, Berlin/Heidelberg

Long MB, Levin PS, Fourguette DC (1985) Simultaneous two-dimensional mapping of species concentration and temperature in tubulent flames. Opt Lett 10:267

Long MB, Smooke MD, Xu Y, Zurn RM, Lin P, Frank JH (1993) Computational and experimental study of OH and CH radicals in axisymmetric laminar diffusion flames. Proc Comb Inst 24:813

Lovell W (1948) Knocking characteristics of hydrocarbons. Ind Eng Chem 40:2388
Lovett JA, Abuaf N (1992) Emissions and stability characteristics of flameholders for lean-premixed combustion. Proc. International Gas Turbine and Aeroengine Congress, ASME paper no 92-GT-120
Lozano A, Yip B, Hanson RK (1992) Acetone: a tracer for concentration measurements in gaseous flows by planar laser-induced fluorescence. Exp. Fluids 13:369
Lutz AE, Kee RJ, Miller JA (1987) A Fortran program to predict homogeneous gas-phase chemical kinetics including sensitivity analysis. SANDIA Report SAND87-8248, Sandia National Laboratories, Livermore CA
Lutz AE, Kee RJ, Miller JA, Dwyer HA, Oppenheim AK (1989) Dynamic effects of autoignition centers for hydrogen and $C_{1,2}$-hydrocarbon fuels. Proc Comb Inst 22:1683
Lyon RK (1974) Method for the reduction of the concentration of NO in combustion effluents using ammonia. U.S. Patent No 3 900 544
Maas U (1990) private communication. Universität Stuttgart
Maas U (1998) Efficient calculation of intrinsic low-dimensional manifolds for the simplification of chemical kinetics. Comp. and Visual. in Science 1:69
Maas U, Pope SB (1992) Simplifying chemical kinetics: Intrinsic low-dimensional manifolds in composition space. Comb Flame 88:239
Maas U, Pope SB (1993) Implementation of simplified chemical kinetics based on intrinsic low-dimensional manifolds. Proc Comb Inst 24:103
Maas U, Warnatz J (1988) Ignition processes in hydrogen-oxygen mixtures. Comb Flame 74:53
Maas U, Warnatz J (1989) Solution of the 2D Navier-Stokes equation using detailed chemistry. Impact of Computing in Science and Engineering 1:394
Mach JJ, Varghese PL (1998) Velocity measurements using filtered Rayleigh scattering of near-IR diode lasers. AIAA paper no 98-0510
Magre P, Dibble RW (1988) Finite chemical kinetic effects in a subsonic turbulent hydrogen flame. Comb Flame 73:195
Malte PC, Pratt DT (1974) Measurement of atomic oxygen and nitrogen oxides in jet-stirred combustion. Proc Comb Inst 15:1061
Marsal (1976) Die numerische Lösung partieller Differentialgleichungen in Wissenschaft und Technik. Bibliographisches Institut Mannheim/Wien/Zürich
Masri AR, Bilger RW, Dibble RW (1988) Turbulent nonpremixed flames of methane near extinction: probability density functions. Comb Flame 73:261
Mathur S, Tondon PK, Saxena SC (1967) Heat conductivity in ternary gas mixtures. Mol Phys 12:569
Mauss F, Schäfer T, Bockhorn H (1994a) Inception and growth of soot particles in dependence on the surrounding gas phase. Comb Flame 99:697
Mauss F, Trilken B, Breitbach H, Peters N (1994b) Soot formation in partially premixed diffusion flames at atmospheric pressure. In: Bockhorn H (ed) Soot formation in combustion. Springer, Berlin/Heidelberg
McKinnon JT (1989) Chemical and physical mechanisms of soot formation. Ph.D. Dissertation, MIT, Cambridge (Massachusetts)
McMillin BK, Palmer JL, Hanson RK (1993) Temporally resolved two-line fluorescence imaging of NO temperature in a transverse jet in a supersonic cross flow. Appl Optics 32:7532
McMurtry PA, Menon S, Kerstein AR (1992) A linear eddy sub-grid model for turbulent reacting flows: application to hydrogen-air combustion. Proc Comb Inst 24:271
Melius CF, Miller JA, Evleth EM (1992) Unimolecular reaction mechanisms involving C_3H_4, C_4H_4, and C_6H_6 hydrocarbon species. Proc Comb Inst 24:621

Miller JH (1990) The kinetics of polynuclear aromatic hydrocarbon agglomeration in flames. Proc Comb Inst 23:91

Miller JA (1996) Theora and modeling in combustion chemistry. Proc Comb Inst 26:461

Miller JA, Melius CF (1991) The formation of benzene in flames. Am Chem Soc, Div Fuel Chem 36:1440

Mittelbach G, Voje H (1986) Anwendung des SNCR-Verfahrens hinter einer Zyklonfeuerung. In: NO_x-Bildung und NO_x-Minderung bei Dampferzeugern für fossile Brennstoffe. VGB-Handbuch

Mongia RM, Tomita E, Hsu FK, Talbot L, Dibble RW (1996) Use of an optical probe for time-resolved in situ measurement of local air-to-fuel ratio and extent of fuel mixing with application to low NOx emissions in premixed gas turbines. Proc Comb Inst 26:2749

Mongia R, Dibble RW, Lovett J (1998) Measurements of air-fuel ratio fluctuations caused by combustor driven oscillations. ASM paper no 98-GT-304

Morley C (1987) A fundamentally based correlation between alkane structure and octane number. Comb Sci Technol 55:115

Moss JB (1979) Simultaneous measurements of concentration and velocity in an open premixed turbulent flame. Comb Sci Technol 22:115

Moss JB (1994) Modeling soot formation for turbulent flame prediction. In: Bockhorn H (ed) Soot formation in combustion. Springer, Berlin/Heidelberg

Moss JB, Stewart CD, Young KJ (1995) Modeling soot formation and burnout in a high temperature laminar diffusion flame burning under oxygen-enriched conditions. Comb Flame 101:491

Mungal MG, Lourenco LM, and Krothapalli A (1995) Instantaneous velocity measurements in laminar and turbulent premixed flames using on-line PIV. Comb. Sci. Tech 106:239

Nau M, Wölfert A, Maas U, Warnatz J (1996) Application of a combined pdf/finite-volume scheme on turbulent methane diffusion flames. In: Chan SH (ed) Transport pheneomena in combustion, Taylor & Francis, Washington D. C., p 986

Nehse M, Warnatz J, Chevalier C (1996) Kinetic modelling of the oxidation of large aliphatic hydrocarbons. Proc Comb Inst 26:773

Neoh KG, Howard JB, Sarofim AF (1974) Effect of oxidation on the physical structure of soot. Proc Comb Inst 20:951

Nguyen QV, Dibble RW, Hofmann D, Kampmann S (1993) Tomographic measurements of carbon monoxide temperature and concentration in a Bunsen flame using diode laser absorption. Ber Bunsenges Phys Chem 97:1634

Nowak U, Warnatz J (1988) Sensitivity analysis in aliphatic hydrocarbon combustion. In: Kuhl AL, Bowen JR, Leyer J-C, Borisov A (eds) Dynamics of reactive systems, Part I. AIAA, New York, p 87

Onsager L (1931) Reciprocal relations in irreversible processes I. Phys Rev 37:405, Reciprocal relations in irreversible processes II. Phys Rev 38:2265

Oppenheim AK, Manson N, Wagner HG (1963) Recent progress in detonation research. AIAA J 1:2243

Oran ES, Boris JP (1993) Computing turbulent shear flows – a convenient conspiracy. Computers in Physics 7:523

Palmer HB, Cullis CF (1965) The formation of carbon from gases. In: Walker PL (ed), Chemistry and physics of carbon Vol 1, p 265. Marcel Dekker, New York.

Paul P, Warnatz J (1998) A re-evaluation of the means used to calculate transport properties of reacting flows. Proc Comb Inst 27:495

Paul P, van Cruyningen I, Hanson RK, Kychakoff G (1990) High resolution digital flow-field imaging of jets. Exp Fluids 9:241

Penner SS, Bernard JM, Jerskey T (1976a) Laser scattering from moving polydisperse particles in flames I: Theory. Acta Astr. 3:69
Penner SS, Bernard JM, Jerskey T (1976b) Laser scattering from moving polydisperse particles in flames II Preliminary experiments. Acta Astr. 3:93
Perrin M, Namazian N, Kelly J, Schefer RW (1995) Effect of confinement and blockage ratio on nonpremixed turbulent bluff-body burner flames. Poster, 23 Symp (Intl) Comb, Orleans
Peters N (1987) Laminar flamelet concepts in turbulent combustion. Proc Comb Inst 21:1231
Peters N (2000) Turbulent combustion. Cambridge University Press, Cambridge
Peters N, Warnatz J (eds) (1982) Numerical methods in laminar flame propagation. Vieweg-Verlag, Wiesbaden
Pfefferle LD, Bermudez G, Byle J (1994) Benzene and higher hydrocarbon formation during allene pyrolysis. In: Bockhorn H (ed), Soot formation in combustion. Springer, Berlin/Heidelberg
Pitz WJ, Warnatz J, Westbrook CK(1989) Simulation of auto-ignition over a large temperature Range. Proc Comb Inst 22:893
Poinsot T (1996) Using direct numerical simulations to understand premixed turbulent combustion. Proc Comb Inst 26:219
Poinsot T, Veynante D, Candel S (1991) Diagrams of premixed turbulent combustion based on direct simulation. Proc Comb Inst 23:613
Pope SB (1986) PDF methods for turbulent reactive flows. Prog Energy Combust Sci 11:119
Pope SB (1991) Computations of turbulent combustion: Progress and challenges. Proc Comb Inst 23:591
Pope SB (2000) Turbulent flows. Cambridge University Press, Cambridge
Prandtl L (1925) Über die ausgebildete Turbulenz. Zeitschrift für Angewandte Mathematik und Mechanik 5:136
Prandtl L (1945) Über ein neues Formelsystem der ausgebildeten Turbulenz. Nachrichten der Gesellschaft der Wissenschaften Göttingen, Mathematisch-Physikalische Klasse, p 6
Raffel B, Warnatz J, Wolfrum J (1985) Experimental study of laser-induced thermal ignition in O_2/O_3 mixtures. Appl Phys B 37:189
Raffel B, Warnatz J, Wolff H, Wolfrum J, Kee RJ (1986) Thermal ignition and minimum ignition energy in O_2/O_3 mixtures. In: Bowen JR, Leyer J-C, Soloukhin RI (eds), Dynamics of reactive systems, Part II, AIAA, New York, p 335
Raja LL, Kee RJ, Deutschmann O,Warnatz J, Schmidt LD (2000) A critical evaluation of Navier-Stokes, boundary-layer, and plug-flow models of the flow and chemistry in a catalytic-combustion monolith. Catalysis Today 59:47
Razdan MK, Stevens JG (1985) CO/air turbulent diffusion flame: Measurements and modeling. Comb Flame 59:289
Reh CT (1991) Höhermolekulare Kohlenwasserstoffe in brennstoffreichen Kohlenwasserstoff/Sauerstoff-Flammen. Dissertation, TH Darmstadt
Reynolds WC (1986) The element potential method for chemical equilibrium analysis: implementation in the interactive program STANJAN version 3. Dept. of Engineering, Stanford University
Reynolds WC (1989) The potential and limitations of direct and large eddy simulation. In: Whither turbulence? Turbulence at crossroads. Lecture notes in physics, Springer, New York, p 313
Rhodes RP (1979) A probability distribution function for turbulent flows. In: Murthy SNB (ed) Turbulent mixing in non-reactive and reactive flows, Plenum Press, New York, p 235
Riedel U, Schmidt R, Warnatz J (1992) Different levels of air dissociation chemistry and Its coupling with flow models. In: Bertin JJ, Periaux J, Ballmann J (eds), Advances in Hypersonics - Vol. 2: Modeling Hypersonic Flows. Birkhäuser, Boston

Riedel U, Schmidt D, Maas U, Warnatz J (1994) Laminar flame calculations based on automatically simplified chemical kinetics. Proc. Eurotherm Seminar #35, Compact Fired Heating Systems, Leuven, Belgium

Roberts WL, Driscoll JF, Drake MC, Goss LP (1993) Images of the quenching of a flame by a vortex – To quantify regimes of turbulent combustion. Comb Flame 94:58

Robinson PJ, Holbrook KA (1972) Unimolecular reactions. Wiley-Interscience, New York

Rogg B, Behrendt F, Warnatz J (1987) Turbulent non-premixed combustion in partially premixed diffusion flamelets with detailed chemistry. Proc Comb Inst 21:1533

Roshko A (1975) Progress and problems in turbulent shear flows. In: Murthy SNB (ed) Turbulent mixing in nonreactive and reactive flow, Plenum, New York

Rosner DE (2000) Transport processes in chemically reacting flow systems. Dover Publication, Mineola NY

Rosten H, Spalding B (1987) PHOENICS: Beginners guide; user manual; photo user guide. Concentration Heat and Momentum Ltd, London

Roth P, Brandt O, von Gersum S (1990), High temperature oxidation of suspended soot particles verified by CO and CO_2 measurements. Proc Comb Inst 23:1485

Roth P, von Gersum S (1993) High temperature oxidation of soot particles by O, OH, and NO. In: Takeno T (ed), Turbulence and molecular processes in combustion, Elsevier, London, p 149.

Rumminger MD, Dibble RW, Heberle NH, Crosley DR (1996) Gas temperature above a porous radient burner: Comparison of measurements and model predictions. Proc Comb Inst 26:1755

Santoro RJ, Yeh TT, Horvath JJSemerjian HH (1987) The transport and growth of soot particles in laminar diffusion flames. Comb Sci Technol 53:89

Schlatter JC, Dalla Betta RA, Nickolas SG, Cutrone MB, Beebe KW, Tsuchiya T (1997) Single digit emissions in a full scale catalytic combustor. ASME paper no 97-GT-57

Schmidt D (1996) Modellierung reaktiver Strömungen unter Verwendung automatisch reduzierter Reaktionsmechanismen, PhD Thesis, Universität Heidelberg

Schwanebeck W, Warnatz J (1972) Reaktionen des Butadiins I: Die Reaktion mit Wasserstoff-Atomen. Ber Bunsenges Phys Chem 79:530

Seinfeld JH (1986) Atmospheric chemistry and physics of air pollution. John Wiley and Sons, New York

Semenov NN (1935) Chemical Kinetics and Chain Reactions. Oxford University Press, London

Seitzman JM, Kychakoff G, Hanson RK (1985) Instantaneous temperature field measurements using planar laser-induced fluorescence. Opt Lett 10:439

Sherman FS (1990) Viscous flow. McGraw-Hill, New York

Shirley JA, Winter MA (1993) Air mass flux measurement system using Doppler-shifted filtered Rayleigh scattering. AIAA paper no 93-0513

Shvab VA (1948) Gos Energ izd Moscow-Leningrad; see also Zeldovich (1949), Williams (1984)

Sick V, Arnold A, Dießel E, Dreier T, Ketterle W, Lange B, Wolfrum J, Thiele KU, Behrendt F, Warnatz J (1991) Two-dimensional laser diagnostics and modeling of counterflow diffusion flames. Proc Comb Inst 23:495

Sirignano WA (1992) Fluid dynamics of sprays – 1992 Freeman scholar lecture. J Fluids Engin 115:345

Smith JR, Green RM, Westbrook CK, Pitz WJ (1984) An experimental and modeling study of engine knock. Proc Comb Inst 20:91

Smith DA, Frey SF, Stansel DM, Razdan MK (1997) Low emission combustion system for the Allison ATS engine. ASME paper no 97-GT-292

Smoluchowski MV (1917) Versuch einer mathematischen Theorie der Koagulationskinetik kolloider Loesungen. Z. Phys. Chem. 92, 129

Smooke MD ed (1991) Reduced kinetic mechanisms and asymptotic approximations for methane-air flames. Lecture notes in physics 384, Springer, New York

Smooke MD, Mitchell RE, Keyes DE (1989) Numerical solution of two-dimensional axisymmetric laminar diffusion flames. Comb Sci Technol 67:85

Smoot LD (1993) Fundamentals of coal combustion. Elsevier, Amsterdam/Oxford/New York

Solomon PR, Hamblen DG Carangelo RM, Serio MA, Deshpande, GV (1987) A general model of coal devolatilization. ACS paper 58/ WP No 26

Spalding DB (1970) Mixing and chemical reaction in steady confined turbulent flames. Proc Comb Inst 13:649

Speight JG (1994) The chemistry and technology of coal. Marcel Dekker, Amsterdam/New York

Stahl G, Warnatz J (1991) Numerical investigation of strained premixed CH_4-air flames up to high pressures. Comb Flame 85:285

Stapf P, Maas U, Warnatz J (1991) Detaillierte mathematische Modellierung der Tröpfchenverbrennung. 7. TECFLAM-Seminar „Partikel in Verbrennungsvorgängen", Karlsruhe, p 125. DLR Stuttgart

Stefan J (1874) Sitzungsberichte Akad. Wiss. Wien II 68:325; see also: Curtiss, Hirschfelder (1949)

Stein SE, Walker JA, Suryan MM, Fahr A (1991) A new path to benzene in flames, Proc Comb Inst 23:85

Strehlow RA (1985) Combustion fundamentals. McGraw-Hill, New York

Stull DR, Prophet H (eds) (1971) JANAF thermochemical tables. U.S. Department of Commerce, Washington DC, and addenda

Subramanian VS, Buermann DH, Ibrahim KM, Bachalo WD (1995) Application of an integrated phase Doppler interferometer/rainbow thermometer/point-diffraction interferometer for characterizing burning droplets. Proc Comb Inst 23:495

Tait NP, Greenhalgh DA (1992) 2D laser induced fluorescence imaging of parent fuel fraction in nonpremixed combustion. Proc Comb Inst 24:1621

Takagi Y (1998) A new era in spark ignition engines featuring high pressure direct injection. Proc Comb Inst 27:2055

Takeno T (1995) Transition and structure of jet diffusion flames. Proc Comb Inst 25:1061

Takeno T, Nishioka M, Yamashita H (1993) Prediction of NO_x emission index of turbulent diffusion flames. In: Takeno T (ed), Turbulence and molecular processes in combustion, p 375. Elsevier, Amsterdam/London

Thévenin D, Behrendt F, Maas U, Przywara B, Warnatz J (1996) Development of a parallel direct simulation code to investigate reacting flows. Computers and Fluids 25:485

Thorne AP (1988) Spectrophysics, 2nd ed, Chapman and Hall, London/New York

Thring RH (1989) Homogeneous charge compression ignition (HCCI) engines. SAE paper 892068

Tien CL, Lienhard JH (1971) Statistical thermodynamics. Holt, Rinehart, and Winston, New York

Tom HWK, Mate CM, Zhu XD, Crowell JE, Heinz TF, Somorjai GA, Shen YR (1984) Surface studies by optical second harmonic generation: The adsorption of O_2, CO, and sodium on the Rh(111) surface. Phys Rev Lett 52:348

Tsuji H, Yamaoka I (1967) The counterflow diffusion flame in the forward stagnation region of a porous cylinder. Proc Comb Inst 11:979

Tsuji H, Yamaoka I (1971) Structure analysis of counterflow diffusion flames in the forward stagnation region of a porous cylinder. Proc Comb Inst 13:723

Turns SR (1996) An introduction to combustion. McGraw-Hill, New York

Vagelopoulos CM and Egolfopoulos FN (1998) Direct experimental determination of laminar flame speeds. Proc Comb Inst 27:513

Vandsburger U, Kennedy I, Glassman I (1984) Sooting Counterflow Diffusion Flames with Varying Oxygen Index. Comb Sci Technol 39:263

von Gersum S, Roth P (1992) Soot oxidation in high temperature N_2O/Ar and NO/Ar mixtures. Proc Comb Inst 24:999

von Karman Th (1930) Mechanische Ähnlichkeit und Turbulenz. Nachrichten der Gesellschaft der Wissenschaften Göttingen, Mathematisch-Physikalische Klasse, p 58

Wagner HG (1979) Soot formation in combustion. Proc Comb Inst 17:3

Wagner HG (1981) Mass growth of soot. In: Siegla DC, Smith GW (eds), Particulate carbon formation during combustion. Plenum Press, New York

Waldmann L (1947) Der Diffusionsthermoeffekt II. Z Physik 124:175

Warnatz J (1978a) Calculation of the structure of laminar flat flames I: Flame velocity of freely propagating ozone decomposition flames. Ber Bunsenges Phys Chem 82:193

Warnatz J (1978b) Calculation of the structure of laminar flat flames II: Flame velocity of freely propagating hydrogen-air and hydrogen-oxygen flames. Ber Bunsenges Phys Chem 82:643

Warnatz J (1979) The structure of freely propagating and burner-stabilized flames in the H_2-CO-O_2 system. Ber Bunsenges Phys Chem 83:950

Warnatz J (1981a) The structure of laminar alkane-, alkene-, and acetylene flames. Proc Comb Inst 18:369

Warnatz J (1981b) Concentration-, pressure-, and temperature dependence of the flame velocity in the hydrogen-oxygen-nitrogen mixtures. Comb Sci Technol 26:203

Warnatz J (1981c) Chemistry of stationary and instationary combustion processes. In: Ebert KH, Deuflhard P, Jäger W (eds) Modelling of chemical reaction systems, Springer, Heidelberg, p 162

Warnatz J (1982) Influence of transport models and boundary conditions on flame structure. In: Peters N, Warnatz J (eds), Numerical methods in laminar flame propagation, Vieweg, Wiesbaden

Warnatz J (1983) The mechanism of high temperature combustion of propane and butane. Comb Sci Technol 34:177

Warnatz J (1984) Critical survey of elementary reaction rate coefficients in the C/H/O system. In: Gardiner WC jr. (ed) Combustion chemistry. Springer-Verlag, New York

Warnatz J (1987) Production and homogeneous selective reduction of NO in combustion processes. In: Zellner R (ed) Formation, distribution, and chemical transformation of air pollutants. DECHEMA, Frankfurt, p 21

Warnatz J (1988) Detailed studies of combustion chemistry. Proceedings of the contractors' meeting on EC combustion research, EC, Bruxelles, p 172

Warnatz J (1990) NO_x Formation in high-temperature processes. Eurogas '90, Tapir, Trondheim, p 303

Warnatz J (1991) Simulation of ignition processes. In: Larrouturou B (ed) Recent advances in combustion modeling. World Scientific, Singapore, p 185

Warnatz J (1993) Resolution of gas phase and surface chemistry into elementary reactions. Proc Comb Inst 24:553

Warnatz J, Bockhorn H, Möser A, Wenz HW (1983) Experimental investigations and computational simulations of acetylene-oxygen flames from near stoichiometric to sooting conditions. Proc Comb Inst 19:197
Warnatz J, Allendorf MD, Kee RJ, Coltrin ME (1994) A model of hydrogen-oxygen combustion on flat-plate platinum catalytist. Combust. Flame 96:393
Warnatz J, Chevalier C, Karbach V, Nehse M (2000) Survey of reactions in the C/H/O system, publication in preparation
Weinberg FJ (1975) The first half-million years of combustion research and today's burning problems. Proc Comb Inst 15:1
Weinberg FJ (1986) Advanced combustion methods. Academic Press, London/Orlando
Wersborg BL, Howard JB, Williams GC (1973) Physical mechanisms in carbon formation in flames. Proc Comb Inst 14:929
Westblom U, Aldén M (1989) Simultaneous multiple species detection in a flame using laser-induced fluorescence. Appl Opt 28:2592
Westbrook CK, Dryer FL (1981) Chemical kinetics and modeling of combustion processes. Proc Comb Inst 18:749
Williams A (1990) Combustion of liquid fuel sprays. Butterworth & Co, London
Williams FA (1984) Combustion theory. Benjamin/Cummings, Menlo Park
Williams WR, Marks CM, Schmidt LD (1992) Steps in the reaction $H_2 + O_2 = H_2O$ on Pt: OH desorption at high temperature. J Chem Phys 96:5922
Wilke CR (1950) A viscosity equation for gas mixtures. J Chem Phys 18:517
Wieschnowsky U, Bockhorn H, Fetting F (1988) Some new observations concerning the mass growth of soot in premixed hydrocarbon-oxygen flames. Proc Comb Inst 22:343
Wolfrum J (1972) Bildung von Stickstoffoxiden bei der Verbrennung. Chemie-Ingenieur-Technik 44:656
Wolfrum J (1986) Einsatz von Excimer- und Farbstofflasern zur Analyse von Verbrennungsprozessen VDI Berichte 617:301
Wolfrum J (1998), Lasers in combustion – From basic theory to practical devices. Proc Comb Inst 27:1
Woods IT, Haynes BS (1994) Active sites in soot growth. In: Bockhorn H (ed) Soot formation in combustion. Springer, Berlin/Heidelberg
Xu J, Behrendt F, Warnatz J (1994) 2D-LIF investigation of early stages of flame kernel development after spark ignition. Proc. COMODIA, p 69. JSME, Yokohama
Yang JC, Avedisian CT (1988) The combustion of unsupported heptane/hexadecane mixture droplets at low gravity. Proc Comb Inst 22:2037
Zeldovich YB (1946) The oxidation of nitrogen in combustion and explosions. Acta Physicochim. USSR 21:577
Zeldovich YB (1949) K teorii gorenia neperemeshannykh gazov. Zhur Tekhn Fiz 19:1199; English: On the theory of combustion of initially unmixed gases. NACA Tech Memo No 1296 (1951)
Zeldovich YB, Frank-Kamenetskii DA (1938) The theory of thermal propagation of flames. Zh Fiz Khim 12:100
Zhang QL, O'Brien SC, Heath JR, Liu Y, Curl RF, Kroto HW, Smalley RE (1986) Reactivity of large carbon clusters: Spheroidal carbon shells and their possible relevance to the formation and morphology of soot. J Phys Chem 90:525

20 Index

D

N

O

P

Q

R

T

U

V

W

Z